CHEMICAL VAPOR DEPOSITION

CHEMICAL VAPOR DEPOSITION

Thermal and Plasma Deposition of Electronic Materials

S. Sivaram

VNR SPRINGER SCIENCE+BUSINESS MEDIA, LLC

Originally published by Van Nostrand Reinhold in 1995

I(T)P™ A division of International Thomson Publishing Inc.
The ITP logo is a trademark under license.

1 2 3 4 5 6 7 8 9 10 QEBFF 01 00 99 98 97 96 95

Library of Congress Cataloging-in-Publication Data

Sivaram, S.
Chemical vapor deposition : thermal and plasma deposition of electronic materials / S. Sivaram.
p. cm.
Includes bibliographical references and index.
ISBN 978-1-4757-4753-9 ISBN 978-1-4757-4751-5 (eBook)
DOI 10.1007/978-1-4757-4751-5
1. Microelectronics industry. 2. Chemical vapor deposition.
3. Microelectronics—Materials. I. Title
TK7836.S54 1994 94-39696
621.3815'2—dc20 CIP

To my family,
starting with my father,
whose dreams I am living.

Contents

Preface

In early 1987 I was attempting to develop a CVD-based tungsten process for Intel. At every step of the development, information that we were collecting had to be analyzed in light of theories and hypotheses from books and papers in many unrelated subjects. These sources were so widely different that I came to realize there was no unifying treatment of CVD and its subprocesses. More interestingly, my colleagues in the industry were from many disciplines (a surface chemist, a mechanical engineer, a geologist, and an electrical engineer were in my group). To help us understand the field of CVD and its players, some of us organized the CVD user's group of Northern California in 1988. The idea for writing a book on the subject occurred to me during that time.

I had already organized my thoughts for a course I taught at San Jose State University. Later Van Nostrand agreed to publish my book as a text intended for students at the senior/first year graduate level and for process engineers in the microelectronics industry,

This book is not intended to be bibliographical, and it does not cover every new material being studied for chemical vapor deposition. On the other hand, it does present the principles of CVD at a fundamental level while uniting them with the needs of the microelectronics industry. As a materials scientist, I have always been bothered by the lack of attention paid in the microelectronics industry to the relationship between the structure and properties of thin films to the conditions during deposition. I have attempted to bridge the gap and looked at CVD from an applications viewpoint.

In dealing with broad topics, such as reaction thermodynamics, kinetics, reactor design, and plasma fundamentals, I have attempted to keep the treatment and terminology similar to well-established textbooks. I have also included many of these texts as recommended reading at the end of the relevant chapters. However, I have simplified the treatment with a narrow focus aimed at CVD and kept at a level that can be understood by researchers with a basic background in science and engineering.

Close to my heart is the chapter on manufacturability. Too many processes are being developed and integrated into manufacturing only to be re-engineered on line. Cost of ownership and quality need to be remembered even in the fervor of novel processes development.

Librarians at three different institutions have provided me with immeasurabe assistance. I thank the people at the libraries of SEMATECH in Austin, Texas; Matsushita Electric Industrial in Moriguchi, Osaka; and Intel in Santa Clara, California, for being there when I needed help.

I also acknowledge the original inspiration to write the book from Rama Shukla of Intel and Shyam Murarka of Rensselaer Polytechnic Institute. The encouragement and help from Ken Cadien, Joe Schoenholtz, Xiao Chun Mu, Dave Fraser, Paolo Gargini, Robert Tolles, and Hubert Bath deserve my gratitude.

Time to write a book in the midst of Intel's explosive growth did not come cheap. More than anyone else, my wife Ranjana paid the price during my lengthy absences. To her and to my children, Varun and Uttara, thank you for being there even if I was not always there!

Chapter 1

Introduction

Chemical vapor deposition (CVD) is a technique for synthesizing materials in which chemical components in vapor phase react to form a solid film at some surface. The occurrence of a chemical reaction is central to this means of thin film growth, as is the requirement that the reactants must start out in the vapor phase. Ability to control the components of the gas phase, and the physical conditions of the gas phase, the solid surface, and the envelope that surrounds them determines our capability to control the properties of the thin films that are produced.

CVD is a sequential process which starts from the initial vapor phase, progresses through a series of quasi steady-state subprocesses, and culminates in the formation of a solid film in its final microstructure. This sequence is illustrated schematically in Figure 1.1.

a. Diffusion of gaseous reactants to the surface.
b. Adsorption of the reacting species on to surface sites, often after some migration on the surface.
c. Surface chemical reaction between the reactants, usually catalyzed by the surface.
d. Desorption of the reaction by-products.
e. Diffusion of the by-products away from the surface.
f. Incorporation of the condensed solid product into the microstructure of the growing film.

Each of the individual steps will be studied in detail in various chapters throughout the book. However, this sequence also serves to illustrate the multidisciplinary nature of the subject. Reaction chemistry is central to an understanding of the process, but flows and temperatures have to be

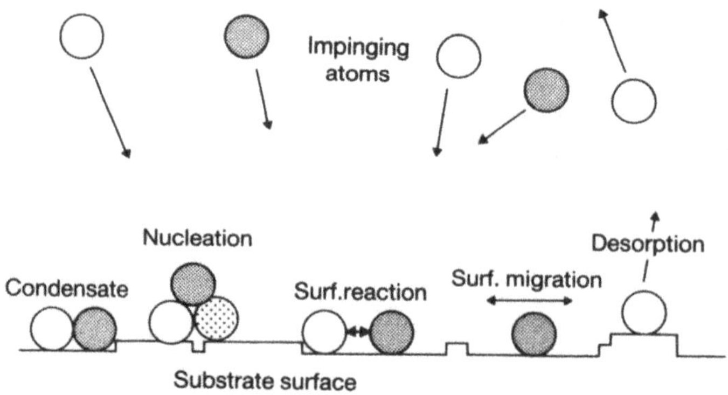

FIGURE 1.1. Schematic of events that occur around the substrate surface. Summation of these events leads to deposition on the wafer.

engineered to facilitate the desired reaction at the desired location. Together, chemistry and engineering have to produce a thin film with the required structure and properties.

The objective of the book is to enable the reader to understand these processes in order to produce thin films with desired properties. Thus in the sequence described above, step (f) is the end result of this effort. We are, however, forced by the nature of the CVD process to understand steps (a) through (e), and to be aware of their influence in producing the final film.

Chemical vapor deposition is only one of many methods available for thin film synthesis. However, we will learn of its unique capabilities for tailoring and controlling film properties to suit the requirements of microelectronic components. Whereas the science underlying the CVD process itself is universal to all CVD applications, microelectronic processing imposes its own constraints. It is in providing an understanding of the definitions, procedures, and applications of CVD to microelectronics that this book will differ from most other thin film treatises.

1.1. BASIC ASSUMPTIONS

Thin solid films are a class of materials which have one of their dimensions much smaller than the other two. The result is a dramatic increase in surface to volume ratio, as illustrated in Figure 1.2. The two bounding surfaces are so close to each other as to have a decisive influence on the physical, mechanical, chemical, electrical, and other properties of the material.

Commonly used units for measuring thin film thicknesses are angstroms (Å), nanometers (nm) and microns (μm). Even though nanometers, the SI unit,

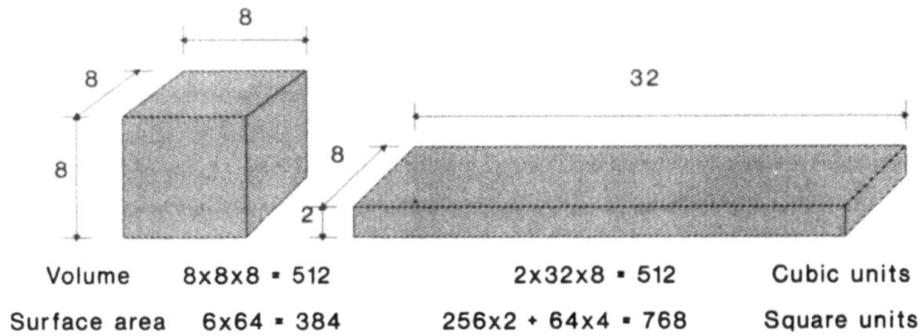

FIGURE 1.2. Surface to volume ratio in thin films is very high. As one dimension of a cube is reduced keeping the volume constant, one can see the rapid increase in surface area.

are preferred, all three units are commonly used, both in the literature and in everyday practice. These units are related to a meter as follows:

$$1 \text{ nm} = 10 \text{ Å} = 1 \times 10^{-3}\,\mu\text{m} = 1 \times 10^{-9}\text{ m}$$

We will restrict ourselves in our discussion to thin films in the region 10 nm to 10 000 nm during the course of this book. The only situation where we will have to concern ourselves with films thinner than 10 nm is in understanding nucleation of thin films and the effect of nucleation on film properties. We will consider properties of films thicker than 10 000 nm to be essentially similar to bulk materials. A class of materials known as *thick films* are industrially important, especially in hybrid electronics. These films, often as thin as 10 μm, would still be operationally classified as thin films by the definitions used in this book.

In all their applications in microelectronics, thin films are bonded to substrates, which are three-dimensional systems. The dimension of the substrate in the direction normal to the film surface is very large compared to the thickness of the film. We will see in subsequent chapters that the nature of the bond between the thin film and the substrate is also an intrinsic property of the system.

Thin films on substrates can be broadly categorized into two types: diffused and overlaid.[1] Diffused films involve active participation of the substrate during film growth, resulting in a chemical reaction between the substrate and the growth ambient. An example is the formation of silicon oxides through the thermal oxidation of silicon. Similarly, growth of metal silicides through the reaction between substrate silicon and the metal film involves a diffusive process. But an overlaid film requires only a passive role from the substrate; it might provide mechanical or catalytic support during thin film growth while remaining chemically intact. Before any reaction occurs, an evaporated metal film on a substrate can be considered overlaid. All of the CVD films in this book fall into the category of overlaid films.

1.2. CVD IN MICROELECTRONICS

In microelectronic applications, semiconducting films are patterned to form electronic devices. Conductive films act as the conduits through which electrical signals are transmitted to and away from these devices. Dielectric films are used to isolate devices from each other and to insulate the conductive wires. The application is enough to deduce many of the properties required by dielectric films used in microelectronics. For instance, a dielectric film should have very low electrical conductance, high breakdown strength, and should be physically void-free. Conversely, a conductive film should exhibit high electrical conductance and should have good long-term reliability under conditions of high current density. Since many electrical devices are formed in monocrystalline semiconductive films, crystallographic perfection is essential. These are the generic requirements for the three kinds of films in microelectronics.

Additionally, due to the fine geometry of features in microelectronic circuits, overlaid films are often required to reproduce exactly the underlying topography. Depending on the device and circuit demands, the dielectric may need to have a high or a low dielectric constant; and the semiconductor may need dopants of a specific kind. It is essential to determine all the properties required by the application before starting to design a process, whether the film is produced by CVD or otherwise. Considerable savings in

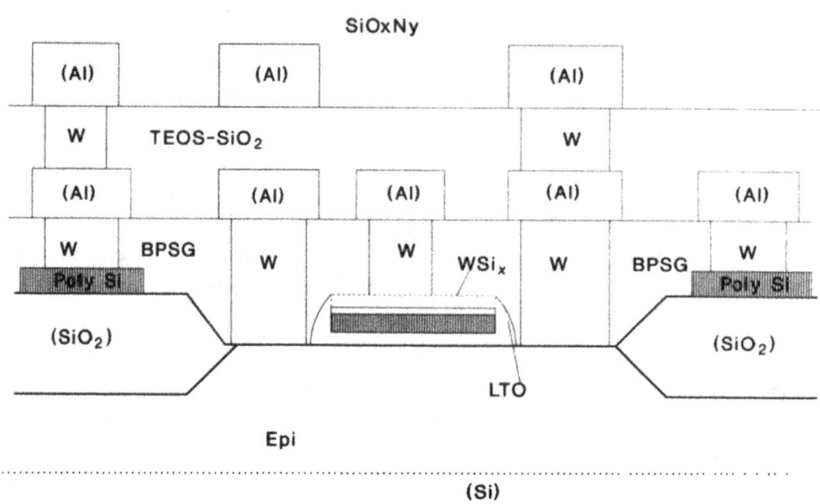

FIGURE 1.3. Schematic cross section of a two metal MOS structure shows the number of deposited layers which utilize CVD techniques. Non-CVD layers shown in paretheses.

time and money can be realized by working out beforehand an appropriate production path and the constraints it places on the processing conditions.

During the past several years, CVD has become an essential part of very large scale integrated circuit (VLSI) manufacture. Conductor, dielectric, and semiconductor films are now routinely deposited using various CVD techniques. More and more film layers are attempted using CVD due to its often superior conformality (see Section 2.4). Figure 1.3 shows a schematic cross section of a typical metal oxide semiconductor (MOS) device with two levels of metallic interconnects. One can already see the number of layers in the cross section that are being deposited by CVD. Newer CVD techniques are being developed for the deposition of more layers. The importance of understanding the fundamentals of chemical vapor deposition are therefore crucial to the development of an integrated process for manufacturing large-scale ICs.

1.3. ORGANIZATION OF THE BOOK

This book is arranged to provide a logical course through two of the major avenues of chemical vapor deposition: thermal and plasma assisted CVD (see flowchart, Figure 1.4). After presenting their fundamentals, we apply them to the development of deposition processes for particular materials.

Chapters 1 and 2 are overview chapters from which the rest of the material is derived; do not bypass them. Chapter 2 describes the general principles of thin film phenomena and relates film properties to growth conditions. Chapter 3 deals with process design for manufacturability and contains essential material for scientists and engineers developing processes for commercial applications. It also provides an overview of thermal and plasma assisted CVD applications. Students of CVD science need not go through this chapter to follow the rest of the book. However, a sound understanding of the principles of experimental design and data analysis are invaluable to any experimental scientist, for the proper setup of experiments and the interpretation of results.

Chapters 4 and 5 introduce the reader to fundamentals of reaction thermodynamics and kinetics. They are followed by principles of reactor design, which concentrate on a narrow section of a much larger field to provide emphasis on the ability to change thin film properties through changes in reactor configuration.

Introduction to the physics of plasmas and the role of plasmas in supplying energy to chemical reactions is dealt with in Chapters 6 and 7. We also study modifications that conventional reactors require, due to the presence of the

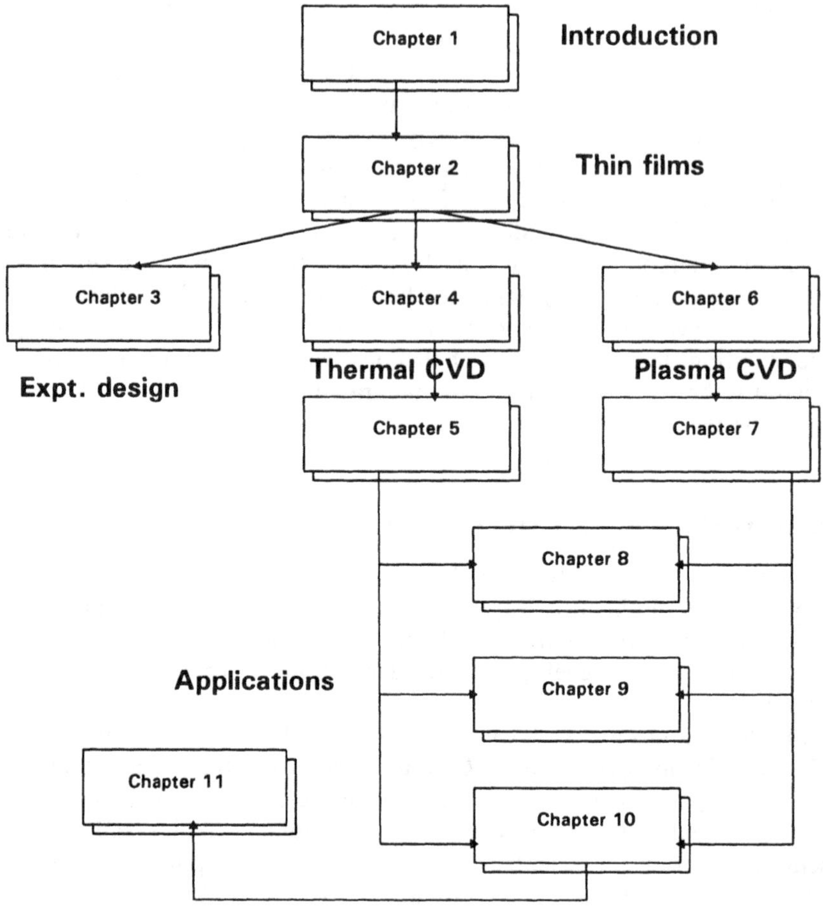

FIGURE 1.4. Flowchart for the sequence of chapters in the book. Thermal and plasma CVD techniques are both illustrated in the applications section.

plasma. This is followed by the deposition of conducting, insulating, and semiconducting films in Chapters 8 through 10. The book concludes by introducing the reader to novel CVD concepts and other narrowly focused branches of CVD. These are application-specific topics and are of interest to researchers looking for newer and more efficient ways of producing thin films.

Scattered throughout the text are illustrative examples and problem sets relating to the discussions immediately preceding them. They pertain to analyses of commonly encountered practical situations and will be useful in developing similar processes. They also provide a bridge from theoretical treatments of the science to troubleshooting and problem solving. Where data is available, tables of properties of the materials being discussed are included.

Reference

1. R. F. Bunshah, *Deposition Technologies for Films and Coatings*, Noyes, Park Ridge, N.J., 1982.

Chapter 2

Thin Film Phenomena

Before we begin to address each of the subprocesses that constitute CVD, it is worthwhile to understand some of the general concepts of thin film growth. Conditions during film growth affect many of the properties of thin films. The importance of early stages of thin film growth on the final film properties cannot be overemphasized. In this chapter, we describe the processes that occur during this stage of growth and correlate the growth conditions to physical, mechanical, and electrical properties of the film.

We begin with a thermodynamic treatment of condensation and nucleation of atoms from the gas phase to form the condensed matter on the substrate. We then proceed to the development of structure in the condensate, and use the conditions of growth to study thin film properties of conducting, insulating, and semiconducting films. Most of the discussion is pertinent to any thin film growth technique, such as evaporation, physical vapor deposition, or CVD. However, factors that are germane to CVD are highlighted throughout the discussion.

2.1. EARLY STAGES OF THIN FILM GROWTH

Regardless of the growth technique used to deposit thin films, initial stages of thin film growth are characterized by three major phenomena: (a) condensation and nucleation, (b) diffusion-controlled island coalescence, and (c) steady-state growth on top of the first layer. Considerable research has occurred to characterize the early phase of thin film growth during such deposition techniques as evaporation and physical vapor deposition. Chemical vapor deposition, however, adds an additional feature to the normal sequence of thin film growth: a chemical reaction occurs on the surface, often catalyzed by the surface, after the incident reactant molecules are adsorbed on the surface. Most of the discussion in this chapter is based on a model of simple condensation from a gas phase, ignoring the chemical reaction. However, the following specific differences between CVD and simple condensation have to be kept in mind:[1]

a. Whereas simple condensation is exothermic, most useful CVD reactions are endothermic. This allows us to delay the reaction effectively till the gases reach the heated substrate surface. Among the rare exceptions is the growth of GaAs using GaCl.
b. While most condensation processes occur at very low pressures, of the order of 1 millitorr or less, CVD is operated at higher pressures, 100 millitorr to 1 atmosphere. Reactants have to diffuse through reaction products in order to reach the surface.
c. It is only the condensing species that needs to be considered in the vicinity of the substrate during simple condensation. In contrast, CVD has to deal with *intrinsic impurities*, which relate to the reaction products.

In order to base our discussion on the three phases of thin film growth on a simple condensation model, two fundamental assumptions will be made.[2] We will assume that decomposition on the surface occurs before significant surface diffusion takes place. And we will assume the chemical reaction is much faster than the adsorption process (see Figure 1.1). While there are many CVD situations where these assumptions are tenuous, they allow us to make use of the extensive evidence that has been gathered using other methods of thin film growth in illustrating the growth phenomena. The next three sections treat the processes of condensation from gas phase, nucleation, and island coalescence in sequence.

2.1.1. Condensation from the Gas Phase

By our earlier definition of CVD (Chapter 1) the reactants for CVD films start out in the vapor phase. The reactant molecule or atom impinging from the vapor phase is attracted to the substrate surface by the instantaneous dipole moment of the substrate surface atoms.[3] If the incident kinetic energy is not large, the component of the impinging velocity normal to the surface is lost within a short time. The velocity parallel to the surface provides the energy for the atom to migrate on the surface. The probability that an impinging atom will be incorporated into the surface is the sticking coefficient. It is measured as the ratio of the amount of material condensed on the surface to the total amount impinged. Figure 2.1 illustrates the processes that occur close to the surface when an atom impinges on a surface.

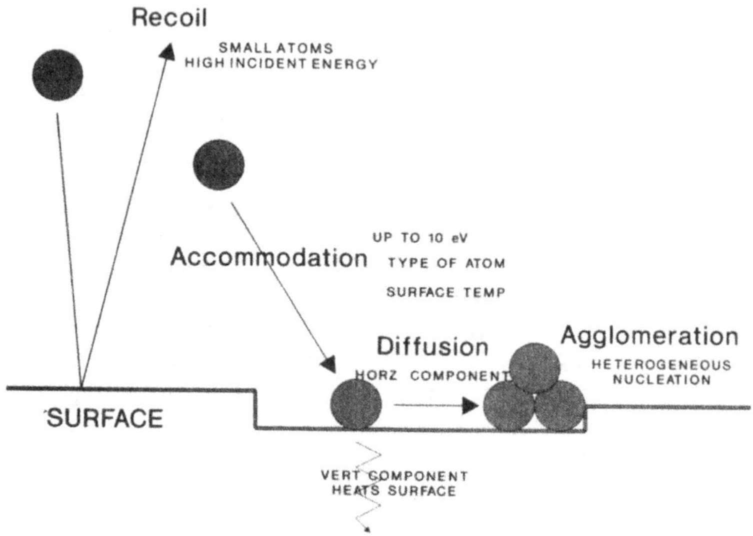

FIGURE 2.1. When an atom impinges on a surface, any of the illustrated processes can occur, depending on the energy of the atom and the substrate temperature.

In a CVD system which has one or more species in the gas phase, the rate of impingement on a unit area of the substrate from the gas phase is given by

$$I = \tfrac{1}{4} n \bar{v} \tag{2.1}$$

where n is the number density of the molecules in the gas phase and $\bar{v}$ is their average molecular velocity. From the kinetic theory of gases, we can substitute for the number density of molecules in terms of pressure P, and for the average velocity in terms of temperature T. This results in

$$I = \frac{P}{\sqrt{2\pi mkT}} \tag{2.2}$$

where m is the mass of a molecule and k is Boltzmann's constant. When several species are present in the gas phase, the rate of impingement of each species is proportional to its partial pressure.

Two factors play important roles in determining whether an incident atom is accommodated on the substrate: the energy of the impinging species and the temperature of the substrate. For equal masses of incoming and substrate atoms, almost all of the incoming atoms are accommodated, for gas phase temperatures as high as 10^6 Kelvin. If the incoming atom is considerably smaller than the substrate atom and if it has high incident energy, the sticking coefficient can often be appreciably less than 1. In an extreme situation where a small incoming atom has enough energy to overcome electronic and nuclear repulsion, it can get implanted at a considerable depth below the surface. We will not consider implantation in this discussion.

The effect of the substrate temperature is tied to the atoms already on the surface, i.e., desorption of atoms adsorbed on the surface. These atoms have to overcome a surface binding energy, Q_{des}, before they can leave the surface.[4] An adsorbed atom or molecule stays on the surface for a length of time, τ_s, given by:

$$\tau_s = \frac{1}{\nu} \exp(Q_{des}/kT) \tag{2.3}$$

where ν is the surface vibrational frequency of the adsorbed atom. When the binding energy is high, the adsorbed atom spends a long time on the surface so the probability of the atom being accommodated is higher. But when the surface binding energy is only on the order of kT, the kinetic energy of the surface atoms, the atom is considered *hot* and has a high probability of being desorbed without being accommodated on the surface.

During its stay on the surface the equilibrated atom migrates on the surface, jumping from one lattice position to another. The average distance the atom can migrate depends on the time it spends on the surface τ_s and the activation energy for surface diffusion, Q_{dif}.[5] The concept of activation energy for surface diffusion is illustrated in Figure 2.2. Often, we observe empirically that Q_{des} is between 0.25 and 0.33 of the surface binding energy, especially for metals. Using this empirical relationship between the binding energy and the activation energy for surface diffusion, we can express the

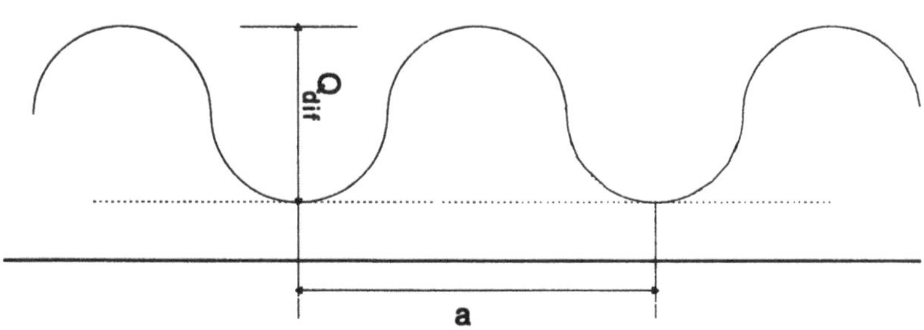

FIGURE 2.2. Potential distribution on the surface. The energy needed for an atom to overcome the potential barrier is termed the activation energy for surface diffusion.

TABLE 2.1 Residence Time on Tungsten

Gas	Q_{des}	Residence time at 300 K (seconds)
He	0.1	1.3×10^{-13}
H_2	1.5	1.3×10^{-12}
Organics	10–15	3.2×10^{-6} to 1.8×10^{-2}
F	30	4×10^{9} (> 100 yrs)
O	150	10^{1100} (1 s at 2 500°C)

average migration distance $\bar{x}$ as,

$$\bar{x} = \sqrt{2}a \exp[(Q_{des} - Q_{dif})/2kT] \simeq 1.4a \exp(0.35Q_{des}/kT) \qquad (2.4)$$

where a is the distance between adjacent lattice sites.

To illustrate the point regarding the surface binding energy and the temperature of the substrate, Table 2.1 shows the binding energies of various gases on a tungsten surface. Notice that when the bond between the adsorbed atom and the substrate is weak, as in helium on tungsten, the residence time is very small and the probability of the atom staying on the surface is very low. As the binding energy increases, the residence time and the migration distances are large. More interestingly, a strong bond, such as that between oxygen and tungsten, can prevent the adsorbed atom from ever leaving the surface. During the discussion on the evolution of film properties later in this chapter, we shall see the effects of residence time and surface migration on film structure.

A feature unique to CVD is the presence of reaction by-products and impurities in the close vicinity of the substrate. Let us assume the pressure in the reactor is made of three components in the gas phase: a reactant with a partial pressure of p_r, the by-products with a partial pressure of p_b, and impurities, such as oxygen, with a partial pressure of p_i. The impingement rates scale with the partial pressures and hence the impingement rate of the reactants will be given through equation (2.2), as

$$I_r = \frac{p_r}{\sqrt{2\pi mkT}} \qquad (2.5)$$

and so on for the other two components. Often it is the partial pressure of the impurities that is most important to thin film properties, since the reaction by-products have very low binding energies to the condensate. Impurities such as oxygen have very high binding energies, as shown in Table 2.1. For ease of calculation if we assume the sticking coefficients of the reactants and the impurities are close to 1, then we can directly correlate the partial pressure in the gas phase to the impurity concentration in the condensate.

As an example, let us take the case of the growth of tungsten film from tungsten hexafluoride. The reaction occurs at a pressure of 1 torr, with $p_{WF_6} = 50$ millitorr. If there is a leak in the reactor contributing to an oxygen partial pressure of 0.05 millitorr, the ratio of the impingement rates is given by

$$\frac{I_{WF_6}}{I_{O_2}} = \frac{p_{WF_6}}{p_{O_2}} \sqrt{\frac{m_{O_2}}{m_{WF_6}}} \tag{2.6}$$

The molecular weight of WF_6 is 298 and the molecular weight of O_2 is 16. For every 230 molecules of WF_6 impinging on the surface, one molecule of oxygen also impinges on the surface. When the sticking coefficient is unity, the oxygen concentration can be as high as 0.4 at%.

This example serves to illustrate the reason why reactors are often pumped to very low pressures prior to flowing any reactants. However, the calculation is simplistic in the use of direct molecular impingement, without considering other flow effects.

2.1.2. Nucleation

Aggregation of the condensed species to form clusters of atoms possessing a specific volume (i.e., bounded by a defined surface) is called nucleation. The energy to create a unit area of surface is always positive. Hence a larger surface area necessarily means increasing the energy of a system. The concept of minimization of total energy through a reduction in total surface area will be a common thread through most of the discussion that follows. In the absence of a substrate, it would be very difficult to nucleate a condensed phase from the vapor, since the very small particles that would form would have a high surface area and a large surface free energy. The total energy of a small spherical particle has two major components: a positive surface free energy equal to $4\pi r^2 \gamma$, where r is the radius of the particle and γ is the specific surface free energy; and a negative free energy of formation ΔG_V with a volume V.[6] Hence

$$\Delta G = 4\pi r^2 \gamma + \frac{4}{3}\pi r^3 \Delta G_V \tag{2.7}$$

The volumetric free energy term is given by

$$\Delta G_V = -\frac{kT}{V} \log S \tag{2.8}$$

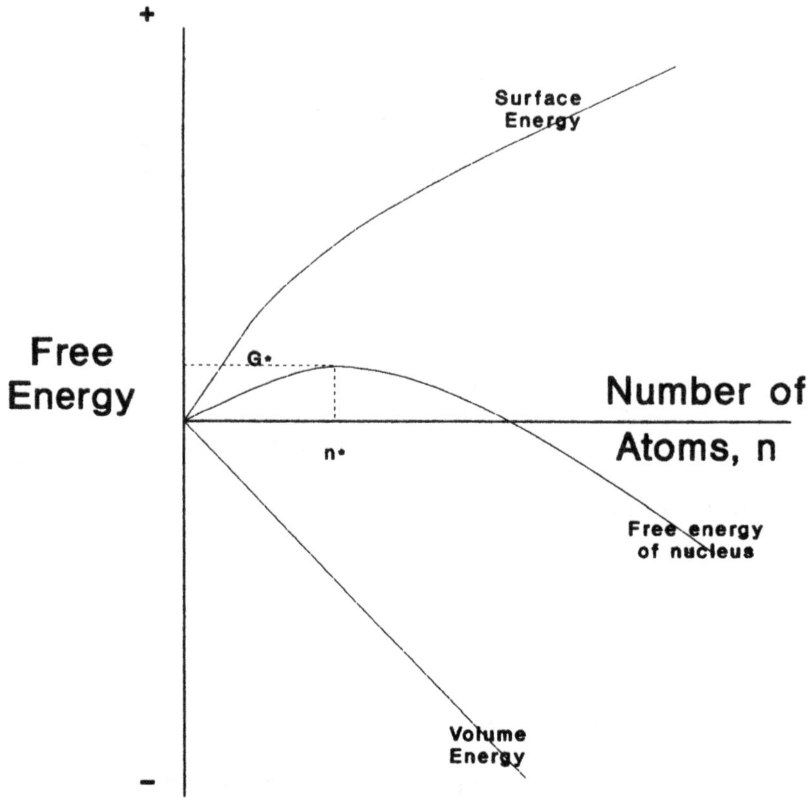

FIGURE 2.3. Change in free energy as the nucleus size grows is the sum of two energies: the increasing surface energy and the decreasing volume free energy. Over the critical nucleus size, increase in nucleus size is energetically favorable.

where $S = P/P_{eq}$ is the degree of supersaturation of the system. Supersaturation refers to the ratio of the system pressure P to the equilibrium vapor pressure of the system P_{eq}. For a more detailed discussion of P_{eq}, refer to Section 4.2. Initially the atomic aggregate is very small and the surface free energy is the larger of the two terms. In this situation, the condensate particle cannot grow even at high degrees of supersaturation ($P \gg P_{eq}$). Figure 2.3 illustrates the change in the total free energy as a function of the radius of the nucleus. The total free energy passes through a maximum at a critical nucleus radius r^*, which can be calculated by setting $d(\Delta G)/dr = 0$.

The critical radius r^* is then given by

$$r^* = -\frac{2\gamma}{\Delta G_V} = \frac{2\gamma V}{kT \log(P/P_{eq})} \tag{2.9}$$

When the radius of the nucleus is smaller than r^*, there is a high likelihood of the particle disintegrating back to the gas phase. If the radius is larger than the critical nucleus r^*, the particle is highly likely to be a stable nucleus. Further growth occurs because the addition of another atom decreases the total energy.

The substrate on which the thin film is growing allows heterogeneous nucleation, in contrast to homogeneous nucleation discussed above. Instead of being spheres, the nuclei now have a cap shape on the substrate. The angle of the cap is determined by

$$\gamma_{CV} \cos\theta = \gamma_{SV} - \gamma_{SC} \tag{2.10}$$

where the terms in equation (2.10) are illustrated in Figure 2.4. The effect of the heterogeneous surface is to lower the critical nucleus size and act as a catalyst for the nucleation process. The presence of the substrate effectively lowers the total surface energy making r^* smaller. In the limiting case, when the angle θ in Figure 2.4 tends to zero, complete wetting of the surface occurs and the situation is most favorable for nucleation. On the other hand, when θ tends to 180°, the situation is the same as homogeneous nucleation and the substrate plays no part in the nucleation process.

The concentration of critical nuclei (i.e., those with $r = r^*$),and the rate at which molecules join the critical nuclei by surface diffusion determines the nucleation rate J. The concentration of critical nuclei N^* is given by

$$N^* = n_0 \exp(\Delta G^*/kT) \tag{2.11}$$

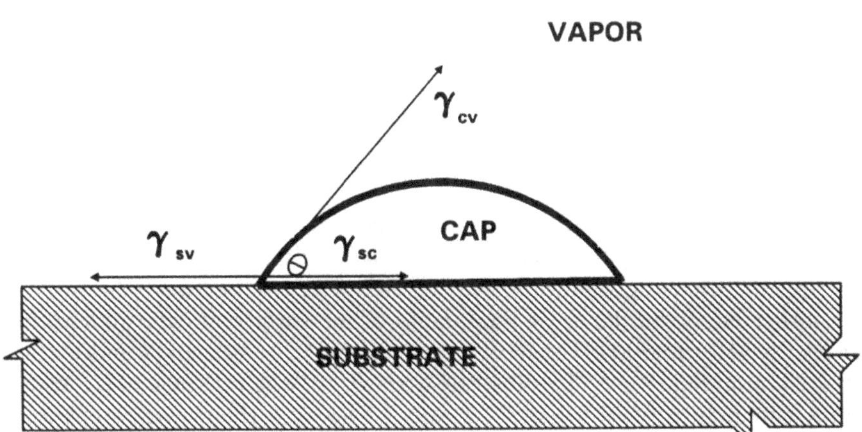

FIGURE 2.4. During heterogeneous nucleation, the nucleus is in the shape of a cap, balancing the surface tension forces between the nucleus, the substrate, and the ambient.

where n_0 is the concentration of surface adsorption sites available for nucleation and ΔG^* is the free energy of the critical nucleus. The higher the free energy of the critical nucleus, the easier it is to form more of them.

The surface diffusion term is similar to the condensation term. The surface diffusion rate F is given by

$$F = n_1 a\nu \exp(-Q_{\text{dif}}/kT) \tag{2.12}$$

where n_1 is the surface concentration of adsorbed molecules.

The nucleation rate J can be determined by combining equations (2.9) through (2.12), resulting in

$$J = Kr^* n_0 a \sin\theta \frac{P}{\sqrt{2\pi mkT}} \exp[(Q_{\text{des}} - Q_{\text{dif}} - \Delta G^*)/kT] \tag{2.13}$$

where K is a proportionality constant.

The nucleation rate has the direct effect of determining the initial microstructure of the growing film. Examining equation (2.13), the factors over which one has direct control to affect J are: (a) the temperature, (b) the supersaturation at the critical nucleus radius and hence ΔG^*, (c) pressure, and to some extent (d) the angle of the cap, through surface cleanness.

Temperature is a crucial factor in controlling J. Both the size and the rate of growth of the critical nucleus are affected by temperature. At very low temperatures, the critical nucleus size could be as low as a single atom. We have already seen the effect of heterogeneous nucleation, through the presence of the substrate, in reducing the size of critical nucleus.

Nucleation rate is exponentially dependent upon supersaturation. Supersaturation has different connotations in CVD and simple condensation.[7] In the case of condensation, supersaturation is the ratio of actual pressure in the vapor phase to the equilibrium vapor pressure of the condensed phase. In the presence of a chemical reaction, such as

$$\text{MX(g)} \rightleftarrows \text{M(s)} + \text{X(g)} \tag{2.14}$$

supersaturation S_{cvd} is defined as $p_{\text{M}}/P_{\text{eq}}$ where p_{M} is the partial pressure of the solid as dictated by the chemical reaction equilibrium. This point, which may appear confusing right now, will be elaborated in Section 4.2.

Another important factor is the quality of the surface in determining the free energy of formation of the critical nucleus (ΔG^*). Ledges and kinks on the surface, contamination of the substrate surface, and the presence of electrical charge can reduce ΔG^*. Figures 1.1 and 2.1 show surfaces with steps and kinks. In particular, the presence of *intrinsic impurities*, i.e., reactant and product gases, can affect the surface free energies and the angle of the cap-shaped nucleus.

The thin film growth process continues from the formation of the critical nucleus on through to the formation of islands. If the critical nucleus consists of at least two atoms and the free energy of its formation from the vapor is positive, an energy barrier prohibits the formation of a continuous film on the substrate. If the barrier is high (large ΔG^*) the radius of the critical nucleus is large, so relatively few large aggregates are formed. For small ΔG^*, the barrier is low and a large number of small aggregates are formed; the film becomes continuous even at a relatively small thickness.

Metals with a high boiling point, such as tungsten, have high values of supersaturation, small critical nuclei, and tend to form continuous films at smaller thicknesses. Strong adhesion between the substrate and the film also tends to lower the size of the critical nucleus.[8]

The effect of the substrate surface on determining nucleation characteristics leads to an interesting phenomenon called selectivity in CVD. The surface, by virtue of its surface charge or its adsorption/desorption characteristics to certain types of atoms, can dramatically modify nucleation. When different surfaces are present on the substrate, each having different nucleation characteristics, the film can be discontinuous, growing only on the favorable surfaces. For instance, in the case of CVD of copper and tungsten, dielectric surfaces do not favor nucleation whereas conductive and semiconducting surfaces nucleate readily. In the case of copper, different gaseous CVD sources nucleate on different surfaces. By prepatterning the substrate surface we can grow the film only on specific locations, but to ensure reproducibility we need to understand the cause of the nucleation behavior. This is not always easy and despite their abundance, selective growth phenomena are not always sufficiently well understood for commercial exploitation.

Exercise 2.1

1. The boiling point of copper is 2 600°C, the surface energy of liquid copper is approximately 1 100 erg/cm^2, and the atomic diameter of copper is 2.55 Å. What is the equilibrium number of nuclei containing 10 atoms in 1 mole of copper vapor just above its boiling point? How many nuclei contain 20 atoms?
2. For liquid silver, the surface energy against another metallic surface is 190 erg/cm^2, and against air is 227 erg/cm^2. If the contact angle is 5°, what is the free energy of the other metal surface to air?

2.1.3. Coalescence of Islands

The randomly formed, three-dimensional nuclei rapidly reach a saturation density with a small amount of deposit. The saturation density is determined when the internuclear distance reaches the mean surface diffusion length. At

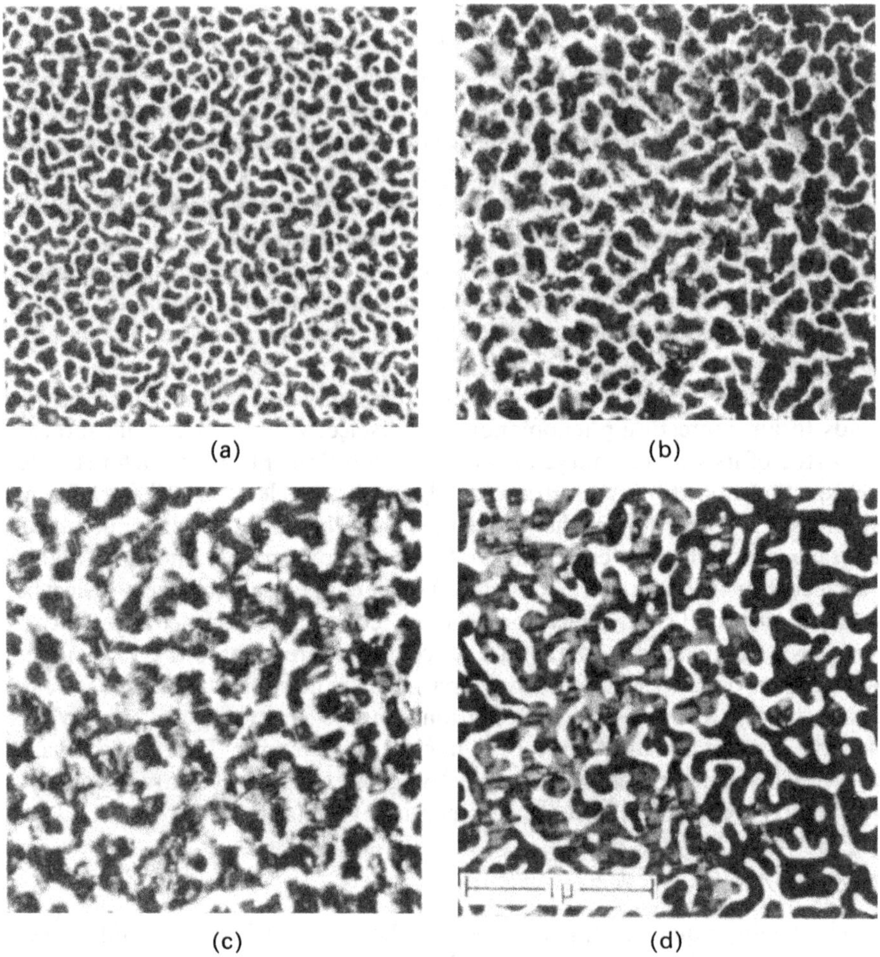

FIGURE 2.5. Sequence of transmission electron micrographs with increasing film thickness in a thin silver film. (a) to (d) indicate increase of 30 Å. Notice the progressive increase in the connected network feature of the film. Reprinted from Ref. 13 with permission from McGraw-Hill Company.

this point the nuclei grow by capture of adsorbed atoms (or adatoms) through diffusion.[9] The depletion of adsorbed atoms, captured by the growing nucleus, prevents any further nucleation. Similar to the nucleation theory, the main driving force for island coalescence is still the reduction in surface energy. The larger islands represent a net reduction in surface energy over many smaller islands.

The islands increase their size by capture of adatoms and other subcritical nuclei. As the islands grow, they get closer together and the larger islands coalesce with the smaller ones. Considerable mass transfer between the islands forms a connected network structure and the islands become flattened out to decrease surface energy. Figure 2.5 shows a planar TEM section of a connected network structure in a silver film. As a rule, when small and large islands coalesce, each having a different crystallographic orientation, the result has the orientation of the large island. This may substantially modify the structure of the film. For example, if the growth conditions favor [1 1 1] orientation, even if [1 0 0] oriented nuclei are greater in number, the larger [1 1 1] nuclei will coalesce with the smaller [1 0 0] nuclei resulting in dominant [1 1 1] grains in the film.

2.2. STEADY-STATE GROWTH AND THE EVOLUTION OF ORDER

Once the islands have coalesced, further film growth is a reenactment of earlier nucleation and growth processes but under steady-state conditions at the substrate surface. The growing film can evolve into any one of the following three types of atomic arrangements: amorphous, polycrystalline, and monocrystalline. The majority of physical properties of thin films depend to a considerable degree on the structure of the film and the structure is most often determined by the conditions during film growth, so it is essential to understand the effect of growth parameters on structure.

2.2.1. Amorphous Films

Amorphous films by definition exhibit no short-range or long-range order in the arrangement of the atoms in the film, i.e., no translational periodicity over several atomic spaces. All glasses used in microelectronic applications, such as SiO_2, Si_3N_4, SiO_xN_y, are amorphous. Oxides of metals are often observed to be amorphous. Amorphous silicon has found widespread applications in solar cells and optoelectronics.

Since structural order is determined largely by the mobility of the adatoms, the formation of amorphous films is promoted by all factors that limit surface mobility during the condensation and nucleation stages of growth. Without surface mobility, incident atoms or molecules would be accommodated at the point of impingement. Since incidence is a random statistical process, atomic arrangement in the film also becomes random. Some of the factors that limit mobility are:

a. Impurity stabilization: the presence of "mobility-inhibiting" gaseous impurities, such as oxygen-bearing species in the residual gas, results in fine-grained or amorphous films. The supercritical nuclei get covered with an oxide phase, which prevents coalescence and further growth.[10]

b. Substrate temperature: atomic mobility is reduced by a condensing the vapor on to a substrate at temperatures cold enough to reduce thermal diffusion completely. The temperature needed to produce amorphous films depends on the type of material being deposited.
c. Very high deposition rate: fine-grained and sometimes amorphous films are produced when the condensation rate is high, especially if the temperature is also low.

In all the discussions pertaining to temperature, the magnitude of the temperature is always relative to the melting point of the material. A useful quantity to remember is homologous temperature T_h, defined as the actual temperature divided by the melting point of the material, all measured in Kelvin. As a rule of thumb, $T_h < 0.3$ can be considered a low temperature, and $T_h > 0.7$ can be considered a high temperature.

For example, for the deposition of aluminum, $T_{melting} = 660°C$, a substrate temperature of 500°C is considered high because $T_h = 0.82$. The same temperature is quite low for the deposition of tungsten, $T_{melting} = 3\,410°C$, because $T_h = 0.2$.

2.2.2. Evolution of Order (Crystallinity)

Two processes that occur simultaneously during film growth result in a transition from a complete lack of order in amorphous films to the fine-grained polycrystalline state. The first is growth of crystallites by accommodation of incident material onto already formed islands. The second is a recrystallization process in which crystallites oriented in nonfavorable growth directions are consumed by grains growing in favorable directions. We alluded to this phenomenon during the island coalescence process. In the limiting case, isolated large crystallites coalesce to form a monocrystalline film resulting in epitaxial growth. We will study epitaxial growth in more detail in Chapter 10 with emphasis on the epitaxial growth of silicon. But now let us examine grain growth from nearly amorphous, fine crystals to films with few large grains. Reduction in the overall free energy the system through a reduction in the surface energy is once again the principal cause of grain growth.

Grain size during film growth is affected by the following parameters: (a) surface temperature, (b) reduced deposition rate, (c) temperature of the gaseous ambient and hence the velocity component parallel to the surface, (d) inertness of the surface, (e) film thickness, and (f) postdeposition annealing temperature, if any. Items (a) through (d) allow for significant atomic motion on the surface and hence larger islands. Film thickness acts as an upper bound for grain size, as we will soon see. To stabilize the film structure and chemical composition, annealing is often performed after deposition. Higher temperatures during annealing promote recrystallization.

2.2.3. Grain Growth Law during Recrystallization

Grain growth with no addition of material (as in the case of recrystallization or annealing) can be modeled after the growth of bubbles in soap froth.[11] As grains grow in size and the grain boundary area diminishes, the total surface energy is lowered. This reduction in energy is the driving force for the recrystallization and grain growth.

Using elementary physics, the pressure difference between the inside and outside of a spherical soap bubble can be expressed as

$$\Delta p = \frac{8\gamma}{D} \tag{2.15}$$

where D is the diameter of the bubble. The equation shows clearly that the smaller the bubble, the greater the excess pressure inside the bubble. The difference in pressure causes gaseous diffusion from the inside to the outside of the bubble. This argument can be extended to soap froth (Figure 2.6a) which closely resembles the cell structure of a polycrystalline film. The froth contains curved walls for individual cells. Figure 2.6b illustrates a single cell bounded by three walls. Across each wall there exists a pressure difference. The difference in internal pressure between adjacent cells is seen through the curvature of the cell walls. Gaseous diffusion causes the cell walls to move towards their respective centers of curvature, i.e., concave walls will move out and convex walls will move towards the center of the cell. The larger the curvature (i.e., the smaller the bubble), the faster the movement of the wall towards its center of curvature.

If we extend this logic to an aggregate of grains, we can write

$$dD/dt = K'c \tag{2.16}$$

where dD/dt expresses the change in the diameter of the grain, c is the curvature of the grain boundary, and K' is a constant of proportionality. If the curvature is inversely proportional to the diameter, then

$$dD/dt = K/D \tag{2.17}$$

where K is another constant of proportionality. Integrating this equation,

$$D^2 = Kt + c \tag{2.18}$$

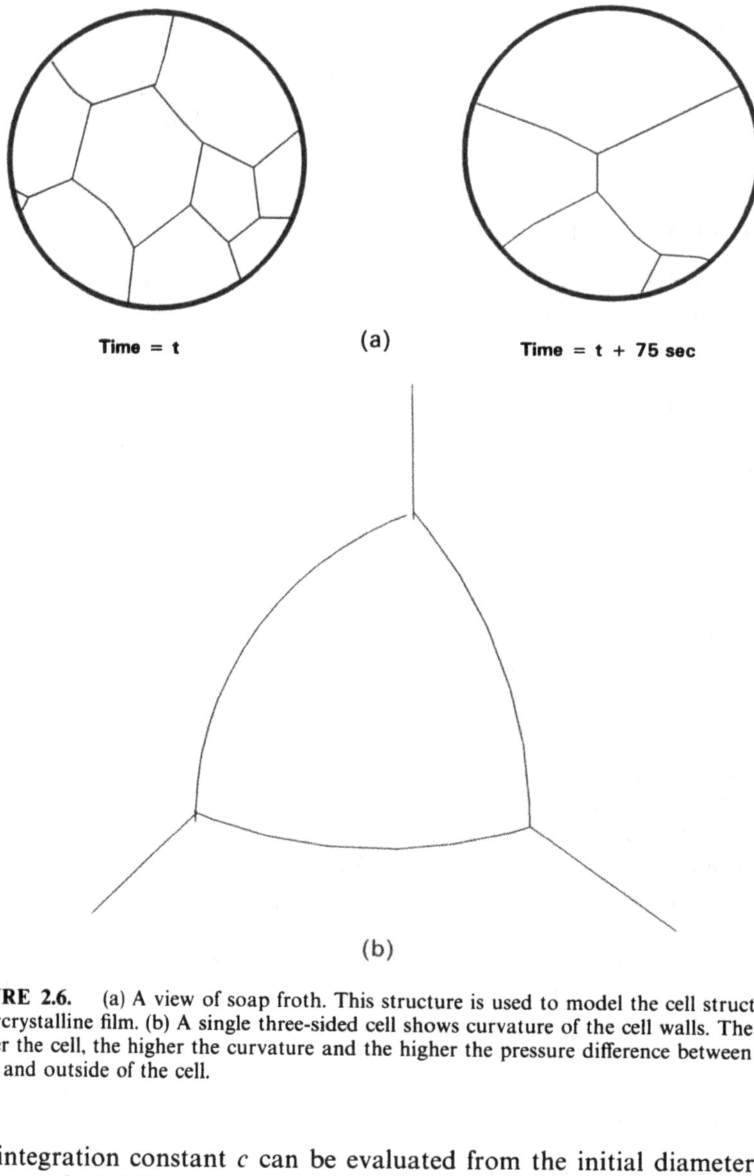

FIGURE 2.6. (a) A view of soap froth. This structure is used to model the cell structure of a polycrystalline film. (b) A single three-sided cell shows curvature of the cell walls. The smaller the cell, the higher the curvature and the higher the pressure difference between the inside and outside of the cell.

The integration constant c can be evaluated from the initial diameter and can be neglected if the initial grain size is very small. If the constant K, which arises due to a diffusive process, can be considered thermally activated with an activation energy Q, then equation (2.18) can be rewritten in terms of time and temperature as

$$D^2 - D_0^2 = k_0 t \exp(-Q/RT) \tag{2.19}$$

This expression has been experimentally confirmed by many researchers. For an isothermal situation with grains initially small, the grain growth law can be simply expressed as

$$D = kt^n \tag{2.20}$$

where n is approximately equal to 0.5.

2.2.4. Free Surface Effects on Grain Growth

Grain boundaries near the surface of the film tend to lie perpendicular to the surface, which results in a net reduction in their curvature. Since thin films might contain only one grain spanning the distance from the substrate to the free surface, this means the grains are more often cylindrical (or columnar) than spherical. Going back to the soap bubble analogy, cylindrical surfaces have a net pressure difference equal to $4\gamma/D$, as opposed to $8\gamma/D$ for spherical surfaces. Hence grain growth during recrystallization of the cylindrical grains is slower than the corresponding spherical grains.

A more important feature associated with grain boundaries and a free surface is called thermal grooving. Surface tension effects result in the formation of grooves or channels on the surface where the grain boundaries intersect the free surface. Transport of atoms along the free surface results in the removal of material to form the groove.[12] Figure 2.7 illustrates the surface tension forces that cause thermal grooving. Thermal grooves tend to

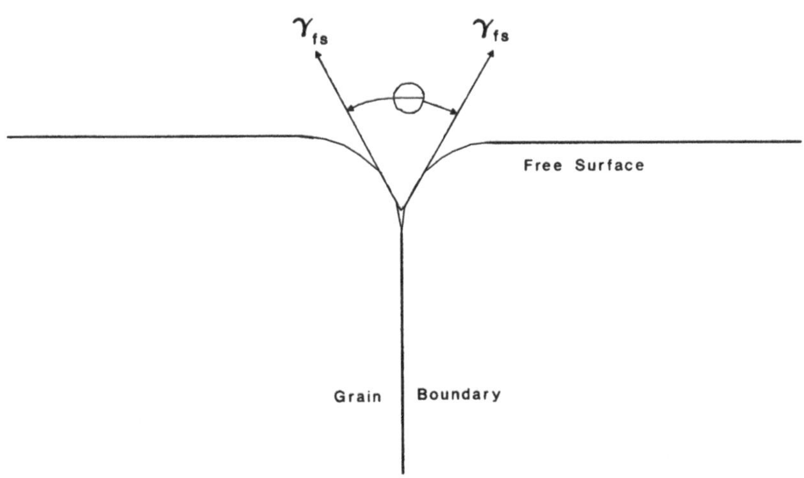

FIGURE 2.7. Surface tension equilibrium at a free surface between the grain boundary, and free surface free energies. Grooving of the boundary is caused by diffusion away from the point of high curvature.

anchor the ends of the grain boundaries, and in grain sizes that approach film thickness, retard further growth. In effect, the thermal grooving plays a strong part in determining the final grain size of the film. Deviations from the grain growth law are observed when the film grain size approaches film thickness.

Exercise 2.2

(1) The surface energy for copper grain boundaries in copper is about $650\,\mathrm{erg/cm^2}$. What is the effective internal pressure for a hypothetical spherical grain of diameter 1 m and a cylindrical grain of diameter 1 μm?
(2) For a metal following the ideal grain growth law with very small initial grain size, how does the driving force for grain growth change with time?

2.2.5. Other Structural Properties of Crystalline Films

Texture, or preferred orientation of grains, occurs when certain crystallographic planes are oriented in a preferred manner with respect to the surface normal of the substrate. Intensities of X-ray diffraction peaks can be used to detect texture readily, by contrasting the peak intensities from the film with a random powder diffraction pattern. The direct result of preferred growth directions is texture. Thin films usually exhibit some texture and it can be critical in determining the anisotropy of their diffusive, electrical, and magnetic properties.

Crystallographic defects, such as vacancies, twins, and dislocations, are commonly observed in thin films, significantly affecting their mechanical properties. Equilibrium concentration of vacancies in any material follows a simple exponential rule:

$$\frac{n}{N} = \exp(-E/kT) \tag{2.21}$$

where n is the number of vacancies and N is the number of lattice sites per unit volume. The activation energy E for the formation of vacancies in most materials is on the order of 1 eV. In thin films, n can be significantly larger, due to the presence of the free surface at close proximity. Conversely free surfaces also act as sinks for vacancies, and well-annealed films will exhibit very low vacancy concentrations, close to bulk equilibrium.

Dislocations in thin films arise due to many sources, including (a) island coalescence and the resultant misfit between two relatively large islands; (b) misfit dislocations between the substrate and the film, where the lattice spacing of the substrate slowly relaxes to the equilibrium film spacing; (c) stacking faults in the islands that result in partial dislocations, and (d) aggregation of vacancies. Dislocations get pinned at the surfaces and often result in a structure which looks work-hardened. Minimization of dislocations is especially important to epitaxial film growth where they can possess electrical activity. Other crystallographic defects such as twins and stacking faults are also of relevance when growing epitaxial films and hence will be treated in Section 10.2.

Exercise 2.3

The activation energies for the formation of vacancies in Al, Ag, and Au are respectively 0.76, 1.09, and 1.1. Their melting points are respectively 660°C, 961°C, and 1 063°C. Calculate the ratio of the equilibrium number of vacancies between room temperature and 5°C below their melting points. What conclusion can you deduce regarding order in films close to their melting points?

2.3. PROPERTIES OF THIN FILMS

Having introduced the phenomena that occur during the growth of thin films, let us explore the properties of thin films, from a structural viewpoint. We will examine the way physical, mechanical, electrical, and optical properties are affected by the film structure and microstructure. We will also examine deviations in material properties on going from bulk to thin film.

2.3.1. Physical Properties of Thin Films

Density or specific gravity of thin films can be determined through weight gain measurements using microbalances. We often observe an increase in density in thin films with increasing thickness, with the bulk density as the upper limit (Figure 2.8). This phenomenon has been attributed to the smaller grain size and the increased grain boundary area, which is normally less dense than the grain itself. More often, thinner films tend to contain microscopic voids that decrease as a percentage of the film volume as the film grows thicker. Tendencies that lead to crystallographic perfection generally lead to an increase in film density. Density of the film often correlates to resistivity and other film properties affected by the presence of voids. Amorphous films are less dense than their crystalline bulk counterparts; their density increases as they become more crystalline.

Surface roughness arises from the random nature of nucleation and coalescence. Deviation from the average thickness Δt for films grown at

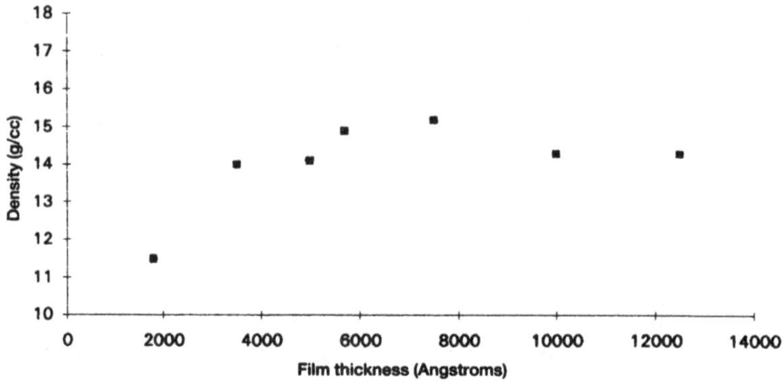

FIGURE 2.8. Density change in a thin tungsten film as the film thickness is increased.

relatively low temperatures and at limited surface mobility can be modeled according to a Poisson distribution.

$$\Delta t \propto \sqrt{t} \tag{2.22}$$

In characterizing roughness, both Δt and the periodicity of the peaks and valleys need to be accounted for. Various optical scattering and surface profilometric techniques have been developed to characterize roughness.

Another contributor to surface roughness is the presence of surface grooves. Since higher temperatures often result in larger grooves, a direct relationship can be observed between roughness and temperature of growth. In this situation, measurement of surface roughness yields a direct measure of the film grain size distribution. Surface roughness acts as an excellent indicator of surface contamination. Often deviation in film roughness can be directly attributed to leaks in the deposition system and the presence of gaseous impurities such as oxygen.

2.3.2. Mechanical Properties of Thin Films

Thin films exhibit remarkable levels of stress, often much higher than the bulk tensile strength of the material. Slip is restrained by the pinning of dislocations at the two surfaces; propagation of microcracks is similarly restrained. However, excessive tensile stress can destroy the film by producing cracks and excessive compressive stress can produce buckling. Stresses can also cause delamination from the substrate and alter the band gap of semiconductors.

Stresses in thin films consist of an intrinsic component and a thermal component. The thermal component is easy to model; it arises from the difference in thermal expansion between the film and the substrate. During cooling from the growth temperature, the film assumes a tensile stress state if the film wants to contract more than the substrate will allow. Conversely, the film assumes a compressive stress state if the substrate contracts more than the film wants to.

The magnitude of this component of stress, S_{th}, can be expressed as[13]

$$S_{th} = \frac{E_f}{1 - \nu_f} \int_{T_1}^{T_2} (\alpha_s - \alpha_f) dT \qquad (2.23)$$

the subscripts f and s denote the film and the substrate respectively. $E/(1 - \nu)$ is the compliance term for thin films where E is the modulus of elasticity and ν, the Poisson's ratio.

The intrinsic component of the stress is complex in origin; intrinsic stress arises probably due to one or more of the following factors:

a. Incorporation of intrinsic CVD impurities in the lattice, or chemical reactions after the atoms have been accommodated.
b. Difference in lattice spacing between the film and the substrate (particularly important for epitaxy).
c Variation of interatomic spacing with crystallite size.
d. Recrystallization and grain boundary movement.
e. Microscopic voids and special arrangements of crystallographic defects.
f. Phase transformation in the film.

Any of these might occur during film growth or during subsequent annealing. Both compressive and tensile stresses have been observed in CVD films. We will discuss means of modulating stress levels in individual film types in Chapters 8 through 10.

Stresses in thin films can be modeled as simple composite plates. This is true only for continuous films without holes and spaces (in contrast to the way they are often found in microelectronic applications). Detailed relationships near edges are not well known. Figure 2.9 shows a general schematic of the stress state of thin films. Shear stresses occur near the film edges and are assumed to be negligible further away. Stresses in thin films are considered biaxial, present only in the plane of the film. Imagine this to be similar to an inflated balloon. All of the stresses in the skin of the balloon are only in the plane of the balloon surface at every location, even though there is a hydrostatic gas pressure inside the balloon. Stresses are assumed to be uniform along the thickness of the film; this may not be rigorously true but it is useful as a mathematical simplification.

Structural imperfections, such as grain boundaries, vacancies, and dislocations, inhibit the movement of other dislocations, increasing the yield stress

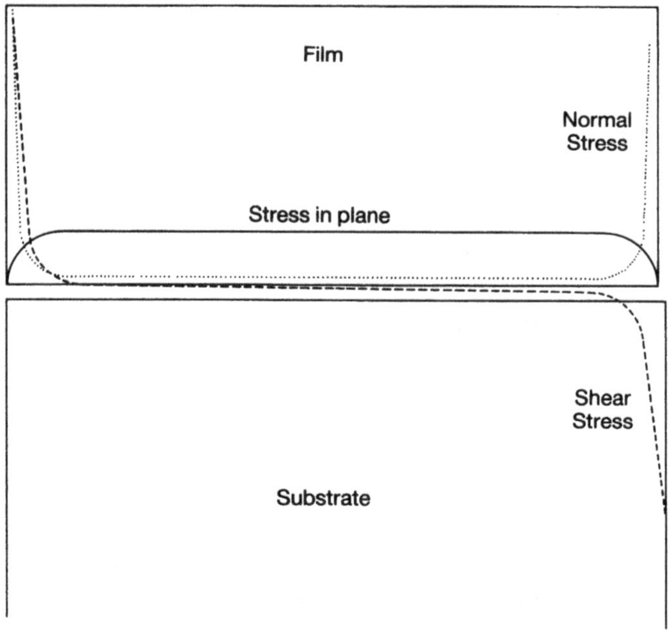

FIGURE 2.9. Schematic representation of stress in thin films attached to substrates. Shear stresses decay away from the edges. Normal stresses are considered uniform along film thickness.

of bulk materials. In particular, Hall and Petch have empirically established the relationship between tensile yield stress and grain size in bulk materials.

$$\sigma_0 = \sigma_i + kD^{-0.5} \tag{2.24}$$

where σ_0 is the yield stress and σ_i is normally interpreted as the frictional stress opposing motion of a dislocation.[14] D is the grain diameter and k is a constant of proportionality related to dislocation pile-up. This empirical relationship holds for thin films, even though the exponent can vary between 0.3 and 0.66.[15]

Measurement of stress in the thin film can often be accomplished by measuring the curvature of the substrate, either as a disk or as a strip. Interference rings, laser holography, traveling microscopes, and optical curvature measurement techniques have been used to measure curvature of disks and deflection of the strips.[16]

Central to all microelectronic applications of thin films is the assumption that the thin film is strongly adherent to the substrate. The level of stresses in the film strongly affects adhesion. Despite complex tests that have been developed to test adhesion of films, the most effective method to measure adhesion has been the adhesive tape test. Applying and peeling off a strip of adhesive tape on a freshly scratched film surface can provide a semi-quantitative measure of adhesion.

There are various theories on the origin of adhesion. Two components to the adhesion of the film arise from mechanical bonding and chemical bonding between the film and the substrate. Mechanical bonding is relatively weak and does not provide sufficient adherence. Chemical bonding is promoted by (a) van der Waals' forces, (b) diffusion of the film into the substrate forming a transition region, (c) chemical reaction between the film and the substrate, and (d) ion bombardment of the surface, which increases surface defect density and enhances the binding energy.

Surface cleanness is often the key player in the promotion of adhesion. Depending on the nature of the impurity, adhesion might degrade or improve but, in general, the cleaner the surface, the better the adhesion.

Exercise 2.4

The yield stress of titanium measured on samples of different grain sizes are summarized below. Calculate the parameters in the Hall–Petch relationship.

Grain size (μm)	Stress (kg/mm^2)
1.1	32.7
2	28.4
3.3	26
28	19.7

2.3.3. Electrical Properties of Thin Films

Electrical properties of very thin discontinuous films can be very complex and are of little value to microelectronic applications. We will concentrate only on thicker films, once the films are continuous. And we will consider the properties of conductor, semiconductor, and dielectric films separately, since each type has its peculiarities.

Conductive Films

Conductivity, σ_B, of a material can be derived from first principles to be equal to

$$\sigma_B = \frac{Ne^2\lambda_0}{mv} \tag{2.25}$$

where N is the number density of conduction electrons, e is the electronic

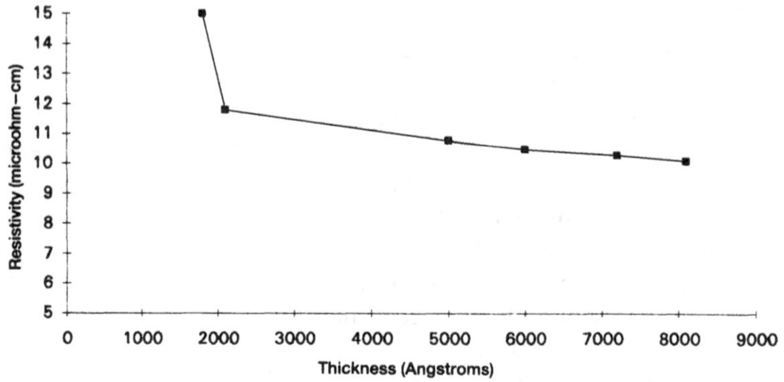

FIGURE 2.10. Change in resistivity of a CVD tungsten film as the film thickness is increased.

charge, m is the electronic mass, and v is its mean thermal velocity.[17] Significant changes in λ_0, the mean free path of the electron in a bulk single crystal material, produce significant changes in conductivity. At film thicknesses smaller than λ_0, scattering of the electrons results in increased resistance. This phenomenon is known as *size effect* or *thin film effect*. Figure 2.10 shows the resistivity of a CVD film as a function of film thickness. The increase in resistivity of the thinner films is very evident.

According to Matthiessen's rule,[18] various electron scattering processes and hence their contribution to resistivity, ρ_f, are additive, i.e.,

$$\rho_f = \rho_B + \rho_S + \rho_I \tag{2.26}$$

where the subscripts B, S, and I refer to contributions to the resistivity from ideal bulk lattice, surface scattering, and imperfections, including impurities, respectively. Each of these components can be calculated for idealized situations fairly accurately.

In particular, the effect of columnar grains on resistivity can be modeled by considering the grain boundaries as internal, specular, electron reflecting surfaces. The result of incorporating grain size into the resistivity of a pure film can be written as

$$\rho = \rho_0\left(1 + \frac{3}{2}\frac{R}{1-R}\frac{\lambda_0}{D}\right) \tag{2.27}$$

where D is the average grain diameter and R is a scattering coefficient equal to 0 for bulk ideal material and 1 for a very thin film with no specular scattering.

A useful quantity to remember in the case of thin films is sheet resistance

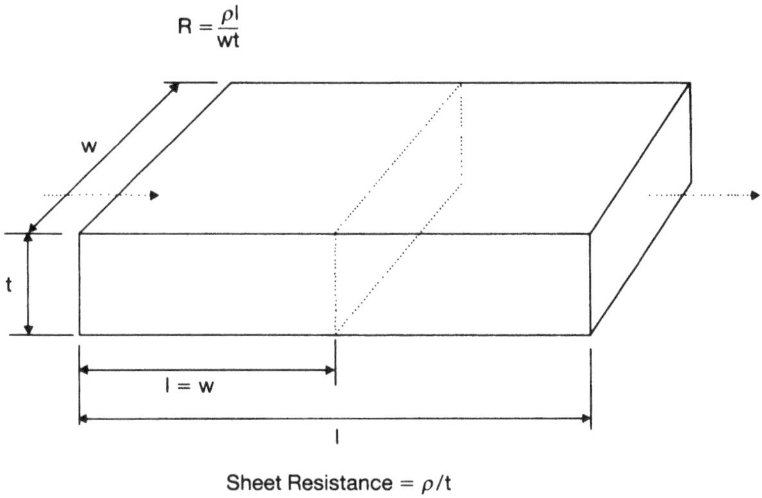

FIGURE 2.11. Squares of any size with identical thickness have the same electrical sheet resistance.

R_s (Figure 2.11). In the fundamental resistance equation

$$r = \rho l/th \tag{2.28}$$

where l, t and h are the length, height, and thickness of a rectangular wire. If two of the dimensions, say l and h, are equal (forming a square), then

$$r = \rho/t = R_s \tag{2.29}$$

Formally, sheet resistance is defined as the resistance of a square of arbitrary sides with a finite thickness. Since all squares, irrespective of their dimensions will have the same resistance for a given thickness, R_s has the units of Ω/square. Sheet resistance is easy to measure, and scales inversely as thickness. In most situations where measurement of thickness is cumbersome, sheet resistance is often used as a direct indicator of thickness.

Another phenomenon that affects the reliability of conductive films is electromigration. Electromigration is the transport of lattice atoms in the direction of electron flow in the conductor, due to momentum transfer from the electrons to the lattice atom. At large current densities and at higher temperatures, the effect can be very pronounced, resulting in void formation at one end of the conductor and material accumulation at the other. This produces opens and shorts in the electrical circuit, and is an important failure mechanism for lighter metals such as aluminum. Detailed discussions regarding the mechanism and failure modeling are beyond the scope of this book; the reader is referred to some excellent reviews on this subject.[19]

Semiconducting Films

The effect of the surface in modifying electrical properties is even more pronounced in semiconducting films than in metallic films. Small changes in impurity concentration or defects in the crystal lattice can affect conductivity significantly. From first principles, the conductivity of a semiconductor is given by

$$\sigma = ne\mu \tag{2.30}$$

where n is the carrier concentration, e is the electronic charge, and μ is the electron mobility.[20] Carrier mobility in a semiconductor can be formally defined as

$$\mu = e\bar{t}/m^* \tag{2.31}$$

where $\bar{t}$ is the average time between scattering events in the lattice and m^* is the effective mass of the carrier in the band. Size effects in the semiconducting film contribute to an additional scattering component to the resistivity. If we define a constant ξ such that

$$\xi = t/\lambda \tag{2.32}$$

where t is the thickness of the film and λ is the electron mean free path, then carrier mobility in a thin film can be written as

$$\mu_f = \frac{\mu_0}{1 + \dfrac{1}{\xi}} \tag{2.33}$$

Thus the carrier scattering at the surface serves to effectively reduce carrier mobility.

Another important effect in semiconducting films is due to the bounding surfaces. The presence of the surfaces results in a surface charge which in turn causes bending of energy bands closer to the surface.[21] Detailed discussion on band bending is beyond the scope of this book. In the chapter on epitaxial silicon growth, we will be mentioning band bending again.

More often the conductivity of the semiconducting films in microelectronics is modulated by the presence of p-type or n-type dopants. For silicon films, boron serves as the p-type dopant, while phosphorus, arsenic, and antimony are used as n-type dopants. The dopant concentrations are very small but they play a dramatic role in increasing carrier concentrations and hence the conductivity of the semiconductor.

Dielectrics

As a class of materials, dielectrics exhibit large energy gaps in their band structure, with few free electrons to participate in electrical conduction. Band gaps in dielectrics can be on the order of a few electron volts. The important characteristics of the dielectric that affect its usefulness in microelectronic application are the dielectric constant, the breakdown strength, and the dielectric loss.

Dielectric constant or permittivity is a measure of the amount of electrical charge a material can withstand at a given electrical field strength, not to be confused with dielectric strength. For a nonmagnetic, nonabsorbing material, the dielectric constant is the square of the index of refraction. A vacuum, the perfect dielectric, has a dielectric constant of unity.

The capacitance C in farads of a parallel-plate capacitor shown in Figure 2.12 with surface area A and a dielectric of thickness t in centimeters is given by

$$C = \frac{\varepsilon_0 \varepsilon A}{4\pi t} = 8.85 \times 10^{-14} \varepsilon A/t \qquad (2.34)$$

A high dielectric constant is required to obtain high values of capacitance for a storage capacitor in a DRAM. Interconnect applications require low capacitance between adjacent metal lines. Even though smaller thicknesses

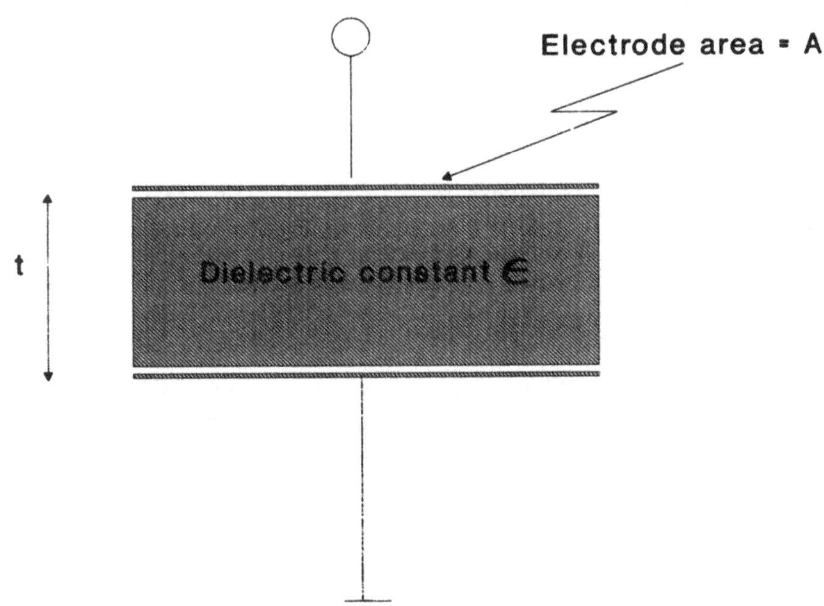

FIGURE 2.12. A parallel-plate capacitor.

can result in a high value of the capacitance, difficulty in obtaining continuous and structurally stable insulators limits the practical lower limits in thickness to about 100 Å.

Dielectric strength or breakdown strength is a measure of the resistance of the dielectric to electrical breakdown under the influence of strong fields (usually expressed in V/cm). Structural integrity of the insulator, the presence of pinholes and metallic contaminants reduce the dielectric strength.

Dielectric loss is a measure of frictional loss, dissipated as heat, in the presence of a varying electric field. The loss occurs because electric polarization in a dielectric is unable to follow the electric field.

Continuous dielectric thin films have remarkably similar properties to bulk materials down to very small thicknesses (on the order of 100 Å). Mechanisms of charge transport and dielectric breakdown behavior in thin films are similar to bulk effects and the reader is referred to many excellent books on the general properties of dielectrics.[21] The few exceptions that arise are in the case of avalanche breakdown, where irreversible dielectric damage in thin films can be contained due to the thickness of the film being smaller than the mean free path of the electrons. Hence the electron is unable to be accelerated in the field to cause the avalanche phenomenon.

2.4. SPECIAL PROPERTY REQUIREMENTS FOR MICROELECTRONICS

Even though we have introduced general properties of thin films, such as mechanical characteristics and electron transport, there are certain unique requirements for thin films in microelectronics. These requirements are extensions of the properties discussed in the preceding sections. However, they possess industry-specific definitions and means of measurement that need to be explicitly stated.

2.4.1. Conformality and Step Coverage

Conformality of a thin film refers to its capability to exactly reproduce the surface topography of the underlying substrate. Conditions during growth and subsequent annealing, along with intrinsic properties of the material determine conformality. In some cases, geometrical constraints of the substrate topography preclude conformality. These concepts are illustrated in Figure 2.13. The need for conformality arises because microelectronic processing proceeds by successively depositing and patterning features on thin films. If successive films do not follow the patterns created on the previous layers, voids in deposited layers begin to form. Etching these layers may

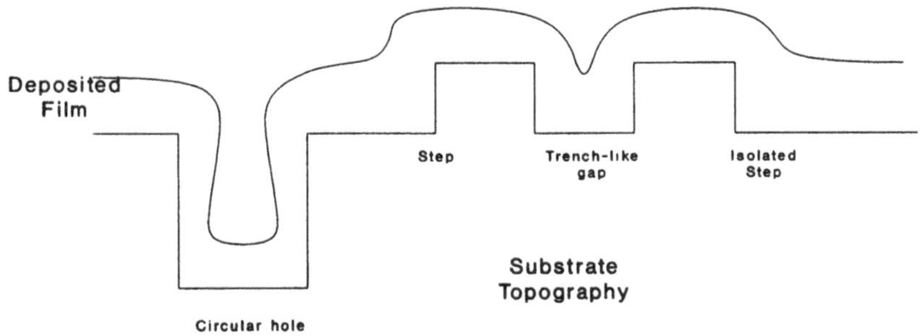

FIGURE 2.13. Microelectronic films need to follow the contours of substrates faithfully.

result in stringers. These can lead to electrical shorts and opens, or to failures caused by trapped material in the voids.

CVD's ability to produce conformal coatings is largely responsible for its widespread use in the microelectronics industry. Physical vapor deposition (PVD) by evaporation or sputtering produces films with thicknesses dependent on the angle of exposure to the incident flux; CVD films have no such dependency, due to the pressure regime of operation.

It is often assumed that surface mobility can contribute significantly to conformality. However, even though surface mobility is a necessary condition for good step coverage, it is far from being sufficient. For example, very heavy atoms with relatively little lateral mobility often exhibit excellent conformality. The rate-determining step in the sequence of subprocesses for CVD plays a key role in determining conformality. As a rule of thumb, if the rate of arrival of reactant species at all locations exceeds the rate of reaction on the surface at all locations, good conformality can be expected. This is not an easy condition to satisfy. We will revisit this condition during the discussion on reactor design.

Conformality over a right-angled step is termed step coverage. It refers to the thinning of the film when covering a surface normal to the substrate plane. The situation is exacerbated in a cylindrical hole with a relatively small hole diameter. Step coverage terms are defined in Figure 2.14. The electron micrographs in Figure 2.15 show aluminum films grown by PVD and CVD techniques on the same underlying topography. Notice how well the CVD film conforms to the underlying topography.

2.4.2. Planarity

A related film property to conformality is planarity. Lithographic imagers are used in microelectronic manufacturing to pattern very small features on the substrate. These imaging tools have limited depths of focus and hence

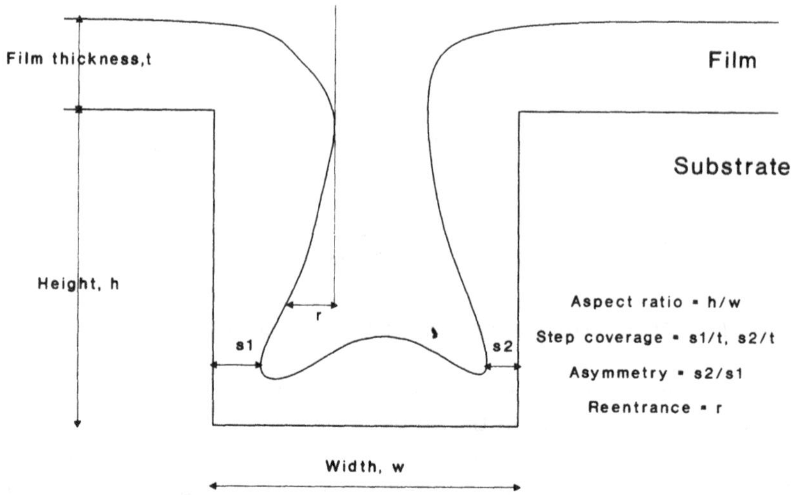

FIGURE 2.14. Step coverage and the related terms.

require that each successive layer is sufficiently planar. Planarity, as explained in Figure 2.16, can be achieved by many means, some of which are unrelated to CVD. However, one of the techniques used in planarization is the thermal flowing of doped oxide glasses deposited by CVD under elevated temperatures. Figure 2.17 shows a cross-sectional electron micrograph of a thermally deposited phosphosilicate glass deposited by CVD before and after flowing at a temperature of 800°C. Notice the improvement in planarity.

2.4.3. Particles

Since the feature size in a microelectronic device is very small, constraints on contaminant particles falling on the substrate are stringent. Densities of particles are expressed in terms of defects per square centimeter and can be as low as 0.01 for particles of diameter 0.05 μm. The sources of particles, even in the clean environment of a microelectronic manufacturing facility can be classified into human, environmental, equipment related, and process related. It is the last two that we will be repeatedly referring to in the rest of the book. Equipment related particles arise from the handling of the wafers in the process equipment. Process related particles are produced along with the film, and in CVD they can result from delamination of film, previously deposited on chamber walls, or from homogeneous nucleation in the gas phase.

Particles present on the surface tend to grow in size as further conformal

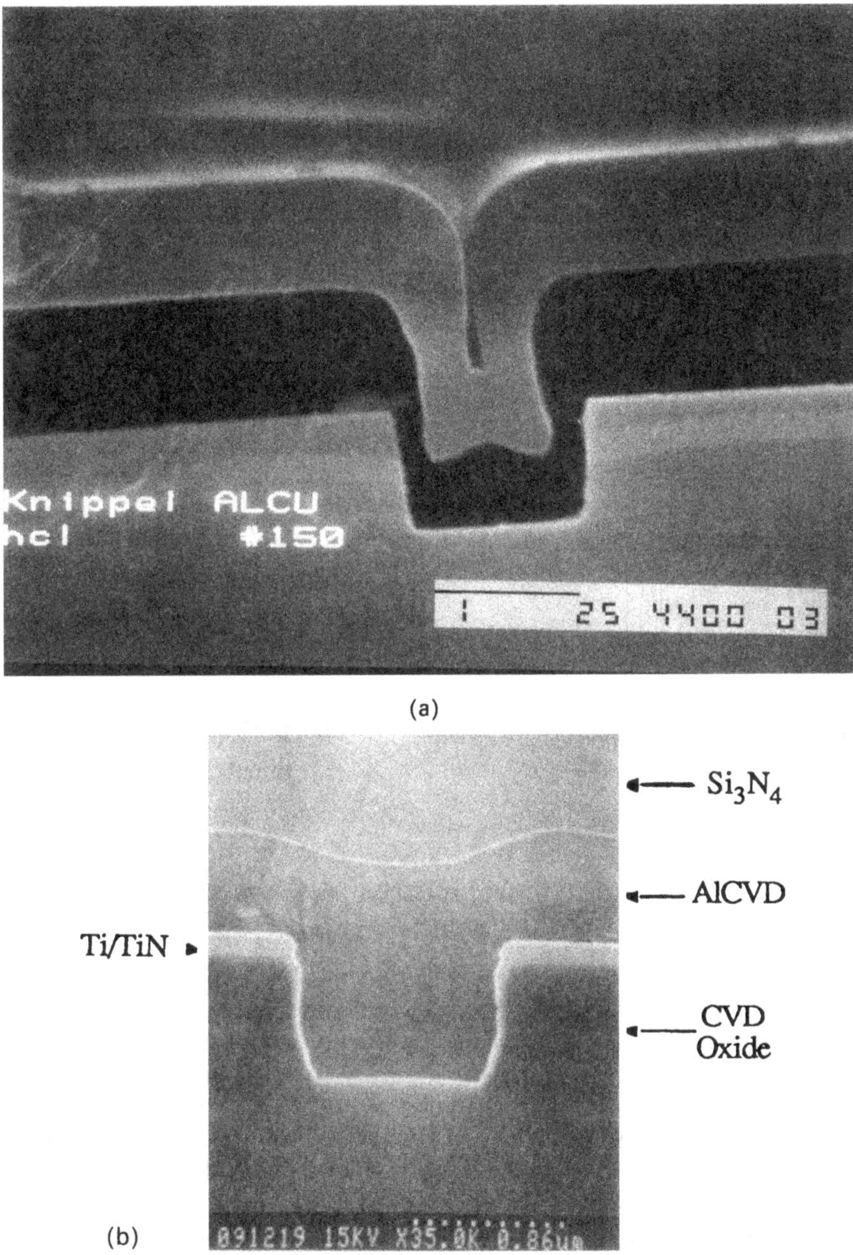

FIGURE 2.15. Aluminum films grown by (a) chemical and (b) physical vapor deposition techniques. The superior conformality of CVD films is evident. Reprint from the 1989 Proceedings of the VLSI Multilevel Interconnection Conference, p. 128.

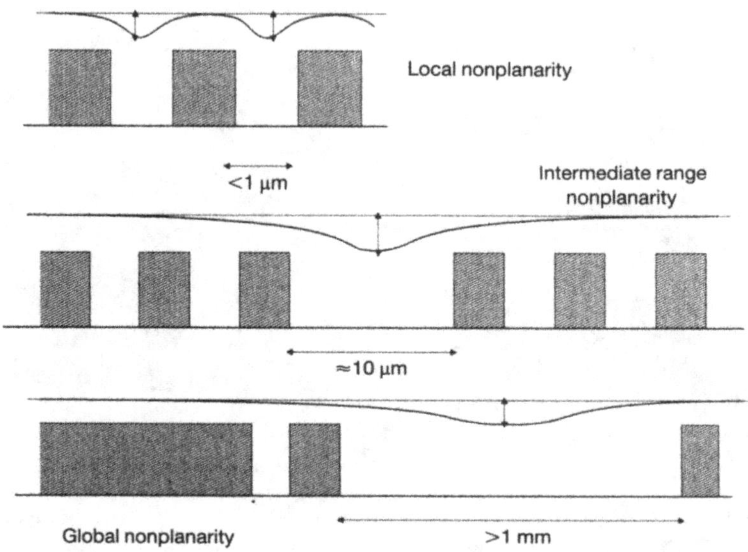

FIGURE 2.16. Planarity in the local, intermediate, and global senses required in microelectronic processing.

films are deposited; this greatly aggravates the problem. During process qualification, particle reduction is probably the single most time-consuming exercise for a CVD development engineer. Identifying and reducing the source of particles is central to any microelectronic processing step.

2.5. REVIEW

In this chapter we followed the growth of a thin film, starting with atomic impingement from the gas phase. We covered accommodation of the species on the surface, nucleation, and the eventual coalescence of the nuclei and islands into a continuous film. Addition of further material to the thin film produced steady-state growth, which could be amorphous or crystalline, based on the growth conditions. We also studied the evolution of order and the laws of grain growth.

Next we examined thin film properties, physical, mechanical, and electrical, and how they are affected by finite film thickness. We went on to define the requirements of microelectronics, such as step coverage, planarity, and defects, and we considered how they might be measured.

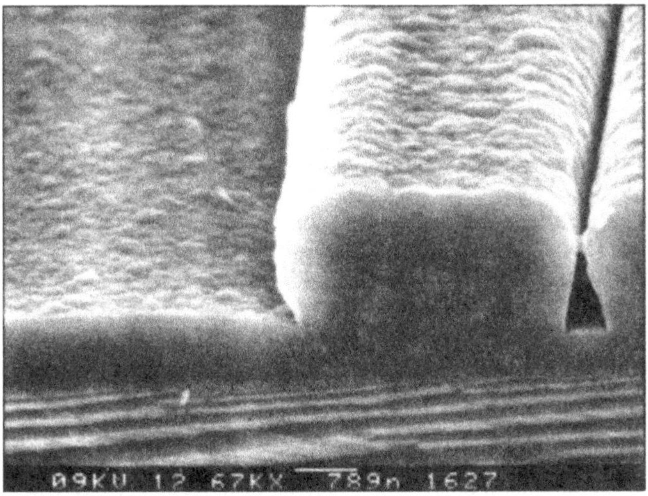

(a)

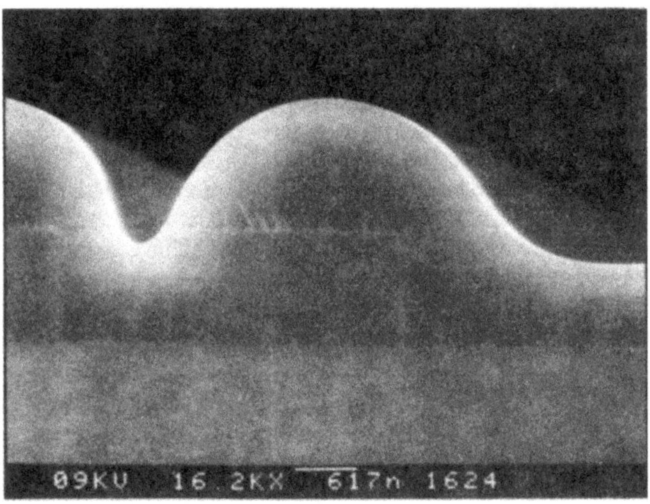

(b)

FIGURE 2.17. Reflow achieves planarity in doped oxide films: (a) before and (b) after reflow. Reprinted with permission from Watkins Johnson, Inc.

References

1. J. M. Blocher, Jr., "The role of intrinsic impurities in CVD," paper presented at the Gordon Research Conf. on Thin Films, New Hampton, N.H., Aug. 22 to Sep. 2, 1966.
2. J. P. Hirth, in *Vapor Deposition* (Powell, Oxley, and Blocher, eds.), Ch. 5, John Wiley, New York, 1966.

3. H. S. W. Massey, and E. H. S. Burhop, *Electronic and Ionic Impact Phenomena*, Ch. 9, Oxford University Press, London, 1952.
4. B. McCarroll, and G. Ehrlich, in *Condensation and Evaporation of Solids* (Rutner et al., eds.), p. 521, Gordon and Breach, New York, 1964.
5. L. Eckertova, *Physics of Thin Films*, 2nd Ed., Plenum, New York, 1986.
6. J. P. Hirth, and G. M. Pound, *Condensation and Evaporation*, Macmillan, New York, 1963.
7. J. P. Hirth, and K. L. Moazed, in *Physics of Thin Films* (G. Hass and R. Thun, eds.), Vol. 4, p. 97, Academic Press, New York, 1967.
8. R. D. Gretz, and J. P. Hirth, Nucleation and growth processes in CVD, in *Proc. Conf. Chemical Vapor Deposition of Refractory Metals, Alloys and Compounds*, 1967, p. 73.
9. E. N. Andrade, *Trans. Faraday Soc.* **31**, 1137 (1935).
10. H. L. Caswell, *J. Appl. Phys.* **32**, 105, 2641 (1961).
11. J. E. Burke, and D. Turnbull, *Prog. Metal Phys.* **3**, 220 (1952).
12. W. W. Mullins, *J. Appl. Phys.* **28**, 333 (1957).
13. K. L. Chopra, *Thin Film Phenomena*, p. 272, Robert E. Krieger Publishing Company, Malabar, Fla., 1985.
14. E. O. Hall, *Proc. Phys. Soc. London* **B64**, 744 (1951); N.J. Petch, *J. Iron and Steel Inst.* **174**, 25 (1953).
15. R. W. Hoffman, in *Physics of Thin Films* (G. Hass and R. Thun, eds.), Vol. 3, p. 211, Academic Press, New York, 1966.
16. L. I. Maissel, and R. Glang, *Handbook of Thin Film Technology*, Ch. 12. 3, McGraw-Hill, New York, 1983.
17. J. M. Ziman, *Electrons and Phonons*, Oxford University Press, Fair Lawn, N.J., 1962.
18. A. H. Wilson, The *Theory of Metals*, Cambridge University Press, New York, 1958.
19. H. B. Huntington, and A. R. Grone, *J. Phys. Chem. Solids* **20**, 76 (1961).
20. B. Streetman, *Solid State Electronic Devices*, 3rd Ed., Prentice-Hall, Englewood Cliffs, 1990.
21. C. Kittel, *Introduction to Solid State Physics*, John Wiley, New York, 1966.

Chapter 3

Manufacturability

This chapter is different from the rest of the book. It is not directly pertinent to the art or science of CVD; instead it deals with the development process. It provides a framework to increase the efficiency with which processes can be made manufacturable. I believe this framework should be a prerequisite for all process/equipment engineers who hope to work in the microelectronic industry.

The chapter begins with an overview of the model of a process engineering environment and proceeds to provide a basic understanding of the statistical methods applied to the different components of the model. There follows a hypothetical example of process development using a CVD reactor and finally an overview of properties of thermal, plasma, and photo CVD processes.

3.1. QUALITY AND MANUFACTURABILITY OF PROCESSES

The term *manufacturability* has many connotations; perhaps it implies all of the following properties:

a. Reduced time to market a product.
b. Reproducibility to predefined specifications.
c. Least cost to make the product to specification.
d. Wide margins for error.
e. Safe for people and the environment.

Given the stringent requirements for ensuring manufacturability, it is essential that the process used to produce the thin film be defined properly when it is developed the first time. Refining or redeveloping a process that has already been implemented into a manufacturing line for reasons that it is inadequate in any of the above properties is often uneconomical. The term *quality* is applied to this concept of getting the process right the first time. This chapter deals with the principles of statistics, especially with reference to statistical design of experiments and statistical process control, such that the process developed can be robust, and right the first time!

Manufacturing involving large number of units is a typical feature of the microelectronics industry. For instance, integrated circuit manufacturing involves fabrication facilities producing 5 000 or more wafers a month, each wafer containing more than 100 dies and each die consisting of millions of transistors. Such large numbers are well suited to the application of the theories of probability and statistics that can predict the performance of the products. Hence ensuring the quality and manufacturability of a thin film process used in the microelectronics industry is facilitated by the application of the concepts of statistics.

3.2. DEFINING THE ENVIRONMENT

Before actual development of a process for depositing the CVD film, it is essential to define the entire *environment* in which the process is going to be developed. Environment means (a) the objective of the development process, (b) the constraints on the process development with respect to resources and time, (c) the application of the thin film, (d) the film specifications, and (e) the constraints that exist on the film properties.

In a typical process development situation, delivering the performance of a thin film has to be balanced against the time schedule and the resources available to complete the project. Hence these three factors—cost, schedule and performance—provide the constraining envelope for process development.[1] The researcher or a team of engineers works within this envelope as shown in Figure 3.1. One of the first roles of the researcher/process engineer is to decide which of these three constraints will be the deciding concern. At any given instant in the process development cycle, all three of these constraints cannot be made limiting. Efficient project management requires at the outset a decision on whether cost, schedule, or performance takes precedence. Having said that, we will concentrate only on the performance aspect of development effort, since cost and resources are outside the scope of this book. Process and equipment responsibilities should not be considered as distinct; the performance of one affects the performance of the other. It is only to emphasize the two roles that Figure 3.1 shows them as separate entities.

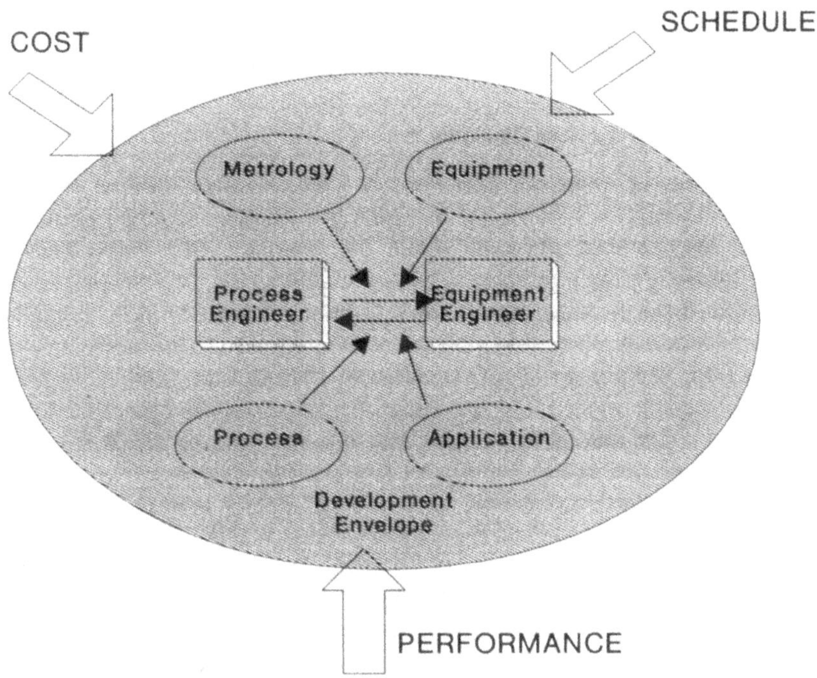

FIGURE 3.1. A process development envelope constrained by performance, cost, and schedule.

3.2.1. Defining the Process Objective

A concise statement of the objective of the CVD film often assists with the design and execution of experiments. This is applicable to all phases of process development: during the exploratory research phase, and while improving the manufacturability of a process whose feasibility has been demonstrated. The objective statement (or charter) should explicitly state the scope of the CVD process development. It should also mention the application of the thin film. An example of an objective statement is

> To deliver by July 2001, a dielectric passivation film capable of protecting 0.5 micron devices from moisture and sodium contaminations.

The objective statement does not necessarily specify the composition of the film, or all its specifications. For instance, the researcher is allowed the choice of using Si_3N_4, PSG, silicon oxynitride, or any other film as long as it meets the moisture and sodium requirements. It mentions specifically the time constraint on the project, after which it is not economically viable to use the process to make a product. Such an objective statement allows the development team to clearly focus its resources on an explicitly stated goal.

3.2.2. Determining the Specifications

Accompanying the objective statement is a list of specifications (or specs). These are the success criteria for the film, the process, and the equipment, and often they depend on each other. For instance, film stress might be reduced by using a high temperature process. However, deposition temperature is often fixed by the process specs; for the passivation film it might be less than 400°C. Similarly the use of certain process conditions might be precluded by equipment specs; equipment throughput might dictate the required process pressure. The specifications list should also provide a range of values on either side of the mean value of a required property, if possible, providing lower and upper bounds for the success criteria.

Film, process, and equipment specs should not be treated lightly. They are the goals on which the success of the process is measured. Specs for the process should not be chosen as convenient round numbers, such as uniformity less than 10%. Specifications are true requirements of the manufacturing process. In a manufacturing line, the specs for each processing step should predict its effect on the downstream processes and incorporate any effects of the upstream processes. Choosing very loose specifications can result in a process that is not manufacturing worthy. Choosing very tight specifications can result in unnecessary wastage of time and resources. Before

TABLE 3.1 A Sample Specifications Table

	Property	Upper spec limit	Lower spec limit	Measurement
1.	Film			
	Thickness (Å)	5 300	4 700	Interference
	Stress (GPa)	0.3	0.5	Optic lever
	Thickness non-uniformity (%)	8	none	Interference
	Moisture permeability	0	none	Device reliability
	Dielectric constant	1.9	2.1	Ellipsometry
2.	Process			
	Temperature	400°C	none	
3.	Equipment			
	Throughput (wafers per hour)	none	20	
	Utilization capability	none	60%	
	Wafer breakage	1/1000	none	

defining specifications, make sure to understand the implications of process integration. Often it is beneficial to arrange the specifications in an order of importance; some might be essential, while others might be only desirable.

Specifications are not stand-alone documents. They have to be accompanied by definitions of how they are to be measured, including the measurement tool and its measurement capability. We will address this issue again in gauge capability studies. Table 3.1 shows the specifications for the hypothetical passivation stated earlier.

3.3. PROCESS DEVELOPMENT SEQUENCE

Allowing for exceptions, our hypothetical process to deposit a passivation film will proceed according to the following general sequence of events, once the objective and specifications have been defined.[2]

a. Initial exploration phase; different film compositions, their advantages, and disadvantages are compared in a paper study based on an available knowledge base. One or two possibilities are isolated, and innovative experimentation on these choices yields feasibility of the process for our film. This phase of the projects is constrained by film performance, not by time or resources.
b. Design of experiments phase; carefully planned experiments result in statistically valid responses as functions of process variables for the process chosen in step (a). A series of experiments yields the response surface in the output parametric space of the items listed in the specifications list. The input parametric space is called the list of variables. An operating point is chosen from the response surface.
c. Measurement of process capability; where an optimized process chosen based on the response surface in step (b) is measured against the specifications. This might include passive data collection (PDC). Based on the results, the process either moves on to step (d) or iterates back to step (b). Figure 3.2 schematically illustrates the concept of process capability.
d. Process qualification; once the process capability is established, the process is qualified by repeating it without any changes so as to obtain statistical process control charts. These establish the natural variability of the process, according to which the process development is either complete or iterates back to step (b). Figure 3.2 also illustrates the difference between process capability and process control.

We will leave the exploration phase to the end, where we give an overview of CVD applications. First we address steps (b), (c) and (d).

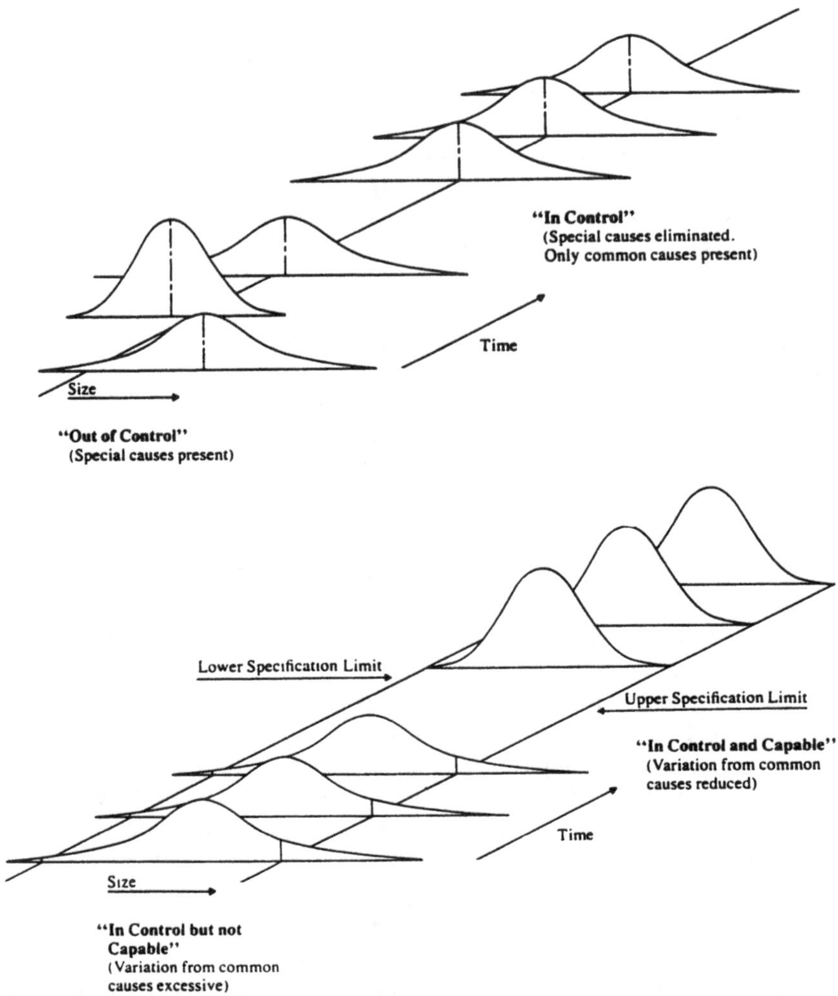

FIGURE 3.2. Process capability is the ability of a process to deliver a specified performance. Process control ensures a capable process repeatedly delivers that specified performance.

3.3.1. Design of Experiments

Once a clear choice of film and a method of growth have been established through the feasibility studies, the process is refined using statistical design of experiments. A process can be optimized to meet the specifications only if it is technically feasible, hence the exploratory stage is crucial in all the following steps.

The first objective of process optimization is to understand what input parameters significantly affect the performance of the CVD film. Using the example of the passivation film, let us assume our exploration phase has dictated a plasma assisted deposition of Si_3N_4.

There is an extensive list of variables in the deposition of this film: the flow rates of the different gases, the process pressure, the temperature of the substrate, the bias used on the wafer, the input power, etc., maybe more than 15 variables. The responses are the parameters listed in the specification

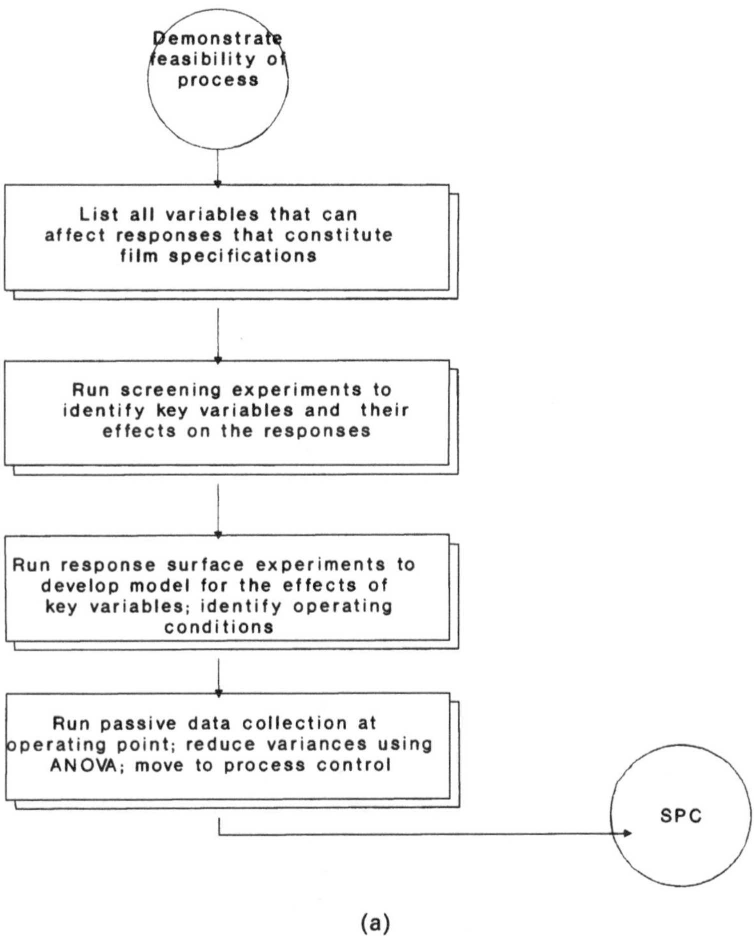

FIGURE 3.3. (a) Overall flow of process development using experimental design principles. The following series of figures illustrate each of the steps.

run	Power (Watts)	Pressure (mTorr)	Temp (DegC)	Si_flow (sccm)	N2O_flow (sccm)
1	0.	0.	0.	0.	0.
2	1.	1.	-1.	1.	-1.
3	-1.	-1.	-1.	1.	1.
4	-1.	1.	-1.	-1.	-1.
5	-1.	-1.	1.	1.	1.
6	-1.	-1.	-1.	-1.	-1.
7	1.	1.	1.	-1.	1.
8	-1.	1.	1.	1.	-1.
9	1.	-1.	1.	1.	-1.
10	1.	-1.	-1.	-1.	1.
11	1.	-1.	1.	-1.	-1.
12	-1.	1.	1.	-1.	1.
13	1.	1.	-1.	1.	1.
14	0.	0.	0.	0.	0.

ANOVA for Gr_rate - screening experiment

Effect	Sum of Squares	DF	Mean Sq.	F-Ratio	P-value
A:Power	32620518.8	1	32620519	20.80	0.0061
B:Pressure	1264252.1	1	1264252	0.81	0.4196
C:Temp	910252.1	1	910252	0.58	0.4883
D:Si_flow	10706852.1	1	10706852	6.83	0.0475
E:N2O_flow	80852.1	1	80852	0.05	0.8317
F:Ar_flow	159852.1	1	159852	0.10	0.7657
G:Frequency	40252.1	1	40252	0.03	0.8806
H:Bias	252.1	1	252	0.00	0.9905
Total error	7841954.2	5	1568391		
Total (corr.)	53625037.5	13			

R-squared = 0.853763 R-squared (adj. for d.f.) = 0.619784

(b)

FIGURE 3.3. (b) The table on top shows an experimental design for screening the important variables listed on the top of the columns. The 14 combinations, with highs (1), lows (−1) and center points (0), define the individual runs. The lower table shows the analysis of the results. It shows the relative importance of the terms and the R-squared values show how good the derived model fits the data.

sheet: let us say we have five responses (growth rate, uniformity, dielectric constant and stress). The 15 variables can affect these five responses in complex ways, often interacting with one another. For instance, temperature might increase growth rate exponentially while pressure increases it linearly. Pressure and flow rate together produce a higher rate than the sum of their individual contributions (this is called an *interaction*). Higher temperature, while improving growth rate, might degrade uniformity. How do we decide which variables affect which responses and what is the functional dependency of the response on the variable?

To solve this problem we need to use statistics. Books listed at the end of this chapter[3–5] deal exclusively with design of experiments (DOE).

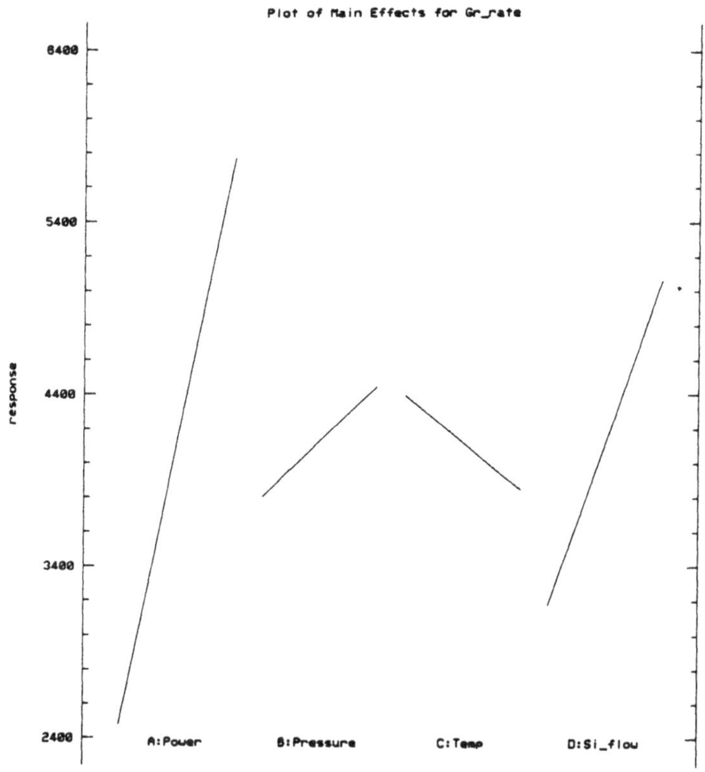

(c)

FIGURE 3.3. (c) A plot of the relative effects of the major variables identified in 3.3b show graphically that power and silane flow have the most influence on growth rate.

Obviously we cannot go into the details of statistical experimental design but we can give a flavor of the kind of problems it can solve.

One of the first steps of DOE is the recognition that we cannot obtain the functional dependence of the responses on the variables by changing the variables one at a time, i.e., keep pressure, flow, etc. constant at one value, change temperature over a range, and measure the responses at each temperature. Not only is this method wasteful, it will not give us the complete information. However, by using sound engineering judgement, developed during the exploration phase, and DOE methods, the results can be obtained by following the steps below. Figure 3.3 follows the same sequence and illustrates the experimental hierarchy.

a. Screening experiments identify the main variables for each response and help to eliminate others. They can reduce the number of important

run	Power (Watts)	Si_flow (sccm)	Pressure (torr)	Temperatur (Deg C)
1	300.	175.	300.	325.
2	300.	425.	300.	325.
3	-100.	175.	300.	325.
4	700.	175.	300.	325.
5	500.	300.	100.	400.
6	100.	300.	500.	250.
7	300.	175.	700.	325.
8	500.	300.	500.	250.
9	100.	50.	500.	400.
10	100.	50.	100.	250.
11	300.	175.	300.	175.
12	100.	50.	500.	250.
13	500.	300.	500.	400.
14	300.	175.	300.	475.
15	300.	175.	-100.	325.
16	500.	50.	100.	250.
17	100.	300.	100.	250.
18	300.	-75.	300.	325.
19	500.	50.	100.	400.
20	500.	300.	100.	250.
21	500.	50.	500.	400.
22	100.	50.	100.	400.
23	500.	50.	500.	250.
24	100.	300.	500.	400.
25	100.	300.	100.	400.
26	300.	175.	300.	325.

(d)

FIGURE 3.3. (d) A more detailed experiment to generate the functional dependence of the variables using the four factors that came out of the screening experiment. The combinations still use the high, low, center point format (the actual values used are shown here).

variables from 15 to perhaps 4 or 5 for each response. An example of a screening experiment is a Plackett–Burman design.[4]

b. Response surface experiments develop response functions, functional dependences between variables and responses. They take the 4 or 5 variables from the screening experiment and find their effect on a response function. An example of a response function is: maximize deposition rate/nonuniformity while keeping the dielectric constant over 1.45; rate is twice as important as uniformity.
c. A mathematical model is developed to rate the response function to the variables. An example of a response surface experiment design is a Box–Behnken design.[4]
d. An optimum point of operation is chosen from the response surface.

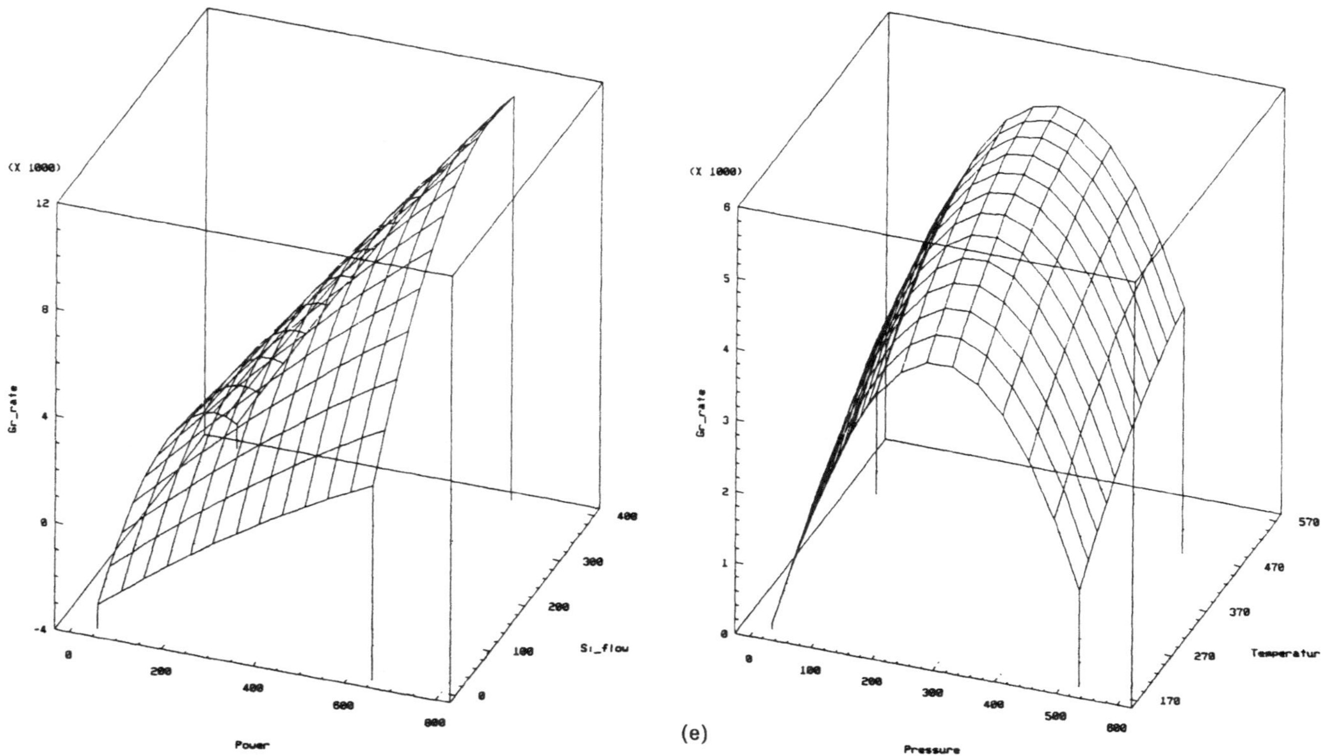

(e)

FIGURE 3.3. (e) Response surfaces developed by the model from the results of the experiment in 3.3d. The plots show the effect of the four variables, power, pressure, temperature, and silane flow, on the response growth rate. A series of such plots can be generated for any combination of settings and used to pick the optimum process condition.

e. Sensitivity measurements evaluate any changes in the responses due to small deviations from the set points of the input variables. For instance, change the input variables by 5% and examine the effect on the responses.

At the end of step (e), we should have a list of settings for each variable that produces a film corresponding to the best response function. We would also be able to obtain a first approximation of the expected natural variance of the process.

3.3.2. Process Capability Measurements

Process capability refers to the ability of the process that we have developed to remain centered between the upper and lower specification limits of our spec sheet (Table 3.1). When a large number of products are being produced, the process suffers natural variances. These variances might be due to many causes: operator related variances, equipment related drifts, environmental variations, etc. In general, these are the variances over which the engineer does not have control. For instance, Figure 3.4 shows the distribution of the response deposition rate over 500 different measurements. The response shows a normal (bell-shaped) distribution. The standard deviation σ is the value on either side of the mean between which approximately 66% of the responses occur. It provides an indication of the variance in the response. The standard deviation needs to be minimized for a manufacturable process, as was shown in Figure 3.2.

Two indices provide the necessary information with regard to process capability. C_p measures how tight is the distribution with respect to the specification limits, and C_{pk} measures whether the distribution is tight and centered between the limits. C_p and C_{pk} are defined as follows:

$$C_p = \frac{\text{upper spec limit} - \text{lower spec limit}}{6\sigma} \tag{3.1}$$

$$C_{pk} = \text{minimum of } C_{pku}, C_{pkl}$$

where

$$C_{pku} = \frac{|\text{USL} - \bar{X}|}{3\sigma}$$

and

$$C_{pkl} = \frac{|\bar{X} - \text{LSL}|}{3\sigma}$$

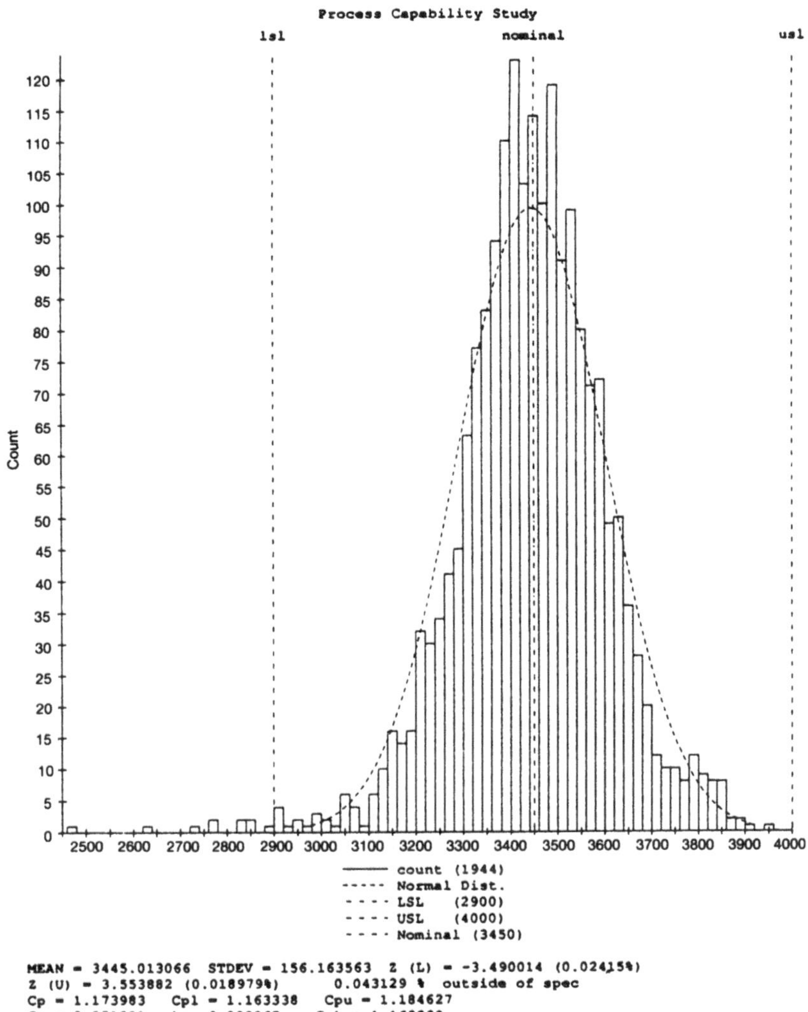

FIGURE 3.4. A set of 500 thickness observations in 2 lots of 24 wafers. There were 9 measurements in each wafer. The lower and upper spcc limits are shown, along with C_p and C_{pk} values.

LSL and USL are respectively the lower and upper specification limits. A C_{pk} index higher than 1.3 is considered essential for a manufacturable process.

Identifying and eliminating systematic (nonrandom) variations helps to improve C_p and C_{pk}. The method of analyzing the variations is called analysis of variance (ANOVA). For instance, the 500 observations shown in Figure 3.4 were measured in the following manner: 9 sites on a wafer, 24 wafers in

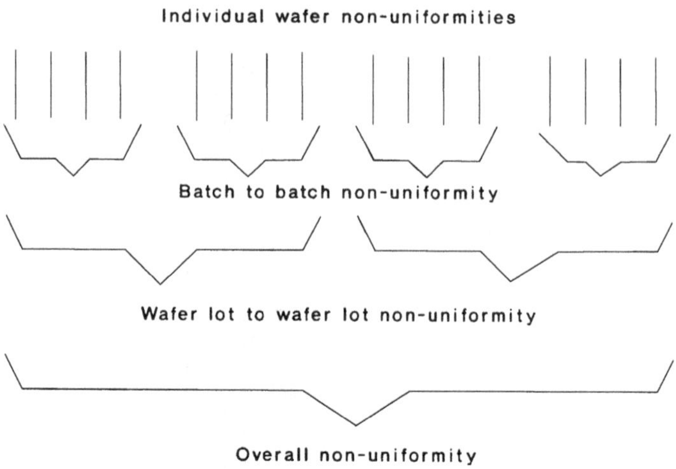

FIGURE 3.5. The nesting of nonuniformities in the set of measurements shown in Figure 3.4. Each of the nonuniformities (lot to lot, wafer to wafer, and within wafer), contributes to the distribution causing the bell shape.

a wafer lot and 2 wafer lots. Figure 3.5 shows how the variances are nested within one another. The biggest component of the variance arises within the wafer, among the nine sites on each wafer. The engineer has to improve nonuniformity within the wafer to reduce the total variance of the process.

Qualification through passive data collection (PDC) and statistical process control charts can begin only if the C_p and C_{pk} indices are acceptable. If they are unacceptable , optimization experiments are continued to identify a more suitable operating point.

3.3.3. Process Control Charts

Statistical Process Control charts keep track of the performance of the process over manufacturing time. For each response two charts are used in tandem, e.g., to graphically track its mean and standard deviation. Figure 3.6 shows a logical sequence in the setup and use of SPC charts. The frequency and sample size can be determined using established formulas as long as the random variances of the process are known.

At least 30 measurements are made before plotting the chart. The measurements should be done sequentially, without in any way adjusting the process. The 30 readings are plotted in the mean and standard deviation

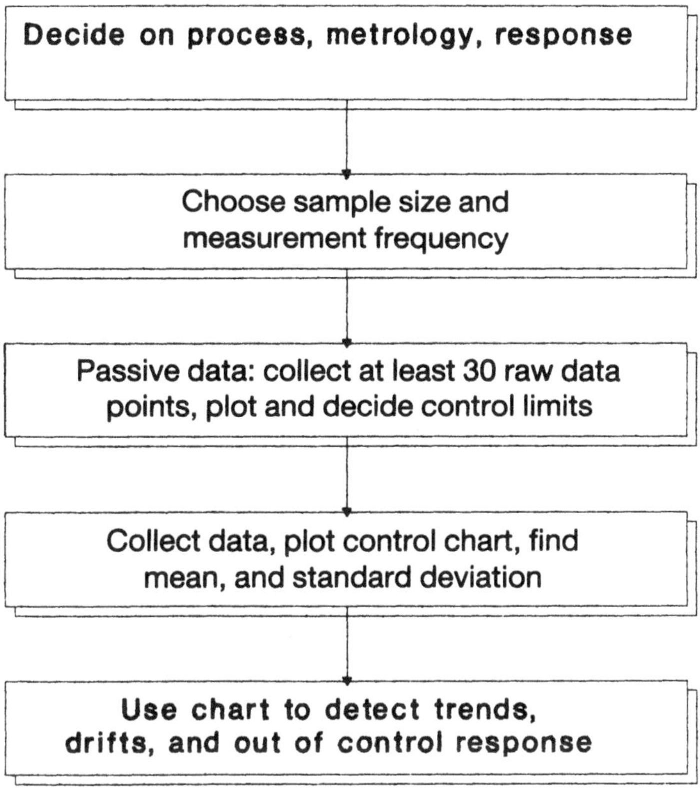

FIGURE 3.6. Flowchart for the establishment of control charts or SPC charts.

formats and a control limit is calculated for them as follows (Figure 3.7):

$$\mathrm{UCL}, \mathrm{LCL} = \mathrm{CL} \pm (3S_u) \tag{3.2}$$

where

$$S_u = \sqrt{\frac{\sum (u_i - \bar{u})^2}{n - 1}}$$

CL represents the centerline placed at the average of the data points u_i. Each u_i in the means chart is the average of the individual data points taken on each wafer. In the standard deviations chart, it is the standard deviation of the individual data points on each wafer. $\bar{u}$ is the average of the u_i's. The control limits represent ± 3 standard deviations of the u_i's, i.e., 99% of all responses should fall within these two lines (Figure 3.7).

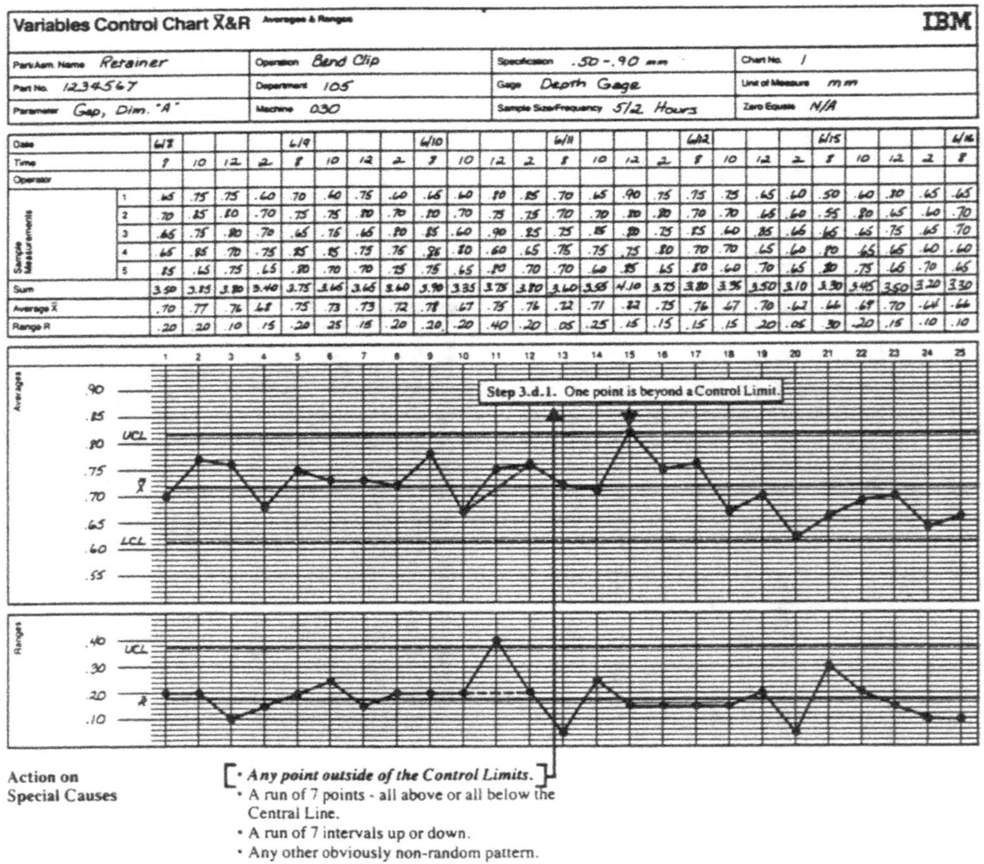

FIGURE 3.7. A typical process chart from a manufacturing line. Reprinted with permission from IBM Corp.[6]

Any response falling outside is considered to be an outlier and needs careful consideration. That response is not part of the natural variation of the process and probably has an external reason for being different from the other data. The control limits lie within the specification limits. How centered and tight the responses are with respect to the specification limits are measured by the C_{pk} index.

Interpretation and use of SPC charts are driven more by common sense than complex mathematics. If one observes a drift in the chart, it implies that the process is not behaving within its natural random variance. Any systematic shifts can thus be visually seen and corrected through the use of SPC charts.

3.4. METROLOGY

Even though we have spent considerable time exploring variances in responses, we have not examined the role played by the measurement tools that provide the responses. The use of specific measurement tools (or gauges) is intimately tied to the list of specifications. Complex specifications without the means to make measurements are not of much use. Measuring devices have to be studied before any experiments are made to develop a process.

Two terms which are often confused in usage with respect to gauges need to be defined: accuracy and precision.

Accuracy of a gauge is its ability to determine a property close to the true value of the property. Accuracy of the tool is determined by measuring a standard. For example, if we know the value of a step in thickness of a standard (provided among others by the National Institute of Standards and Technology) to be 1 000 Å and the gauge measures it to be 950 Å, the accuracy of the gauge is 0.95. Often the tool measures a value consistently different from the true value. This is called the bias of the tool.

Precision is the ability of the gauge to repeat and reproduce the measurement. For instance, if three different operators measure the same thickness 10 times each and all the measurements are 951 ± 10 Å, the precision of the gauge is ± 10 Å. The repeatability component of precision refers to the same operator with the same setup making the measurements. *Reproducibility* refers to the variance introduced by normal procedural variations, such as different operators, different environmental conditions, etc. The smaller the standard deviation of the different measurements, the better the precision of the instrument.

Both accuracy and precision of a gauge are important when choosing a particular gauge to measure a specification. They are combined in the measurement of gauge capability. A commonly used measure of gauge capability is the precision/tolerance ratio (P/T). Tolerance is the difference between the upper and lower spec limits. Precision is the standard deviation of the gauge measurements. P/T shows what component of the spec width is taken up by the gauge capability, before we even develop a process. P/T is the inverse of C_p. To achieve a high C_p, we want P/T to be as small as possible. A proper measurement tool for a specification will have $P/T < 0.1$.

3.4.1. Summary

We have thus far studied the means of optimizing a process of proven feasibility. A designed sequence of statistical experiments was used to converge the complex relationship between the variables and responses to a mathematical model. We were then able to predict an optimized operating point based on the model and to test the robustness of the operating point using sensitivity measurements.

Once the operating point was carefully chosen, we went on to measure the capability of the process through C_p and C_{pk} indices using passive data collection. We also saw the use of process control charts while running the process. Lastly, we defined the metrology requirements and the share of the specification width consumed by metrology alone.

We will now go back to mainstream CVD, so as to consider the feasibility portion of process development. We will examine the conditions that favor CVD over other deposition methods and investigate the strengths and weaknesses of CVD films.

3.5. OVERVIEW OF CVD PROCESSES

A process can be made manufacturable only if it is technically feasible. Despite the best intentions in optimizing and removing random variations, the film will not meet the success criteria of the specification list unless the initial exploration studies show the process is indeed feasible. Hence the first exploratory studies done both on paper and through actual experimentation are the foundation on which the rest of development occurs. A vast bibliography of literature exists with respect to the wide variety of CVD thin films.[7] Make extensive use of the available literature before embarking on any experiments. According to an anonymous quotation, "The best results six months of experimentation can produce are the same as a good day's work in a library!"

CVD methods have been used to deposit many different thin films. Whether CVD is the right choice for the application and what kind of CVD process is suitable for the situation can be determined by knowing the general characteristics of the different processes. The general requirements for using CVD processes to deposit any thin film are as follows:

a. Availability of suitable gaseous precursors: consider the stability of the precursor, its vapor pressure, toxicity, and ability to decompose without leaving residues.
b. Reactions involving the precursor and other gases: endothermic reactions can be controlled at a specific location; lower temperatures are dictated by the IC application, between 350°C and 1 200°C.
c. Residual product species: intrinsic impurities will inevitably be present in the film; high-purity films may need to be deposited by other methods.
d. Step coverage requirements: CVD films generally provide excellent step coverage compared to their PVD counterparts, so where step coverage is an important criterion, CVD films should be the first choice.
e. Epitaxy and other crystalline requirements: CVD is unique in producing epitaxy or large-grained films; usually an advantage, it can be a disadvantage when it results in facetting or high surface roughness.
f. Selectivity: also unique to CVD, some reactions proceed only on certain surfaces, providing a means of depositing films on chosen areas of the substrate.

As we will see in subsequent chapters, there are many classifications of CVD films, particularly based on the source of energy supply to the forward progress of endothermic reactions. In thermal CVD the substrate temperature supplies the heat for the reaction. In plasma enhanced CVD, electrical energy is coupled to the gas phase through a plasma. And in photoenhanced CVD, either laser or ultraviolet irradiation supplies energy to the reaction. The typical characteristics (there are always exceptions!) of the different CVD processes are discussed in subsequent sections.

3.5.1. Thermal CVD

Depending on which of the subprocesses determines the rate of the thermal CVD reaction, the films produced are extremely conformal to the substrate topography.[8] Temperature, as expected, is the key variable in thermal CVD, along with pressure, gas composition, and flow rates. Metallic films are often coarse-grained and contain intrinsic impurities. Amorphous oxide films (in particular SiO_2) have properties, such as dielectric breakdown strength, strongly dependent on the temperature of growth. Excellent homoepitaxy is seen at very high temperatures (above 900°C) in semiconductor films such as silicon.

Common reactor configurations include atmospheric pressure tube reactors, both horizontal and vertical; cold- and hot-walled reduced pressure reactors in single wafer and batch configurations; and continuous belt reactors. Rapid thermal CVD reactors provide rapid temperature cycling capabilities in a single wafer configuration.

3.5.2. Plasma Enhanced CVD

The primary advantage of plasma enhanced CVD (PECVD) is that it removes the temperature constraint from thermal CVD.[9] This is particularly useful for applications near the end of microelectronic processing sequences, such as passivation and interlevel dielectrics. However, other advantages such as metastable reactant species and new radicals have promoted its use in depositing epitaxial films, and dielectric films.

Film properties can often be tailored by the plasma parameters: excellent control in stress, step coverage, dielectric properties, and grain size. Temperature of the substrate is one more variable along with plasma power, frequency, and bias that provide additional capabilities.

By necessity PECVD reactors operate at reduced pressures. Both single and batch configurations that operate at 13.56 MHz or at microwave frequencies are available commercially. Newer plasma sources are being continuously developed.

3.5.3. Photoenhanced CVD

Unique to laser enhanced CVD is a direct write capability.[10] Even though the technique is in its developmental phase, the approach is being actively pursued by researchers for the deposition of metallic films from organometallic precursors. The method is gaining acceptance in wide featured applications such as mask repair.

Similarly ultraviolet (UV) irradiation of patterned substrates in the

TABLE 3.2 Films for Microelectronics by Photoenhanced CVD

UV–CVD	Si	$Si_xH_yCl_z$–H_2	Epi,poly, amorphous Si
	SiO_2	SiH_4–N_2O, PH_3	Undoped, doped
	Si_3N_4	SiH_4–NH_3	Also oxynitrides
	Al, Cu	Organic sources	Nucleation only
Laser CVD	Si	SiH_4, $SiCl_4$	Epi, poly, amorphous Si
	SiO_2	SiH_4–N_2O	Undoped
	Si_3N_4	SiH_4–NH_3	Also oxynitrides
	Al, Cu	Organic sources	Nucleation only
	W	WF_6	Nucleation only
	$TiSi_2$	$TiCl_4$–SiH_4	

presence of CVD gases, allows selective nucleation of the film. Further deposition to full thickness might then be accomplished by thermal CVD;[11] this would reduce lithography and etching. Ultraviolet CVD is also in its infancy and is not a mainstream process currently being used by the microelectronics industry. Table 3.2 lists some of the films relevant to microelectronics that have been deposited by photoenhanced CVD methods.

References

1. "Project Management", Class Notes, Integrated Project Systems, Inc., 1070 6th Ave #112, Belmont, CA 94002.
2. D. C. Montgomery, *Design and Analysis of Experiments,* John Wiley, New York, 1976.
3. G. E. P. Box, W. G. Hunter, and J. S. Hunter, *Statistics for Experimentors,* John Wiley, New York, 1978; R. L. Plackett, and J. P. Burmen, *Biometrika* **33**, 305–325 (1946).
4. W. J. Diamond, *Practical Experimental Designs,* Van Nostrand Reinhold, New York, 1981.
5. A. R. Alvarez, D. J. Walters, and M. Johnson, *Solid State Technol.* **26**(7), 127 (1983).
6. International Business Machines Corp., *Process Control, Capability and Improvement,* The Quality Institute, Southbury, Conn., 1984.
7. C. E. Morosanu, *Thin Films by Chemical Vapor Deposition,* Elsevier, Amsterdam, 1990.
8. W. Kern, and V. S. Ban, in *Thin Film Processes* (Vossen and Kern, Eds.), p. 258, Academic Press, New York, 1978.
9. R. Reif, in *Handbook of Plasma Processing Technology* (Rossnagel, Cuomo, and Westwood, Eds.), p. 260, Noyes, Park Ridge, N. J., 1990.
10. R. Solanki, C. A. Moore, and G. J. Collins, *Solid State Technol.* **28**(6), 220 (1985).
11. J. Peters, F. Gebhardt, and T. Hall, *Solid State Technol.* **23**(9), 121 (1980).

Chapter 4

Chemical Equilibrium and Kinetics

The central feature of chemical vapor deposition is the presence of a heterogeneous reaction on the surface over which we intend to grow the thin film. This reaction is usually endothermic and the energy needed for the forward progress of this reaction can come from many sources: thermal, electrical plasmas, photons, etc. The earliest and most common means of supplying energy to the CVD reaction is by heating the substrate and this technique is commonly referred to as thermal CVD. We will begin our discussion of thermal CVD by concentrating on two constraints placed on the reaction by the economics of a manufacturing process: how much of the reactants are converted to products and how quickly this can be accomplished.

The first question along with the energetics of the reaction is answered by chemical reaction thermodynamics. The rate of the reaction is governed by chemical reaction kinetics. This chapter will introduce concepts of reaction thermodynamics and reaction kinetics, so that useful physical reactors can be constructed to produce thin films. Reactor design for thermal CVD will be discussed in Chapter 5.

4.1. THERMODYNAMICS VERSUS KINETICS

Thermodynamics underlies many of the subprocesses that constitute CVD. Its primary use is to predict the maximum yield of a reaction under a set of controlled variables such as temperature, pressure, or composition. It does not explicitly deal with the time needed for obtaining this yield. For instance, referring to Figure 4.1, even though reaction II ultimately provides a much higher conversion of reactants to products, the time needed for such a conversion is extremely long. In the useful time interval available for a

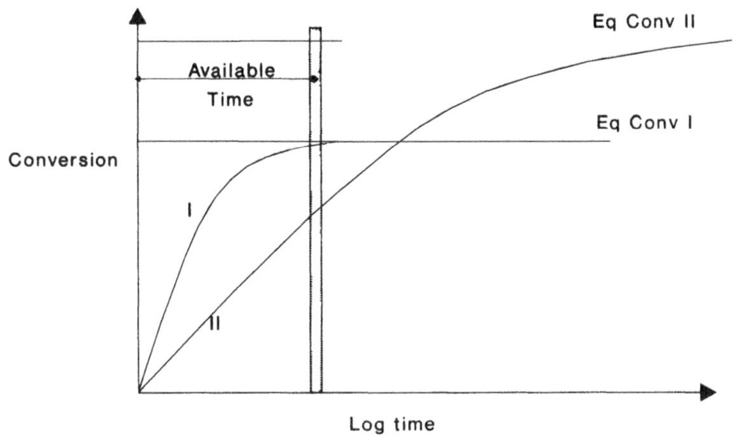

FIGURE 4.1. Yield of two reactions. Even though the yield of reaction II is ultimately higher, in the useful time frame available, reaction I is more practical.

commercial reactor, reaction I provides a higher yield, even though its ultimate yield is lower than reaction II. Hence choice of reactions for CVD has to be judicious, based on both the final yield and the rate at which it is obtained.

Chemical reaction kinetics allows a complex reaction to be broken down into elementary steps for identification of the rate-determining step, and is especially useful in a set of sequential subprocesses such as found in CVD.

4.2. EQUILIBRIUM THERMODYNAMICS OF REACTIONS

This chapter assumes some familiarity with the first and second laws of thermodynamics. For further reference, titles of treatises in related subjects are provided at the end of the chapter.[1,2]

To develop an understanding of thermal CVD, we will first review the fundamentals of chemical reaction thermodynamics and illustrate the calculation of equilibrium constants for relevant chemical reactions. We will then present the principles of chemical kinetics and the effects of reactor operation under different rate limiters on thin film properties. Bear in mind that all reactions considered in CVD are heterogeneous, i.e., they involve at least two phases. We take care during the CVD process to avoid homogeneous reactions in the gas phase, since they do not aid in any way the formation of the film on the solid surface and often have deleterious effect on the particle levels in the reactor.

Another characteristic typical of the conditions in a CVD reactor is ideality of the reactants.[3] Since the operating pressures in CVD are seldom over an atmosphere and the operating temperatures are usually well above the boiling points of the components, most of the vapors and gases in CVD can be treated as following the ideal gas laws. This aids us in making many simplifying assumptions in subsequent discussions. But do remember that ideality of the reactants is a different issue from equilibrium; very often conditions in a CVD reactor are not in equilibrium and equilibrium thermodynamics only provides the limiting case.

4.2.1. The Phase Rule

Under equilibrium or metastable conditions, all systems conform to the Gibbs phase rule, which can be expressed as:

$$\Gamma = N - \Phi + 2 \tag{4.1}$$

The phase rule relates the number of chemical components N required to express the system composition, and the number of phases present Φ through the degrees of freedom for the system Γ. The number of components can be determined as either (a) or (b).

a. The minimum number of chemical species required to produce the system at equilibrium.
b. The number of species in the system minus the number of independent reaction equilibria among these species.

The degrees of freedom are parameters, such as pressure, temperature, or composition, which may be independently changed without causing a change in the number of phases Φ.

Consider a single component system such as an isolated chamber containing pure silicon and its vapor. The number of phases is 2, so by the phase rule the system has a single degree of freedom. If the temperature of the system is changed, the pressure also changes, to return the system to equilibrium. If a pressure different from the equilibrium vapor pressure at a given temperature is imposed on the system, the vapor either condenses or the solid evaporates, so that equilibrium is again reached. Hence for a solid in equilibrium with its vapor, the equilibrium vapor pressure is uniquely determined by the temperature of the system.

Let us now introduce a second inert component into the system, say argon gas. The number of phases remains unchanged at 2. The degrees of freedom have increased to 2. So we can change the partial pressure of argon without upsetting the equilibrium between silicon and its vapor. But if instead we introduce an active component, say oxygen, an independent reaction equilibrium is set up.

Consider the oxidation of silicon:

$$Si(s) + O_2(g) \rightarrow SiO_2(s)$$

The system has three phases: the condensed phases Si and SiO_2, and the gaseous phase. The number of chemical species minus the number of independent reactions is 2 and hence the number of components is 2. By the phase rule, the number of degrees of freedom then, is 1. The temperature may be arbitrarily fixed, in which case the equilibrium pressure of oxygen is fixed. Conversely, if the pressures of oxygen and the equilibrium vapor pressures of silicon and silicon dioxide were arbitrarily fixed, the reaction would move to keep the temperature fixed. Thus the phase rule is useful as a predictor of the importance of controlled variables such as pressure, temperature, and composition, and the effect they have on phase equilibrium. We can predict which way a given reaction will progress when the controlled variables are changed.

Exercise 4.1

Determine the degrees of freedom for the following reactions:

$$SiH_4(g) \rightarrow Si(s) + 2H_2(g)$$

$$SiH_4(Ar) \rightarrow Si + 2H_2(Ar)$$

4.2.2 Criteria for Equilibrium

We have mentioned reaction equilibrium without formally stating what defines equilibrium. One of the fundamental criteria for equilibrium in a multiphase, multicomponent system is that pressure and temperature are constant and uniform throughout the system.[4] This implies that a condensed phase needs to be in equilibrium with the gas phase, and the components in the gas phase should maintain constancy in their partial pressures or concentrations. We will treat these two constraints separately, first through phase equilibrium in a single component system and then through reaction equilibria in the gas phase. The constraints imposed by thermodynamics for internal equilibrium were first deduced by Gibbs.

4.2.3. Single Component Systems

The total Gibbs free energy of a single component system, G^t, can be formally written as

$$G^t = H^t - TS^t \tag{4.2}$$

where the superscript t refers to total value of an extensive property (an extensive property changes with the mass of the substance). H is the enthalpy of the system, which can be written based on the first law of thermodynamics as the sum of two terms, U and PV:

$$H^t = U^t + PV^t$$

U is the internal energy of the system. A change in U is the net result of heat and work interactions between the system and its surroundings, $dU = TdS - PdV$, where V is the volume and S is the entropy. The PV term denotes the effect of external work on the enthalpy. Hence the Gibbs free energy can be written as

$$G^t = U^t + PV^t - TS^t \tag{4.3}$$

and the differential total Gibbs free energy can be expressed as

$$dG^t = -S^t dT + V^t dP \tag{4.4}$$

Through the application of the second law of thermodynamics, it can be uniquely demonstrated that for a system at constant T and P, all irreversible processes within the system will proceed in such a direction as to cause a decrease in the Gibbs free energy of the system. Since the differential terms on the right-hand side of equation (4.4) are both equal to zero at constant T and P, we can write

$$(dG^t)_{T,P} \leqslant 0 \tag{4.5}$$

Thus the equilibrium state of a closed system is that state for which the total Gibbs free energy is a minimum with respect to all possible changes, at the given T and P. Consequently, a general mathematical criterion for a system to be at equilibrium is that

$$(dG^t)_{T,P} = 0 \tag{4.6}$$

Figure 4.2 illustrates the free energy changes on either side of the equilibrium.

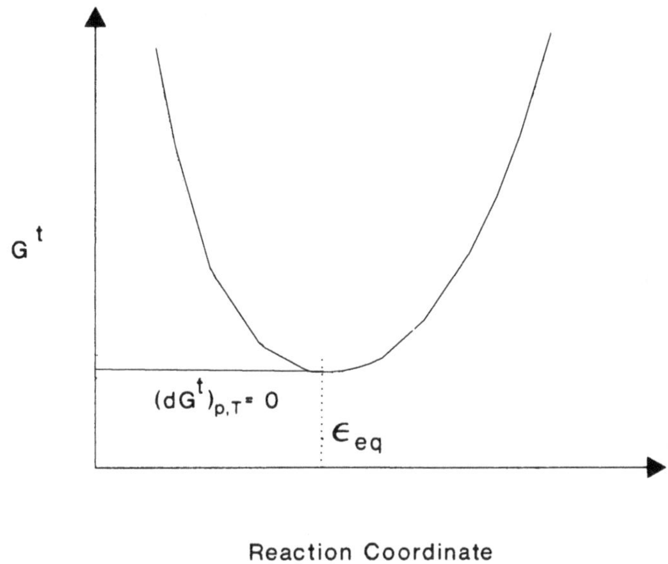

FIGURE 4.2. Free energy change in a reaction. Equilibrium correpsponds to the minimum in free energy. On both sides of equilibrium, the reaction tends toward equilibrium.

4.2.4. Multicomponent Systems

Having discussed the free energy criteria for equilibrium for a single component, multiphase system, now let us turn our attention to equilibrium in the gas phase in a multicomponent, multiphase system. Let us consider the reaction.

$$SiH_4 + 2O_2 \rightleftarrows SiO_2 + 2H_2O \tag{4.7}$$

which is commonly used in the CVD of silicon dioxide. The stoichiometry coefficients that precede the chemical symbols (e.g., the 2 in front of O_2) determine the relationship between the amounts of various components consumed and produced by the reaction. In general, we can rewrite a chemical reaction as

$$\nu_1 A_1 + \nu_2 A_2 + \nu_3 A_3 + \nu_4 A_4 + \cdots = 0 \tag{4.8}$$

where the A_i's represent the chemical symbols and the ν_i's represent stoichiometry numbers. These are numerically equal to the stoichiometry coefficients, but negative for a reactant and positive for a product, by convention.

Considering the silicon dioxide reaction, for every n moles of SiH_4 consumed, $2n$ moles of O_2 are also consumed, forming n moles of SiO_2 and $2n$ moles of H_2O. Using the convention in equation (4.8), this can be generalized as

$$\frac{\Delta n_{A_1}}{\Delta n_{A_2}} = \frac{\nu_{SiH_4}}{\nu_{O_2}} = \frac{1}{-2} \tag{4.9}$$

Applying the same principle for every set of two of the reactants and products,

$$\frac{dn_2}{dn_1} = \frac{\nu_2}{\nu_1}, \frac{dn_3}{dn_2} = \frac{\nu_3}{\nu_2}, \ldots$$

and so on for other components. Comparing all such equations it can be written that

$$\frac{dn_1}{\nu_1} = \frac{dn_2}{\nu_2} = \frac{dn_3}{\nu_3} = \frac{dn_4}{\nu_4} = d\varepsilon \tag{4.10}$$

The general relationship between a differential change in the number of moles of a reacting species and ε is therefore

$$dn_i = \nu_i d\varepsilon \tag{4.11}$$

The variable ε, is called the reaction coordinate, which characterizes the extent to which the reaction has progressed. ε is a quantity we will use both in thermodynamics and kinetics and has been called the reaction coordinate, the degree of progression of the reaction, and several other names. At infinite time, ε reaches equilibrium conversion, as in reaction II in Figure 4.1.

We have shown the criteria for equilibrium in a system with constant T and P in equation (4.6). Since the reaction coordinate ε is the single variable that characterizes the extent of the reaction and hence the composition of the system, we can plot the total Gibbs free energy of the system as a function of ε, as shown in Figure 4.2. At the minimum in the curve, equation (4.6) is satisfied. On either side of the minimum, the reaction proceeds in the direction of the minimum in free energy.

Thus the second criterion for equilibrium in a multiphase, multicomponent system, occurs through reaction equilibrium in the gas phase and can be uniquely characterized by the reaction coordinate. We will shortly define this criterion more precisely. First we need to examine the energy changes in a reaction as it proceeds to equilibrium conversion. We will then attempt to arrive at the criterion for equilibrium by finding the relationship between the composition at equilibrium and the energy change involved.

Exercise 4.2

For the reaction $SiH_4 + O_2 \rightarrow SiO_2 + 2H_2$ determine the number of moles n_i and the mole fractions as a function of the reaction coordinate, if the starting conditions are 2 kmol of silane, 1 kmol of oxygen, 1 kmol of SiO_2, and 4 kmol of hydrogen.

4.2.5. Energy Change in Reactions

All reactions obey the law of conservation of energy. A logical consequence of conservation of energy is Hess's law, which states that the net energy change in a series of reactions is simply the difference in the energies of the final products and the initial reactants, independent of intermediate reactions. Hence heats of even complex multistep reactions can be easily computed by subtracting the heats of formation (ΔH_f) of the reactants from those of the products. The heat of a reaction at any temperature can be calculated from the heat of a reaction at room temperature or at any other temperature where one can obtain heats of formation data, through the following approximation:

[heat absorbed during reaction at temperature T_2] = [heat added

to change temperature of reactants from T_2 to T_1]+

[heat absorbed during the reaction of T_1] +

[heat added to the products to bring them back to T_2 from T_1]

The first and third terms on the right-hand side are dependent on the molar heat capacities of the reactants and products. The molar heat capacity of an ideal gas can be empirically written as

$$C_p = \frac{dH}{dT} = a + bT + cT^{-2} \tag{4:12}$$

Accordingly,

$$\Delta C_p = \frac{d\Delta H}{dT} = \Delta a + \Delta bT + \Delta cT^{-2} \tag{4.13}$$

Integrating equation (4.13) between room temperature T_0 (normally the temperature of input gases in CVD) and the substrate temperature T,

$$\Delta H(T) - \Delta H(T_0) = aT + 0.5bT^2 - c/T - I \tag{4.14}$$

The value of the integration constant I is not zero. If we combine $\Delta H(T_0)$ and I to set it as ΔH^0, the heat of the reaction at any temperature can be written within the validity of equation (4.12) as

$$\Delta H(T) = \Delta H_0 + aT + 0.5bT^2 - c/T \tag{4.15}$$

This equation is quite useful in determining the heat of reaction at any temperature, since the constants a, b, and c are tabulated for many gases. We will again use this expression in the calculation of equilibrium constants later in this section.

By convention an exothermic reaction, a reaction that gives out heat is designated by a minus sign, since the system loses energy. By the same convention endothermic reactions, more relevant to CVD, have a positive energy change.

4.2.6. The Equilibrium Constant

The equilibrium constant relates the criteria for equilibrium, expressed in terms of heats of reaction and hence the free energy change to the equilibrium conversion of reactants and products.

The expression for the total differential free energy of a multicomponent system needs to add in the chemical potentials of each of the components to the single component expression in equation (4.4). The chemical potential of a component μ_i is defined by the following identity:

$$\mu_i \equiv \left.\frac{\delta(nU)}{\delta n_i}\right|_{nV, nS, n_j} \tag{4.16}$$

i.e., the chemical potential is the change in the internal energy of a multicomponent system, through a differential change in the amounts of one of the components, keeping other parameters constant. We can now write the Gibbs free energy equation (4.4) as

$$dG^t = -S^t dT + V^t dP + \sum \mu_i dn_i \tag{4.17}$$

For a chemical reaction similar to the one we considered in equation (4.7), we can rewrite the chemical potential through the reaction coordinate.

$$dG^t = -S^t dT + V^t dP + \sum (v_i \mu_i) d\varepsilon \tag{4.18}$$

At constant temperature and pressure, the expression reduces to

$$\sum (v_i \mu_i) = \left(\frac{\delta G^t}{\delta \varepsilon} \right)_{T,P} \tag{4.19}$$

Equation (4.19) shows that at constant temperature and pressure, the rate of change of the free energy of the system is a function only of the reaction coordinate and the chemical potentials of the components. For a single component system, the situation reduces to the one in equation (4.5). If we again invoke the criterion for equilibrium as the minimum in the free energy curve, the criterion for equilibrium in terms of the chemical potential is

$$\sum v_i \mu_i = 0 \tag{4.20}$$

At this point, we need to review our conventions on the symbols once more. The superscript t, on all expressions refers to the total value of an extensive parameter. We will use the superscript ° to denote standard state properties of a system. Standard states for all components are defined for our discussion as being the pure component at 1 atmosphere pressure, at the system temperature. The concept of a standard state is quite useful, since published data for the properties of components at their standard states is readily available, for instance in the JANAF tables.[6]

Assuming ideal gases, the Gibbs free energy of a component at any given temperature and pressure is related to the standard Gibbs free energy G° by

$$G_i = G_i^\circ + RT \ln p_i \tag{4.21}$$

and since we have shown that the chemical potential of any one component μ_i is identically equal to its free energy G_i from equation (4.16) we can write

$$\mu_i = G_i^\circ + RT \ln p_i \tag{4.22}$$

where p_i is the partial pressure of the ith component expressed in atmospheres. It is implicit that all pressures are divided by 1 atmosphere, our assumed standard state in pressure for all gases.

At equilibrium, equation (4.22) can be combined with equation (4.20) to give

$$\sum v_i G_i^\circ + RT \sum \ln p_i^{v_i} = 0 \tag{4.23}$$

or

$$\ln \prod p_i^{\nu_i} = \frac{-\sum \nu_i G_i^\circ}{RT} \tag{4.24}$$

We can formally define a quantity K as

$$K = \prod p_i^{\nu_i} \tag{4.25}$$

and write equation (4.24) in terms of K as

$$-RT \ln K = \sum \nu_i G_i^\circ = \Delta G^\circ \tag{4.26}$$

K is called the equilibrium constant. It is related to the composition at equilibrium and the stoichiometric numbers of the components in the reaction. We have derived an expression for the energy change of a reaction in terms of the composition at equilibrium, which is what we set out to do at the end of the last section.

Equation (4.26) is an important expression in many ways. This is the expression we were seeking in order to define a criterion for equilibrium for a multicomponent system. ΔG° is the standard Gibbs free energy change of a reaction and equation (4.26) has related it to the composition of the system at equilibrium. Even though K is termed a constant, it is still a function of temperature. The usefulness of this expression comes from the fact that from properties of components at their standard states we can derive the energy change of a reaction. So equation (4.26) provides the equilibrium constant, which is related to composition. The equilibrium constant should not be confused with the rate constant soon to be derived in the kinetics section. K is a thermodynamic quantity which helps us in predicting the yields of reactions at different temperatures.

As it is stated, equation (4.26) deals only with single phase reactions in the vapor phase. CVD reactions, however, are heterogeneous. Equation (4.26) can be made to be valid for solids as well as gases by substituting the partial pressures by activity, which can be thought of as thermodynamic concentration. Since the activity of a pure solid is unity, it often does not directly appear in the expression for the equilibrium constant. However, it is implicitly present. If for any reason, activities in the solid state should change, they have to be included in equation (4.26), since they can have a profound effect on the equilibrium. Thus for the case of a reaction equilibrium involving pure condensed phases and a gas phase, the equilibrium constant can be written solely in terms of those species that occur only in the gas phase.

4.2.7. Effect of Temperature on K

Rewriting equation (4.26) as

$$\frac{\Delta G^\circ}{RT} = -\ln K$$

For a reaction occurring between components in their standard states at constant pressure, from equation (4.4) we can write

$$\Delta S^\circ = -\frac{d(\Delta G^\circ)}{dT}$$

Since

$$\Delta G^\circ = \Delta H^\circ - T\Delta S^\circ$$

we can eliminate ΔS° from the above two equations and write

$$\frac{\Delta H^\circ}{T} - \frac{\Delta G^\circ}{T} = -\frac{d(\Delta G^\circ)}{dT}$$

We can rearrange this expression to give the famous van't Hoff equation, which relates the free energy change to temperature in a differential form.

$$\frac{d(\Delta G^\circ/RT)}{dT} = \frac{-\Delta H^\circ}{RT^2} \tag{4.27}$$

Substituting with K in equation (4.27) gives the dependence of K on temperature:

$$\frac{d\ln K}{dT} = \frac{\Delta H^\circ}{RT^2} \tag{4.28}$$

For small temperature changes, the heat of the reaction (ΔH°) can be assumed constant and equation (4.28) can be easily integrated to give

$$\ln\frac{K}{K_1} = -\frac{\Delta H^\circ}{R}\left(\frac{1}{T} - \frac{1}{T_1}\right) \tag{4.29}$$

Figure 4.3 shows the variation of equilibrium constant over temperature for some reactions of interest using this approximation. Knowledge of the equilibrium constant at each temperature provides one of the answers we were looking for at the beginning of the chapter: prediction of yield of the reaction at the temperature of operation.

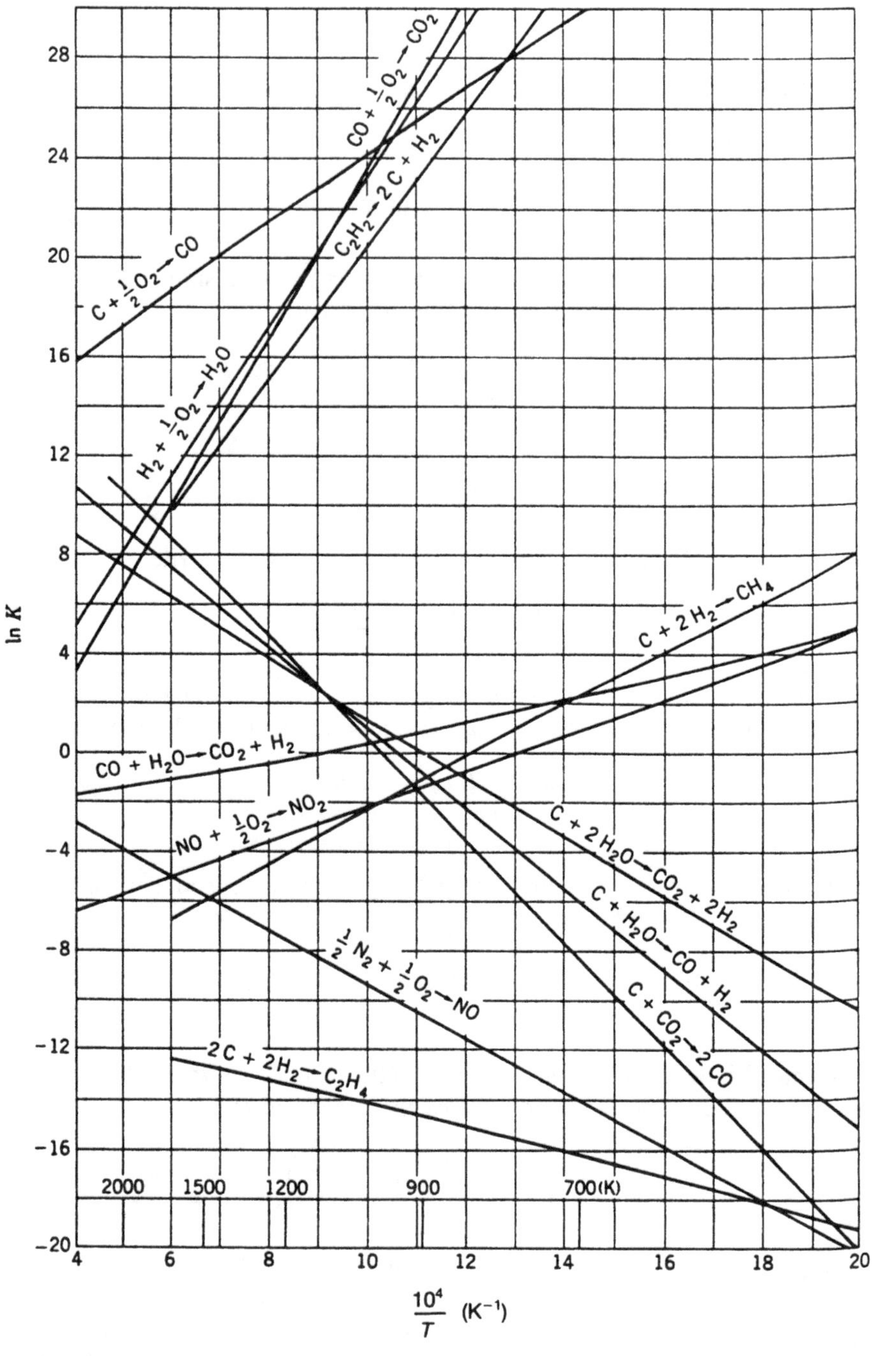

FIGURE 4.3. The variation of equilibrium constant with temperature for some common reactions. Reprinted from Ref. 1 with permission from McGraw-Hill Company.

The following general conclusions can be drawn from the preceding discussion on the equilibrium constant:

a. $K \gg 1$ indicates the possibility of almost complete conversion. Conversely, $K \ll 1$ indicates that the reaction will not proceed to any appreciable extent.
b. The equilibrium constant is unaffected by pressure, the presence of inert gases, or the kinetics of the reaction. It is a function only of the temperature of the system.
c. Equilibrium conversion increases with temperature for endothermic reactions and decreases for exothermic reactions.
d. Even though the equilibrium constant is not a function of pressure or the presence of inert gases, they can influence the equilibrium concentration of materials and hence the equilibrium conversion.
e. Conversion increases with increasing pressure (or decreasing inerts) if the number of moles decreases with the reaction. Conversion decreases with increasing pressure (or decreasing inerts) if the number of moles increases with the reaction.

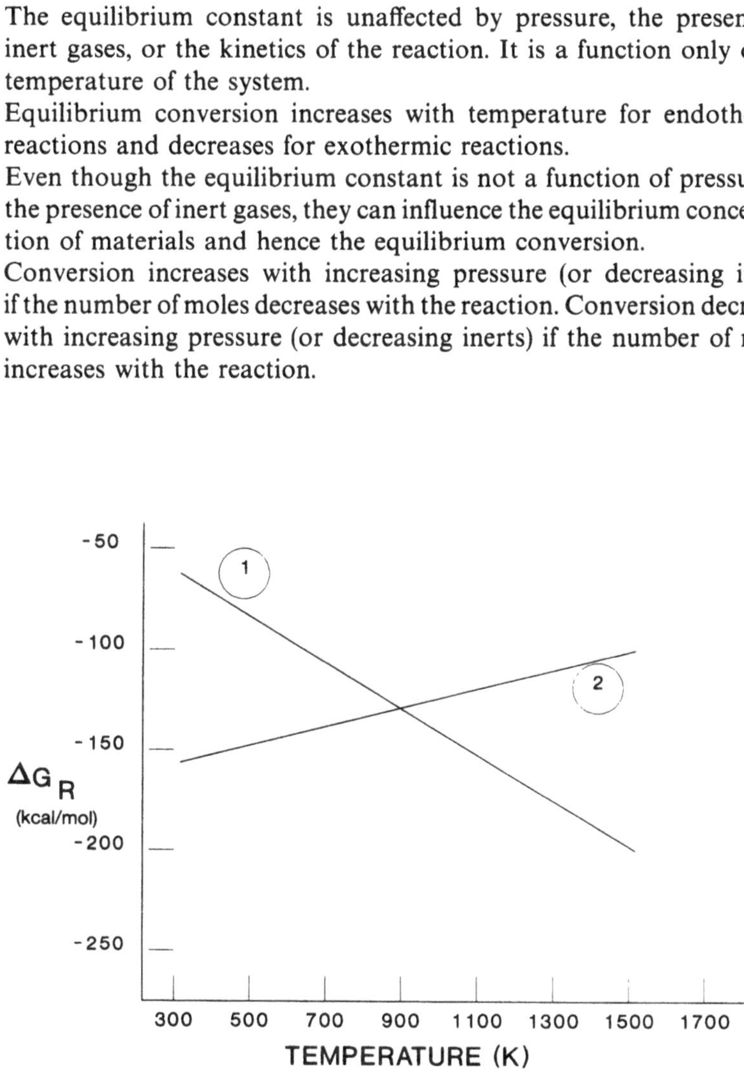

FIGURE 4.4. Free energy change for two reactions that produces silicon nitride from silane: reaction 1 uses ammonia and reaction 2 uses nitrogen. Adapted from C. E. Morosanu and E. Segal, *Thin Solid Films* **88**, 339 (1982) with permission.

Exercise 4.3

1. Figure 4.3 shows the equilibrium constant of several gas phase and heterogeneous reactions of general interest. The figure is intended to provide a feel for the numbers involved. Try to calculate the free energy changes involved at various temperatures and compare them to values taken directly from JANAF tables.[6]
2. The reaction $3SiH_2Cl_2 + 4NH_3 \rightarrow Si_3N_4 + 6HCl + 6H_2$ has a standard Gibbs free energy change of -60 kcal/mol at 300 K and -110 kcal/mol at 700 K (Figure 4.4). Calculate the equilibrium constants at the two temperatures. Using C_p data for the four gaseous compounds and the one solid compound from JANAF tables,[6] plot the change in equilibrium constant and free energy change of the reaction as a function of temperature between 273 K and 900 K.

4.3. CHEMICAL REACTION KINETICS

The situation where a chemical reaction proceeds to equilibrium conversion very fast is the exception rather than the rule. Hence we cannot design reactor systems with thermodynamic information only; we also need to know the *how* and the *how long* of the reaction along with the *how much* information provided by thermodynamics.

Chemical kinetics is the study of the *rate* and the *mechanism* by which one chemical species is converted to another. The rate of a reaction is the mass in moles of a product species produced or reactant species consumed in unit time. It can be expressed as

$$r_i = \frac{1}{W}\frac{dN_i}{dt} = \frac{\text{moles } i\text{th component formed}}{\text{mass of solid} \times \text{time}}. \tag{4.30}$$

By the term mechanism we account for all the individual collisional processes involving the reactant and product atoms that result in the overall rate.[7]

We will introduce concepts of reaction kinetics using homogeneous reactions, reactions involving only one phase, before treating more complex heterogeneous reactions.

4.3.1. Reaction Nomenclatures

An elementary chemical reaction is one in which the rate of disappearance of the reactants or the appearance of the products corresponds directly to the respective stoichiometric coefficients. Consider the reaction

$$A + B \rightarrow C$$

If the formation of C is limited by collisions between molecules of A and molecules of B, the rate of disappearance of A is given by

$$r_A = -kC_AC_B = -kp_Ap_B \tag{4.31}$$

where the C's refer to concentrations and the p's to partial pressures of the gas phase species. Since the rate corresponds directly to the stoichiometry, we call this an *elementary* reaction. For homogeneous, gas phase reactions, the concentrations can be replaced by partial pressures. Consider, however, the formation of tungsten from tungsten hexafluoride.

$$WF_6 + 3H_2 \rightarrow W + 6HF$$

The rate of formation of tungsten is empirically found to be proportional to $[p_{H_2}]^{0.5}$. Since this does not correspond to the stoichiometry directly, this reaction is called *nonelementary*. Nonelementary reactions are thought to be the aggregate of many elementary reactions. One is able to observe only the net reaction in the overall process, which is the summation of the individual elementary reactions.

The *order* of a reaction is defined as follows. Consider the general reaction $A + B + \cdots + D = 0$. The rate of progress of this reaction can often be expressed as

$$r_A = kC_A^aC_B^b\cdots C_D^d \tag{4.32}$$

The exponents $a, b, \ldots, d$ are not necessarily related to the stoichiometry coefficients. These exponents are called the order of the reactions as

ath order in A
bth order in B
nth order overall, where $n = a + b + \cdots + d$.

The constant k relates the concentrations to the overall rate and is called the reaction rate constant.

4.3.2. Kinetic Interpretation of Equilibrium

Even though we discussed equilibrium in the previous sections as a purely thermodynamic concept, it can also be thought of in terms of rates of forward and reverse reactions.[8] Consider the elementary reaction

$$A \leftrightarrows B$$

The rate of formation of B by the forward reaction can be expressed as

$$r_{B,\text{forward}} = k_1 p_A \tag{4.33}$$

Similarly the rate of disappearance of B by the reverse reaction can be expressed as

$$r_{B,\text{reverse}} = -k_2 p_B \tag{4.34}$$

At equilibrium, there is no net formation of B, so

$$r_{B,\text{forward}} + r_{B,\text{reverse}} = 0 \tag{4.35}$$

or

$$\frac{k_1}{k_2} = \frac{p_A}{p_B} \tag{4.36}$$

From equation (4.25), the equilibrium constant K for this situation can be written as

$$K = p_B/p_A$$

Hence at equilibrium, for an elementary reaction,

$$K = k_1/k_2 = p_B/p_A$$

Thus equilibrium can be viewed as (a) for a given temperature and pressure the free energy of the system is at a minimum, and as (b) the rates of forward and reverse elementary reactions are microscopically equal. The latter is the kinetic definition of equilibrium.

4.3.3. Temperature Dependence of Rate

The rate constant k relates the overall rate of the reaction to the concentration-dependent terms. It has been found to be a strong function of temperature and is well represented empirically by the Arrhenius law.

$$k = \chi \exp(-E/RT) \tag{4.37}$$

χ is the collisional frequency term and E is the activation energy. The terms collision frequency and activation energy arise out of the concept of an activated complex.

Consider again the reaction A + B → C. The reaction is hypothesized to occur through a sequence of elementary reactions;

$$A + B \leftrightarrows AB^* \rightarrow C$$

AB* is called the activated complex and is in equilibrium with A and B. The energetics of the reaction are shown in Figure 4.5. The activated complex is formed by the collision of A and B molecules and is incapable energetically of existing by itself. C is formed out of the decomposition of AB*. The rate of formation of C is only dependent on the concentration of AB* and its rate of decomposition, given by statistical thermodynamics as $\kappa T/h$, where κ is Boltzmann's constant and h is Planck's constant.

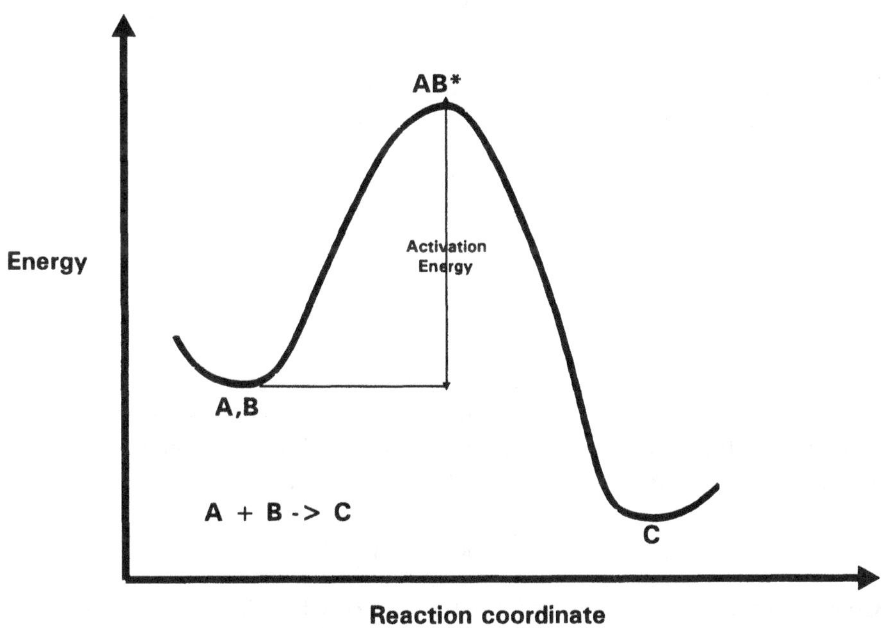

FIGURE 4.5. The change in energy of a reacting species as the reaction proceeds forward. The energy needed for the reactants to overcome the hump in order to form products is known as the activation energy of the reaction.

Let's review our conventions once more. K is the equilibrium constant and is a function only of temperature, k is the reaction rate constant that relates the reaction rate to the concentration terms, and κ is Boltzmann's constant $= 3.297 \times 10^{-27}$ kcal/K $= 8.617 \times 10^{-5}$ eV/K. Since they appear together in the text, take some time to distinguish them.

The rate of the forward reaction, $r_{\text{forward, AB}}$ is then the product of the concentration of AB* and its decomposition rate and can be expressed as

$$r_{\text{forward,AB}} = \frac{\kappa T}{h} K^* C_A C_B \tag{4.38}$$

Since by definition

$$-RT \ln K^* = \Delta G^* = \Delta H^* - T\Delta S^* \tag{4.39}$$

we can split the forward rate into temperature-dependent and temperature-independent terms as

$$r_{\text{forward,AB}} = C_A C_B \left[\frac{\kappa}{h} \exp(\Delta S^*/R) T (\exp(-\Delta H^*/RT) \right] \tag{4.40}$$

By our definition of the rate constant k as the proportionality constant between the rate and the concentration terms, we can write

$$k = \chi T \exp(-\Delta H^*/RT) \tag{4.41}$$

where

$$\chi = \frac{\kappa}{h} \exp(\Delta S^*/R) \tag{4.42}$$

The pre-exponential term χ consists of the entropy term and temperature. Since the linear portion of the temperature dependence is much weaker compared to the exponential term that follows, the linear T is often neglected. The overall rate expression can thus be expressed as

$$r_{\text{forward,AB}} = k C_A C_B = \chi \exp(-\Delta H^*/RT) \tag{4.43}$$

Comparing equation (4.43) and Figure 4.5, we can relate the energy term ΔH^* to the energy required to cross the hump between the reactants and the products. This energy is called the activation energy. The pre-exponential terms determine the frequency of collision between the reactants and the probability of formation of the activated complex and are called the collisional parameters. This model suggests the rate has an exponential dependence on temperature, which is consistent with the empirical Arrhenius relationship proposed earlier.

Measurement of the activation energy of a reaction often helps to identify the reaction mechanism. It also helps to design the reaction chamber to better accommodate the temperature dependence. Evaluation of the activation energy at different temperature regimes also helps to determine whether the reaction mechanism changes with temperature. If the concentration terms are kept constant, the rate of the reaction varies only with the rate constant k.

$$\ln k = E/RT + \text{constant}$$

By plotting in $\ln k$ against $1/T$in what is called an Arrhenius plot, we obtain a straight line; its slope is a measure of the activation energy. Figure 4.6 is an Arrhenius plot showing the temperature dependence of reaction rate. Notice that large activation energies result in high temperature dependencies

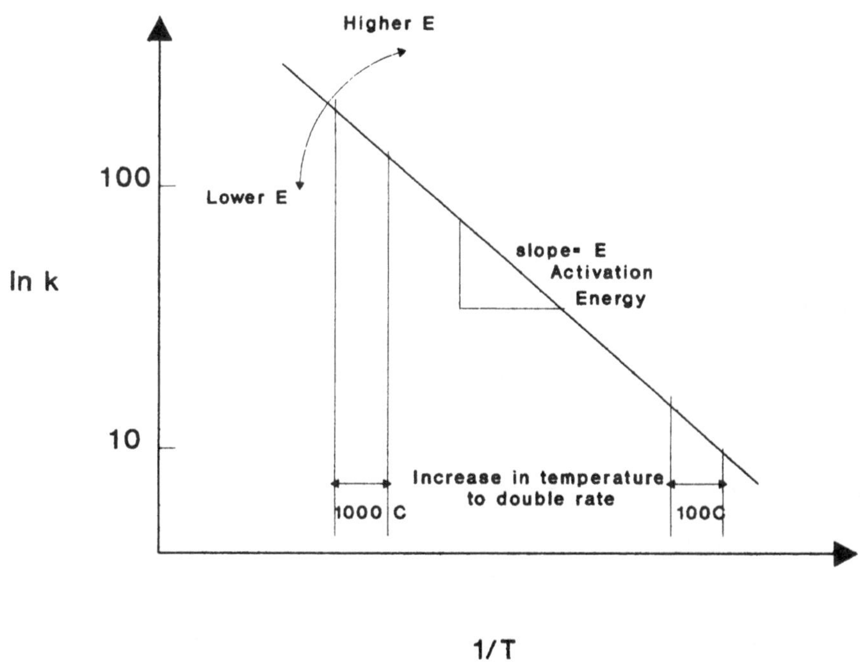

FIGURE 4.6. A plot of the logarithm of reaction rate against inverse temperature is called an Arrhenius plot. Slope of Arrhenius plots indicate the activation energy; notice the rate change is much higher at lower temperatures for a given temperature increase.

and vice versa. At lower temperatures, further to the right-hand side on the horizontal axis, the rate change is more sensitive to changes in temperature than it is at higher temperatures.

Finally, in an Arrhenius plot, the concentration terms are kept constant. This might be easier said than done, especially at high temperatures. The reaction might proceed fast enough for the concentration to change with time, even in a continuous flow reactor. Care must be taken to ensure there are no gradients in the reactor when determining the kinetics of a reaction. We will address this more thoroughly in the section on reactor design, when we discuss the type of reactors most suited to study kinetics.

Exercise 4.4

1. Often it is mentioned that the reaction rate doubles for every 10°C increase in temperature. What activation energy would validate this rule if the reaction were carried out at (a) 298 K and (b) 700 K.
2. The reaction $aA + bB \rightarrow cC + dD$ is elementary as written and the rate expression is $-r_A = kC_A^a C_B^b$. If we write the reaction as $A + (b/a)B \rightarrow (c/a)C + (d/a)D$, will the rate equation become $-r_A = kC_A C_B^{b/a}$.
3. Let us try to get a better understanding of the exponential dependence of rate by comparing some reactions. The activation energies of some typical CVD reactions are

$$SiH_4 = Si + 2H_2 \ (1.6 \text{ eV})$$

$$WF_6 + H_2 = W + 6HF \ (0.7 \text{ eV})$$

$$TiCl_4 + NH_3 = TiN + HCl + H_2 \ (1.2 \text{ eV})$$

Plot these slopes and compare the increase in temperature needed to double the rate at 400°C and 600°C. Conversely, from the same temperature for all three reactions, increase the temperature by 50° and compare the increase in rate.

4.4. HETEROGENEOUS REACTIONS

We introduced the fundamental concepts of reaction kinetics using gas phase, homogeneous reactions. The relative simplicity in the treatment of the kinetics of gas phase, homogeneous reactions results from the absence of any other resistance to the progress of the reaction apart from the activation energy. However, CVD reactions for the production of thin films are by definition heterogeneous, involving a gas phase and at least one solid phase. Thus, for example, all three reactions considered in Exercise 4.4-3 produce

Si, W, or TiN solid films from gaseous sources. In this situation, more of the subprocesses considered in Chapter 1 become critical, and in particular we have to account for mass transfer. Let us examine what happens if we introduce the mass transfer resistance to the reaction. We ignore adsorption and desorption effects until the next section.

4.4.1. Mass Transfer Resistance

Consider the following irreversible reaction:

$$A(g) \Rightarrow B(s) + C(g)$$

The overall reaction can be described in the following three steps:

a. Gas A with a bulk concentration of C_b is transported to the surface, where its concentration is C_s.
b. The reaction occurs on the surface producing gas C.
c. Gas C is transported from the surface to the bulk.

Since the reaction is irreversible, the forward rate of disappearance of A can be written in terms of the concentration of A. At steady state, all the individual steps need to have the same rate to ensure there is no accumulation of any species. Hence the disappearance of A can be written in terms of both steps (a) and (b). The rate according to step (a) is

$$r_p = k_m(C_b - C_s) \tag{4.44}$$

where k_m is the mass transfer coefficient, expressed in terms of a unit area of the surface. Expressing the rate in terms of the reaction rate in step (b),

$$r_p = kC_s \tag{4.45}$$

Note that the rate constant k, in this equation should also be written in terms of unit surface area, not unit mass of reactant disappearing, as used in equation (4.30). The concentration at the interface, C_s, is not an easily measured quantity and hence cannot be left in the final expression. Solving for C_s from the two equations,

$$C_s = \frac{k_m}{k_m + k} C_b \tag{4.46}$$

This expression for C_s can now be substituted into either of the two rate equations to give the overall rate as

$$r_p = \frac{1}{(1/k) + (1/k_m)} C_b \tag{4.47}$$

This equation illustrates how to add the resistances offered to the conversion of reactants to products through each of the subprocesses. Adding these chemical resistances is similar to adding electrical resistances.[9] Thus the net resistance of the reaction consists of the reaction resistance ($1/k$ term) and the mass transfer resistance ($1/k_m$ term).

If the rates are not linearly dependent on concentration, the expressions become very complicated. For example, if the reaction has a second-order dependence on C_s, instead of the first-order dependence treated in equations (4.44) and (4.45), the final expression is[10]

$$r_p = -\frac{k_m}{2k}\left(2kC_b + k_m - \sqrt{k_m^2 + 4k_m k C_b}\right) \tag{4.48}$$

The resistances are no longer additive. The net effect of nonlinear dependence of rate on concentration is to make rate expressions extremely cumbersome, both physically and mathematically.

To circumvent this problem, in a set of serial processes, we introduce the concept of the rate-controlling step. We already know that each of these subprocesses has strong dependence on temperature. The rates of the subprocesses can be very different from one another at any given temperature. Hence it is possible to identify one of the steps in the sequence of processes to be the slowest at a given temperature. Since at steady state all rates have to be the same, this step would then dictate the rate of the overall process. If the rate-controlling process at any temperature can be identified, the prediction of rate becomes much easier. This idea of a rate-controlling step is a powerful concept. Increasing the rates of processes and optimization of reactor designs for economic reasons at any operating temperature often depend on isolating and addressing the slowest link in the chain of processes. We will study the role of the rate-controlling step in the determination of film properties during our discussion on reactor design.

4.4.2. The Kinetics of Adsorption

Let us address one more link in the chain of subprocesses, the kinetics of adsorption of the gaseous molecules onto the substrate surface and the desorption of the products from the surface to the gas phase. Adsorption is the process through which the concentration of molecules attached to surface sites on the substrate becomes different from the interfacial gas phase concentration C_i.[11] Adsorption processes have been studied in extensive detail for specific gas–surface couples.[12] We will limit ourselves to introducing the fundamentals of the rates of adsorptive processes and refer the reader to many fine texts in the area of surface science for further reading.

Consider the equilibrium established between gas A, with a concentration C_i in the gas phase at the interface, and N_a attached to the surface. N_a is similar to C_s, the interfacial concentration of reactants participating in the surface reaction, used in equation (4.45). We use the term N_a to be consistent with literature in surface physics. The adsorption process can be written as

$$A_{gas} \underset{k'}{\overset{k}{\rightleftharpoons}} A_{surface} \tag{4.49}$$

and the net rate of adsorption can be written as

$$\Gamma\left(\frac{\text{molecules}}{\text{cm}^2\ \text{s}}\right) = kC_i - k'N_a \tag{4.50}$$

where k and k' are the rate constants for adsorption and desorption. The surface concentration term N_a has to be distinguished from the other concentration terms, such as C_i, since N_a is expressed in two dimensions. N_a is defined as the number of atoms per unit area of surface, this implies the rate constant k' also has to be expressed in terms of unit area.

The ratio of the number of atoms attached to the surface to the number of available sites on the surface for adsorption is called the surface coverage θ. For low coverages, the rate of desorption is small, so the rate of adsorption can be written as

$$\Gamma = kC_i$$

Using the impingement rate in equation (2.2) the rate constant k can be written as

$$k = \frac{\alpha P}{\sqrt{2\pi mkT}}$$

where α is the adsorption coefficient.

The adsorbed layer concentration, or adlayer concentration, N_a in molecules/cm^2 is governed by the fundamental kinetic relationship of adsorption:

$$N_a = I\tau \tag{4.51}$$

where I is the impingement rate, and τ is the surface residence time, see equation (2.3). We can relate the surface residence time to the surface vibrational frequency $\nu = 1/\tau_0$ and the binding energy on the surface $Q_{des} = \Delta H_d$, the heat of desorption, and using equation (2.3) we can write

$$\tau = \tau_0 \exp(-\Delta H_d/kT) \tag{4.52}$$

If we now substitute for the different terms in the fundamental adsorption equation (4.51) we can write

$$N_a = \frac{\alpha P}{\sqrt{2\pi mkT)}} \tau_0 \exp(-\Delta H_d/kT) \tag{4.53}$$

The surface concentration can thus be estimated from the pressure, the temperature, and the surface binding energy. For the case of Ar on W, under 10^{-6} torr at room temperature, the concentration of Ar on the W surface is about 10^4 atoms/cm^2, compared to a monolayer concentration of 10^{14}/cm^2. If the pressure is raised to 1 atmosphere, the concentration increases to 10^{15} atoms/cm^2.

Since we have related the surface residence time to the binding energy, we can now create an energy diagram similar to the one we used for the activated complex. We can write

$$\Delta H_d = -\Delta H_a = -(E_0 + RT) \tag{4.54}$$

where E_0 is represented energetically in Figure 4.7. Depending on the well depth E_0 in Figure 4.7 the molecule can either be considered to be physically adsorbed (physisorbed with a small E_0, similar to Ar on W) or chemically adsorbed (chemisorbed with a large E_0).

Surface coverage θ can also strongly influence the surface concentration. Since there is only a finite number of surface sites available for adsorption, two simplistic cases can be considered. In the first case, we consider an infinite number of surface sites available for adsorption, hence surface coverage does not determine the surface concentration. In the second case, once the surface site has an atom attached to it, it is no longer available for further adsorption. These two situations are respectively called the Henry's law model[13] and the Langmuir model.[14]

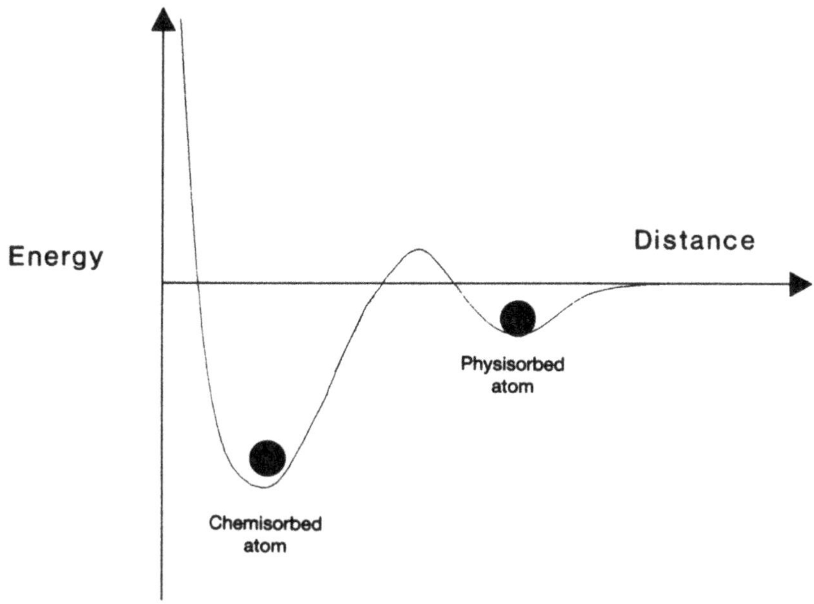

FIGURE 4.7. Schematic of the energy diagram relating to physical and chemical adsorption processes.

The Henry's law model assumes the surface lifetime τ is independent of coverage θ. The adlayer concentration N_a is the product of the impingement rate I and the residence time τ. Hence we can write a relationship between N_a and pressure p for constant T; this is called an adsorption isotherm. For the Henry's law case, the adsorption isotherm is

$$N_a = K_1 P \tag{4.55}$$

As shown in Figure 4.8b, there is a linear dependence of N_a on P, and the gradient of the line increases with decreasing T.

The Langmuir model assumes $\theta \leqslant 1$, i.e., the coverage can never exceed one monolayer. When a molecule is adsorbed onto a site, it effectively removes the site from further adsorption, hence the surface concentration can be written as

$$N_a = I\left(1 - \frac{N_a}{N_0}\right)\tau \tag{4.56}$$

With some algebra, the coverage can be simplified to an expression of the

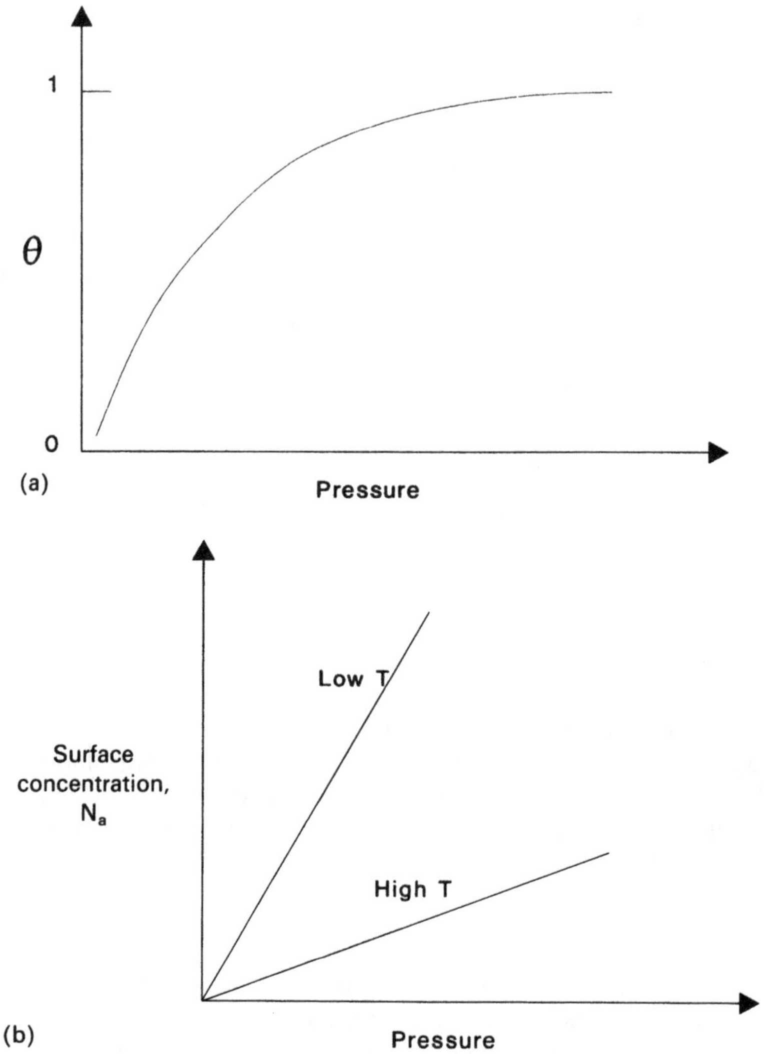

FIGURE 4.8. A plot of surface coverage with pressure at constant temperature is an adsorption isotherm: (a) the Langmuir isotherm and (b) the Henry's law isotherm.

following type for constant temperature:

$$\theta = \frac{K_L P}{1 + K_L P} \tag{4.57}$$

where K_L is a function of temperature. This expression is called the Langmuir adsorption isotherm and is graphically represented in Figure 4.8a. At very small coverages, the Langmuir model approaches the Henry's law model; at high pressures the coverage tends to be constant and equal to unity.

Thus knowing the surface binding energy and the concentration at the interface, we have a relationship to express the surface concentration. We also have rate expressions, so we can determine whether the overall rate of a CVD process is controlled by adsorption or by desorption.

Exercise 4.5

1. For a surface exposed to air (assume its molecular weight is 28), at room temperature, find the time taken to reach a surface coverage of 0.5, if the surface follows (a) Langmuir and (b) Henry's law behavior. The average residence time is 1 s, and the system pressure is 1×10^{-6} torr.
2. A sealed system with an initial pressure of 1 torr, containing nitrogen at room temperature is suddenly cooled to (a) 100 K and (b) 10 K. The average surface residence time is 50 s. What is the final pressure in the two cases? State any reasonable assumptions made. The heat of adsorption is 3 kcal/mol.

4.4.3. Summary

In our discussion of chemical kinetics, we reviewed some notation for the analysis of chemical reactions. We interpreted chemical equilibrium from a kinetic viewpoint and reconciled it with thermodynamic equilibrium. We noted that in a kinetic sense, at equilibrium, forward and reverse reaction rates become microscopically equal.

We explained the exponential dependence of reaction rate on temperature through an Arrhenius equation. We saw how this dependence was modified for heterogeneous reactions that are the basis of CVD. First, mass transfer resistance was modeled for the simple case of first-order equations. When the expression became complicated for higher-order equations, we resorted to modeling only the rate-controlling elementary reaction. We then proceeded to examine adsorption and desorption, two more subprocesses of CVD. Physical and chemical adsorption situations were presented along with Langmuir and Henry's law models of adsorption. Henry's law adsorption assumes no dependence of adsorption on surface coverage of adsorbates; Langmuir adsorption assumes a maximum of monolayer coverage. Now it is time to develop an overall picture to identify the rate-controlling mechanism in CVD processes.

4.5. SEARCHING FOR AN OVERALL MECHANISM

Economic optimization of the reaction necessitates as thorough a knowledge of the overall process as possible, before a reactor can be designed to commercially use the reaction. This helps in extending the operating range of the reaction by increasing its rate or slowing down unwanted side reactions. But above all it helps to obtain the desired film properties for the CVD process.

The search for the mechanism of a heterogeneous reaction begins with an understanding of the stoichiometry of the reaction, and the comparison of the stoichiometric coefficients to the kinetic expression. This is the first indication of whether or not the reaction is elementary. Theoretical calculation of the collisional frequency terms and comparison with experiment can also help to identify elementary reactions. Identification of the mechanism is made considerably easier by understanding a few commonly observed empirical rules.

4.5.1. Principle of Microreversibility

If there are two alternate paths available for a forward reaction, the principle of microreversibility[15] suggests the preferred path in the forward reaction would be the same for the reverse reaction.

Consider for example the thermal decomposition of SiH_4 to silicon.

$$SiH_4 \leftrightarrows Si + 2H_2$$

The forward reaction can be considered as a silane molecule forming an activated complex which then decomposes to the silicon atom and the two hydrogen molecules. By the principle of microreversibility, the reverse reaction has to be the same. This requires two hydrogen molecules to react simultaneously with a silicon atom to form the activated complex. Since such a multibody collision is unlikely, there is probably a simpler forward reaction. In this case, it is a sequence of elementary reactions proceeding as

$$SiH_4 \leftrightarrows SiH_2 + H_2$$

and

$$SiH_2 \leftrightarrows Si + H_2$$

Thus the principle of microreversibility eliminates many possible reaction paths as unlikely.

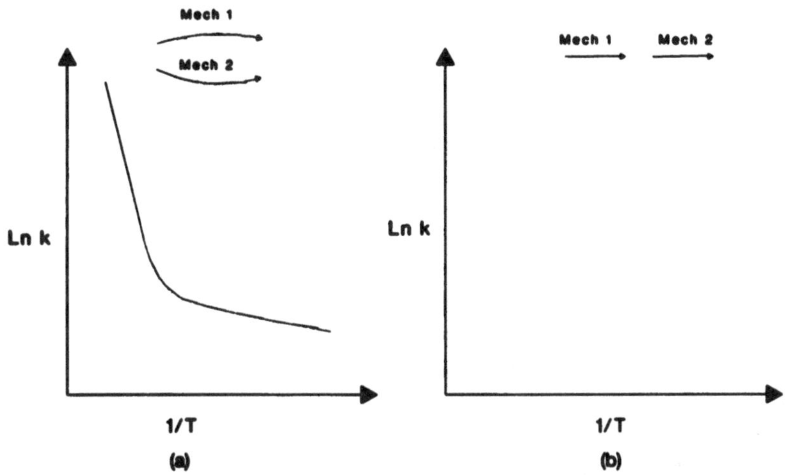

FIGURE 4.9. Identifying reaction mechanism by observing changes in the Arrhenius plot: (a) changes in parallel mechanisms and (b) changes in sequential mechanisms.

4.5.2. Changes in Activation Energy

In an Arrhenius plot of a process, a change in the slope of the line indicates a change in the rate-controlling step or the mechanism itself. If the change is from a high slope to a low slope on increase of temperature, it often indicates change in the rate-controlling step of a sequence of subprocesses. On the other hand, an increase in slope going from low to high temperatures, often indicates a different process in a set of parallel paths becoming the preferred path. The two cases are illustrated in Figure 4.9.

4.5.3. Surface Reaction and Mass Transfer Controlled Growths

Let us again consider a process whose activation energy plot is shown in Figure 4.9b. At lower temperatures, the Arrhenius plot has a steep slope, indicating a strong dependence on temperature. The situation is similar to Section 4.3.3, where the rate of the forward reactions rose very fast when the temperature was increased at relatively low temperatures. Hence the overall process is constrained by the rate of the forward reaction. Thin film growth is *controlled by the reaction rate.* We are implying that all other subprocesses, including transport of the reactant to the surface, adsorption, desorption, and transport away from the surface, are faster than the forward reaction rate, which is now the rate-controlling step.

Under these conditions, films exhibit relatively low growth rates and all the properties that go with them (see Chapter 2). They have strong texture and relatively small grain size, due to the low growth temperature. There is little incorporation of impurities, since desorption is no longer rate-controlling. They exhibit excellent conformality and step coverage, since at all surfaces, both horizontal and vertical, the reaction has the same rate. Growth rate uniformity is a very strong function of temperature uniformity; isothermal reactors exhibit better film uniformity than cold wall reactors.

An increase in the temperature produces a transition area in the Arrhenius plot after which the slope is lower, indicating a different rate-controlling step. The transition area is often indicative of mixed rate-controlling steps in the reactor, with gradients in the reactor causing different rate-controlling steps at different locations. Although we might want to operate at as high a temperature as possible within the rate-controlled regime, we should still avoid the transition area because we cannot be certain its properties are uniform.

At even higher temperatures, the Arrhenius plot shows a greatly reduced slope, indicating a process that is less sensitive to temperature. This regime is often called *mass transfer controlled*; the reaction rate is fast enough so that it does not control the overall rate. Rather, the arrival rate of the reactants and hence the surface concentration controls the rate. All the species adsorbed on the surface react immediately and the overall reaction is constrained by the nonavailability of reactants. Such a situation may arise due to many reasons:[16]

a. The gas feed rate to the reactor is not sufficient to keep up with the consumption of the reactants by the reaction, resulting in a *starved* reactor.
b. Transport of the reactants from the bulk gas stream to the heterogeneous reaction on the surface, as shown in equation (4.46), is not fast enough to keep up with the reaction.
c. Desorption of products of the reaction is slow, resulting in a competition for surface sites such that the adsorption process described in Section 4.4.2 is rate-controlling.

We will examine the effect of reactor configuration on the regime in the next chapter. Each of the above conditions can be tested separately. (a) can be tested by increasing the partial pressures of each of the reactants. (b) can be tested by reducing the overall pressure to increase the gas phase diffusivity. And (c) can be tested by ensuring the pumping speed is large enough to maintain a low residence time for all species in the reactor.

Films grown in this regime exhibit excellent property control and the process is ideally suited for manufacturing. Gas flow rates can be better controlled than substrate temperature, especially under reduced pressures, hence high and uniform growth rates can be achieved simultaneously.

However, an overwhelming problem with this growth regime is poor film conformality. Since mass transfer through various features, such as cylindrical contact holes, can be different for different gases and can be very poor overall, growth rates can vary within the feature.

Microstructurally, the higher temperature allows for larger grain size. However, very fast growth rates combined with high temperatures can result in dynamic recrystallization of the growing film, often resulting in high film stresses.

Exercise 4.6

1. Homogeneous decomposition of $SiCl_4$ proceeds according to the rate equation $r_{SiCl_4} = k[SiCl_4]^2[Cl_2]^{-1}$. What is the overall order of the reaction? Suggest a mechanism for the reaction. How will you test for it?

References

1. For chemical engineering thermodynamics: J. M. Smith, and H. C. Van Ness, *Introduction to Chemical Engineering Thermodynamics,* McGraw-Hill, New York, 1975. For materials engineering: G. W. Gaskell, *Metallurgical Thermodynamics,* McGraw-Hill, New York, 1975.
2. For statistical thermodynamics: F. C. Andrews, *Equilibrium Statistical Mechanics,* John Wiley, New York, 1963; H. C. Van Ness, *Understanding Thermodynamics,* Ch. 7, McGraw-Hill, New York, 1969.
3. D. R. Lide and H. V. Kheiaian, *CRC Handbook of Thermophysical and Thermochemical Data,* CRC Press, Boca Raton, 1994.
4. K. Denbigh, *The Principles of Chemical Equilibrium,* 4th ed., Cambridge University Press, Cambridge, UK, 1983.
5. J. M. Smith, *Chemical Engineering Kinetics,* p. 18, McGraw-Hill, New York, 1981.
6. *JANAF, Thermochemical Tables,* 2d ed. (D. R. Stull and H. Prophet, eds.), NSRDS-NBS37, 1971.
7. R. H. Perry, and C. H. Chilton, *Chemical Engineers' Handbook,* 5th ed., McGraw-Hill, New York, 1973.
8. O. Livenspiel, *Chemical Reactor Engineering,* p. 18, John Wiley, New York, 1972.
9. J. H. Oxley, in *Vapor Deposition* (Powell, Oxley, and Blocher, eds.), Ch. 4, John Wiley, New York, 1966.
10. O. Livenspiel, *Chemical Reactor Engineering,* p. 351, John Wiley, New York, 1972.
11. G. A. Somorjai, *Principles of Surface Chemistry,* Prentice Hall, Englewood Cliffs, N.J., 1972.
12. G. A. Somorjai, *Chemistry in Two Dimensions: Surfaces,* Cornell University Press, Ithaca, N.Y., 1981.
13. D. O. Hayward and B. M. W. Trapnell, *Chemisorption,* 2d ed., Butterworth, London, 1964.
14. I. Langmuir, *J. Am. Chem. Soc.* **40,** 1361 (1918).
15. O. Livenspiel, *Chemical Reactor Engineering,* p. 31, John Wiley, New York, 1972.
16. G. B. Raupp, "CVD Reactor Design," in *Tungsten and Other Refractory Metals for VLSI Applications III* (Wells, ed.), Materials Research Society, Pittsburgh, 1988.

Chapter 5

Reactor Design for Thermal CVD

Classical reactor design requires the application of the principles of heat, mass, and momentum transfer for the efficient conversion of reactants to products. We will treat this subject with a narrower focus: thin films with desired properties from gaseous sources. Such a discussion may be insufficient for the design of a commercial reactor from first principles. However, for the modification of an existing reactor for the purpose of tailoring film properties or for the construction of laboratory prototypes, this discussion will be adequate.

This chapter can be divided into two major sections, the first of which deals with the general understanding required for the design of a conceptual reactor for a thermal CVD process. The second deals with applications of these concepts to commercial atmospheric pressure and reduced pressure CVD reactors. We will discuss the properties of thin films produced in these reactors under various regimes of operations.

5.1. CLASSIFICATION OF REACTORS

While laboratory reactors can be built to suit short-term needs of a researcher, economic necessities require commercial reactors to follow certain standards, classified according to configurations. As we refer to these configurations throughout the chapter, here is a brief overview.

Based on the number of substrates (wafers) a reactor can process at one time, commercial reactors can be classified as either single-wafer or batch type. A batch type reactor offers economies of scale in throughput, while single-wafer reactors frequently offer better process control from one wafer to the next. It is speculated that as substrate sizes get very large (>200 mm

TABLE 5.1 Types of Reactors

	Type	Distinguishing feature	Commonly grown films
1.	Single wafer	Process control	Dielectrics, metals
2.	Batch	High throughput	Epi, oxides, W, WSi_x
3.	APCVD	High rate	Doped oxides, epi
4.	LPCVD, RPCVD	Process control, uniformity	Dielectrics, W, WSi_x, poly
5.	Hot wall	Isothermal	Epi, TiN, poly, nitride
6.	Cold wall	No deposition on walls	W, oxides
7.	Chamber shape	Tube	Poly, doped oxides, nitride
		Pancake	Epi, poly
		Barrel	Epi

in diameter), the two types might converge to a large, single-wafer configuration for certain applications.

We will discuss the role of pressure in reactor design in the next section. Thermal CVD reactors can either operate at atmospheric pressure or at reduced pressures. The choice is based on the ability of the reaction chemistry in delivering film properties such as uniformity and step coverage. Lower pressure reactors require pumping and other vacuum related equipment to maintain reduced pressures. Thus operating pressure is another convenient way of classifying reactors: low pressure CVD (LPCVD), atmospheric pressure CVD (APCVD), subatmospheric CVD (SACVD), etc.

The temperature of the substrate is determined by the requirements on the reaction rate. However, the temperature of the other surfaces in the reactor can be maintained at any optimum temperature for the reaction. The temperature of the chamber walls gives rise to yet another way of classifying reactors: hot or cold wall reactors. Hot wall reactors are isothermal reactors whereas cold wall reactors keep the gas phase at ambient temperatures. Table 5.1 summarizes different reactor types.

Other possible reactor classifications include those based on the shape of the chamber (pancake, tube, barrel, etc.), and method of heating (resistive, inductive, etc.). Since there are multiple reactor configurations available for each type of film, choice of reactor for a particular film is often the most critical step during process development. Figure 5.1 illustrates the different reactor types available for the deposition of epitaxial silicon alone.[1]

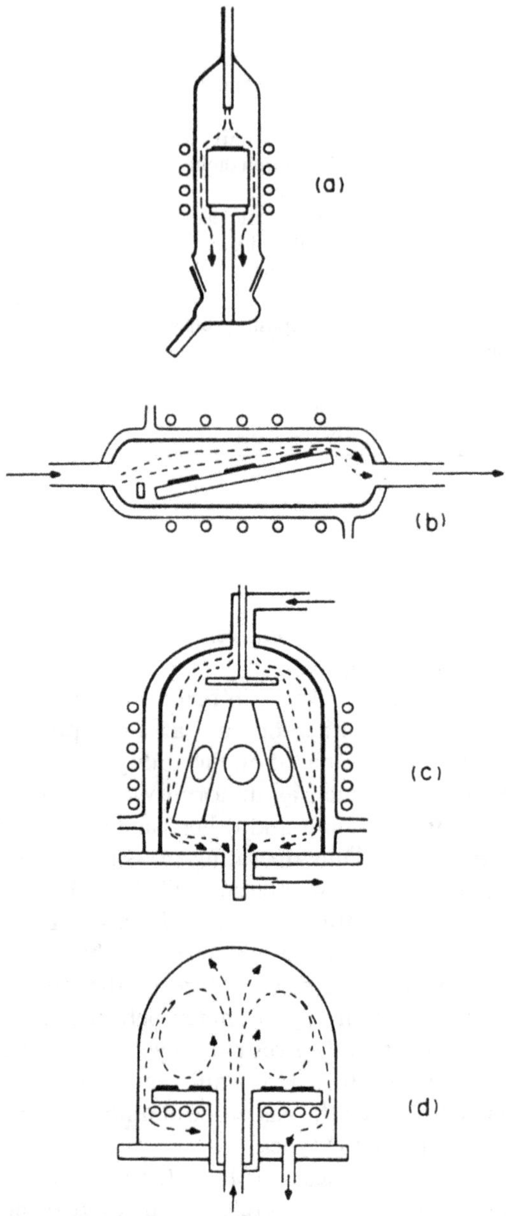

FIGURE 5.1. Reactors for epitaxial film deposition: (a) horizontal, (b) barrel, and (c) vertical. Requirements on film properties have forced this evolution in reactor configurations. Reprinted from W. Kern and V. S. Ban, in *Thin Film Processes*, Academic Press, 1978, p. 284, with permission.

5.2. PRESSURE AND FLOW REGIMES IN CVD REACTORS

Flow of a mass of fluid occurs because of the existence of a pressure gradient. We considered partial pressures in the last chapter only as replacements for gas phase concentration terms. We will now discuss the role of pressure in fluid flow and how it affects reactant transport to the surface. Except for specific deviations, the discussion still assumes ideality in gases. Kinetic theory of gases provides most of the fundamental framework needed for the ensuing discussion and hence a background in its principles is strongly recommended. Reference texts on kinetic theory of gases are provided at the end of chapter.[2]

5.2.1. Continuum and Discrete Treatments of Flow

The ideal gas law can be written in terms of number density of molecules in the gas phase n as

$$P = nkT \tag{5.1}$$

where k is Boltzmann's constant. Note that we are no longer using k to mean reaction rate constant and we are now using k instead of κ for Boltzmann's constant. Given this relationship between the number density and the pressure, we can also express the mean free path of molecules λ in terms of n. λ is the average distance travelled by molecules between collisions and is given by

$$\lambda = \frac{1}{\sqrt{2}n\sigma} \tag{5.2}$$

where σ is the effective collisional cross section between the molecules. From equations (5.1) and (5.2), we can tabulate the mean free path as a function of pressure at constant temperatures, as shown in Table 5.2.

The calculations are for a molecule of mass 30 amu (air) at room temperature. However, the orders of magnitude in Table 5.2 are applicable to a wide range of atomic masses.[3] At a typical operating pressure of 1 torr in LPCVD, an average atom can travel about 0.045 mm before it collides with another atom.

Using Table 5.2 we can define a dimensionless number to predict the pressure at which the flow of a body of vapor can be treated as a continuum (viscous treatment of flow) and where the continuum treatment breaks down and the role of individual molecules needs to be taken into account (discrete

TABLE 5.2 Mean Free Path as a Function of Pressure at Constant Temperature

Pressure (torr)	Number density (molecules/cm^2)	Intermolecular spacing (cm)	Mean free path (cm)	Impingement rate (molecules/cm s)
760	2.7×10^{19}	3.3×10^{-7}	6×10^{-6}	4×10^{23}
1	3.5×10^{16}	3×10^{-6}	4.5×10^{-3}	5×10^{20}
10^{-3}	3.5×10^{13}	3×10^{-5}	4.5	5×10^{17}
10^{-6}	3.5×10^{10}	3×10^{-4}	4.5×10^{3}	5×10^{14}

treatment of flow). A dimensionless parameter, called Knudsen number is defined as

$$K_n = \frac{\lambda}{a} \tag{5.3}$$

where a is a characteristic dimension of the system under consideration. When the Knudsen number is less than 0.01, we can consider the flow as viscous and apply continuum mechanics to explain it. When the Knudsen number is greater than 1, gas–wall collisions become important and the system has to be considered molecular or discrete. The treatment of flow can no longer use continuum mechanics, but has to look into statistical mechanics and kinetic theory. In the region where $0.01 < K_n < 0.1$, both gas–gas and gas–wall collisions become important. Called transition or slip flow, it involves components of molecular migration along the system walls and is extremely difficult to treat mathematically. In this discussion we will ignore transition flow, even though it applies to several LPCVD situations.

Of the two factors that go into the calculation of the Knudsen number, the mean free path, determined by pressure, is easily obtained. But for a given system, the characteristic dimension can depend on what is being evaluated. Consider the deposition of SiO_2 on a wafer at a reactor pressure of 100 millitorr. The reactor is a cylinder with a diameter of 25 cm. The Knudsen number for this situation is much smaller than 0.01. But deposition is actually occurring on the wafer, where we need to fill a 0.3 μm gap, a typical intermetal space. In this case the characteristic diameter is no longer 25 cm but 0.3 μm. The Knudsen number is 1.5×10^4 and the situation is clearly molecular. Viscous laws do not apply.

5.2.2. Continuum or Viscous Flow

Let us begin by considering momentum transfer in fluid flow. When a fluid flows with a velocity u over a stationary surface, the immediate layer of fluid above the surface is also stationary. A gradient in velocity develops normal to the surface till the velocity reaches u at some distance δ above the surface. Through his law of viscosity Newton intuitively defined the shear force exerted in the plane of the surface due to the velocity gradient as

$$F = \eta dA \frac{du}{dy} \tag{5.4}$$

where F is the force acting on the area dA and y is the distance normal to the surface. η is called the coefficient of viscosity. Again from kinetic theory of gases, the coefficient of viscosity can be derived from molecular impingement rates on a unit area, and can be shown to be dependent on the mean free path and mean molecular velocity as

$$\eta = \frac{\lambda n m \bar{v}}{3} \tag{5.5}$$

The velocity gradient du/dy in the expression for the viscous shear force has been calculated exactly for a cylindrical tube. If we assume no gradients in velocity in the axial direction, u varies parabolically in the radial direction as shown in Figure 5.2. For a tube of radius a and with an axial pressure gradient $\Delta P/\Delta x$ (the cause of the flow), u can be written as

$$u = \frac{1}{4\eta} \frac{\Delta P}{\Delta x} (a^2 - r^2) \tag{5.6}$$

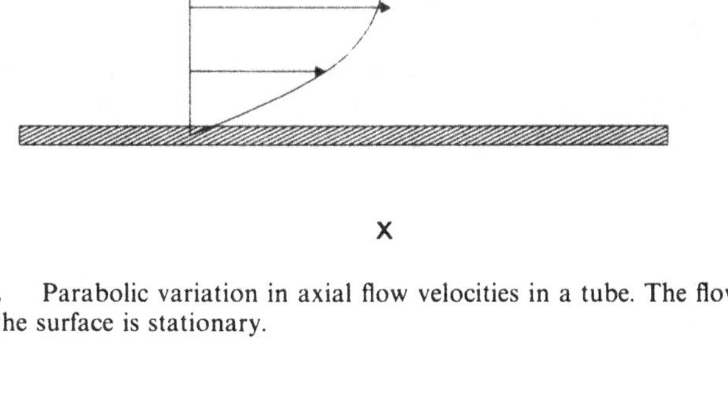

FIGURE 5.2. Parabolic variation in axial flow velocities in a tube. The flow layer in contact with the surface is stationary.

Let us define a new dimensionless number by taking the ratio of inertial to viscous forces acting on a fluid element. Called the Reynolds number, it can be defined as

$$R_e = \frac{\rho u D}{\eta} \tag{5.7}$$

Here ρ is the mass density of the gas and D is the tube diameter. When the Reynolds number is very high ($R_e > 2\,000$), the flow in the tube exhibits an irregular flow pattern and is called turbulent. In general, turbulent flow is observed at high pressures and high flow rates. Turbulent flow causes particulate problems in CVD reactors, so it has to be avoided. At lower Reynolds numbers ($R_e < 1\,000$), the flow occurs in a smooth, layered manner and is termed laminar flow. The occurrence of turbulent flow at high Reynolds numbers is intuitively obvious if we examine the factors involved. For a fluid with low viscosity, high fluid velocities in a large tube lead to a breakdown of smooth layered flow. In commercial reactors, the Reynolds number is very low ($R_e < 100$), so the flow is almost always laminar. The most probable location for the incidence of turbulent flow is at the mouths of pumping ports, especially if the pumping speed is very high and the pressure is close to atmospheric.

The region from the stationary wall to where the fluid reaches the bulk velocity is known as the boundary layer. Its thickness δ delineates a relatively stagnant layer, through which CVD reactants in the bulk of the gas need to diffuse to reach the surface. Hence the thickness of the boundary layer is an important term in determining the mass transfer of reactants from the gas phase to the surface. The thickness of the boundary layer can be expressed as

$$\delta = \sqrt{\frac{x}{R_e}} \tag{5.8}$$

where x is the distance along the tube. In the case of a tubular reactor, the thickness of the boundary layer increases in the axial direction, increasing the resistance to mass transfer. In order to keep the boundary layer thickness constant, we can continuously increase the Reynolds number by increasing the flow velocity. Such a scheme is commonly referred to as a tilted susceptor reactor[4] and is shown in Figure 5.3.

Exercise 5.1

Plot the Knudsen number of the cylindrical reactor of diameter 20 cm and height 25 cm shown in Figure 5.3 as a function of the chamber pressure

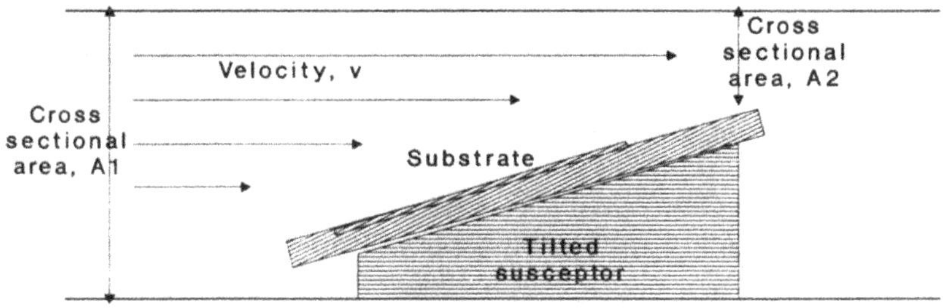

FIGURE 5.3. Tilted suscepptor reactor that tends to maintain constant Reynolds number along the axis of the reactor by varying the flow velocity.

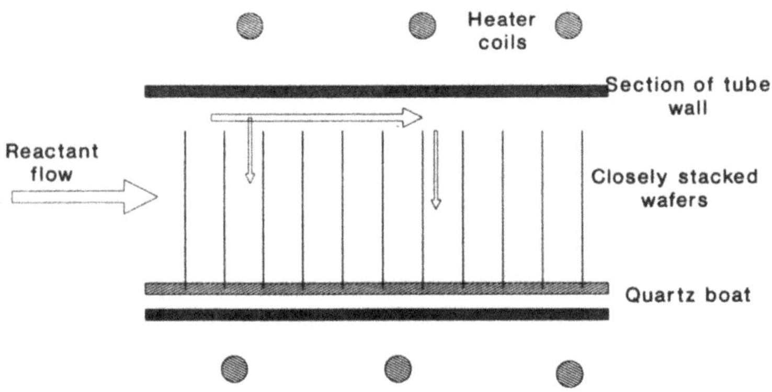

FIGURE 5.4. Diffusivity of gas species between closely spaced, vertically placed wafers in a tube reactor. The diffusivity is enhanced by decreasing the pressure in the tube.

between 760 torr and 10 millitorr. Argon gas is flowed into the reactor at a rate of 200 sccm. Also calculate the boundary layer thickness on the surface of a 15 cm circular wafer placed in the center of the chamber as a function of the distance from the cylinder axis and the axial distance from the edge of the wafer.

5.2.3. Need for Low Pressure CVD

Decreasing the total pressure of a reactor increases the mean free path and increases the diffusion coefficient of the reactants. Often in a batch reactor, when substrates are closely stacked (Figure 5.4), diffusivity of the reactants between the substrate wafers can limit the growth rate, leading to large growth rate differences within the wafer. One solution is to increase the space between the wafers, leading to smaller load sizes; another is to reduce the total pressure. Since the earliest use of low pressure CVD (LPCVD) in batch, tubular reactors, other advantages have come to light and LPCVD has become the deposition of choice for many applications. For instance, control of pressure and temperature allows the user to select the rate-controlling subprocesses. Coupling of electrical plasmas to CVD reactors in order to affect both deposition and reactor cleaning is often contingent on the presence of lower pressures in the reactor. Thus it is common to find even a single-wafer reactor in both LPCVD and atmospheric pressure configurations.

5.2.4. Molecular Flow

We saw earlier that when the mean free path is of the same order of magnitude as the characteristic dimension of the reactor, the flow needs to be considered molecular, and gas–wall collisions become important. Gas molecules strike the wall and are reemitted in random directions. Elastic collisions with the wall result in pressure. Inelastic collisions, including adsorption and surface reaction, result in a cosine distribution of gas molecules desorbing from the surface. There is no velocity gradient (or boundary layer) in the direction normal to the wall. From kinetic theory and the ideal gas law, some important characteristics of the gas phase can be summarized as follows:

1. Gas phase velocities follow the Maxwell–Boltzmann distribution, illustrated in Figure 5.5 and expressed as

$$\frac{dn}{dv} = 4\pi^2 n\left(\frac{m}{2\pi kT}\right)^{3/2} v^2 \exp(-mv^2/2kT) \tag{5.9}$$

   where dn/dv is the number of molecules having velocities between v and $v + dv$.
2. The average molecular velocity in the gas phase is a function only of the temperature and can be written as

$$\bar{v} = \sqrt{\frac{8kT}{\pi m}} \tag{5.10}$$

3. The rate of collision against the walls is equal to the impingement rate and is given by

$$I = \frac{1}{4} n\bar{v} = \frac{P}{\sqrt{2\pi mkT}} \tag{5.11}$$

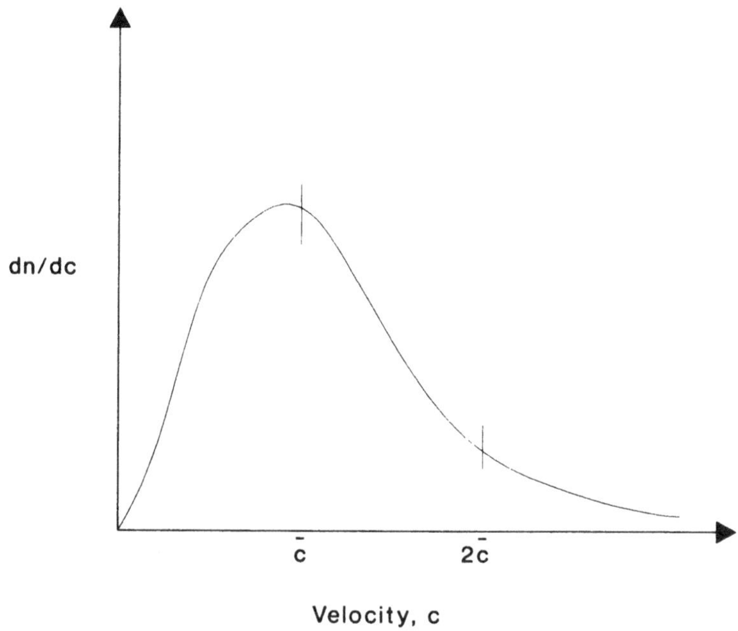

FIGURE 5.5. Maxwell–Boltzmann distribution of gas phase velocities in an ideal gas. Notice the mean velocity $\bar{c}$ is slightly to the right of the most probable velocity, which is the peak of the distribution.

The resistance offered by a tube to fluid flow can also be different for different regimes. We will treat this separately for viscous and molecular flow in the vacuum technology section of the appendix. Molecular transport through orifices can be an effective model for predicting film growth rates inside contact holes.

5.3. RESIDENCE TIMES IN REACTORS

In an ideal CVD reactor, all the reactants introduced at the source spend a fixed time in the reactor, dependent on the reactor volume and the incoming reactant flow rate. The flow rate and the conditions inside the reactor are optimized for conversion of reactant to product and to ensure the solid film has the necessary properties. Such a situation can be idealized into two extremes,[5] illustrated in Figure 5.6.

Figure 5.6a illustrates an ideal, steady-state, stirred tank reactor. Its key feature is that mixing in the reactor volume is complete, so properties such as concentration and temperature are uniform, i.e., there are no gradients within the reactor. Even though reaction occurs only at the solid surfaces

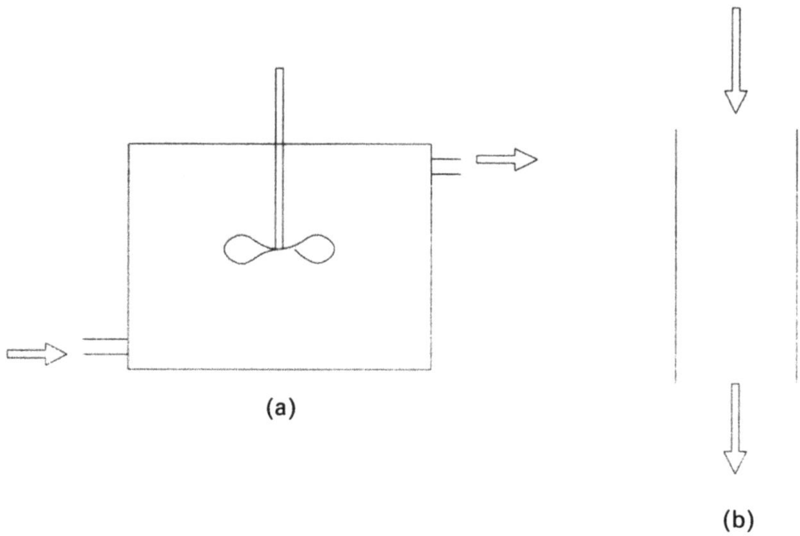

FIGURE 5.6. (a) Ideal stirred tank reactor that ensures no gradients within the reactor. (b) Ideal plug flow reactor where the gas stream shows a concentration gradient along the axis.

and there are no homogeneous reactions, mixing is sufficiently complete to ensure the exit stream truly represents the composition within the reactor and there is no net accumulation of gas phase reactants or products. The mean residence time in this reactor is simply the volume of the reactor divided by the flow rate. Individual fluid elements show a distribution of residence times.

Figure 5.6b illustrates the other extreme ideal situation of a tubular or plug flow reactor. In this situation, there is no mixing in the axial direction, the direction of flow, and complete mixing in the radial direction. Although there is a concentration gradient in the direction of flow, there is none in the radial direction. The reaction rate will vary in the axial direction, unless otherwise compensated. The residence time for every element of fluid is the same and is equal to the volume of the reactor divided by the flow rate.

However, in practical CVD reactors, due to the presence of the heterogeneous reaction, and due to factors other than reactant conversion, nonideality in flow is a common occurrence. Elements of fluid may move through the reactor at different velocities leaving *dead spots* or local eddies. Channelling might cause some elements of fluid to travel faster through the reactor without sufficient time for reaction. Figure 5.7 illustrates some of the nonideal flow situations in reactors.[6] In a CVD reactor, deviations from ideality need to be avoided for many reasons.

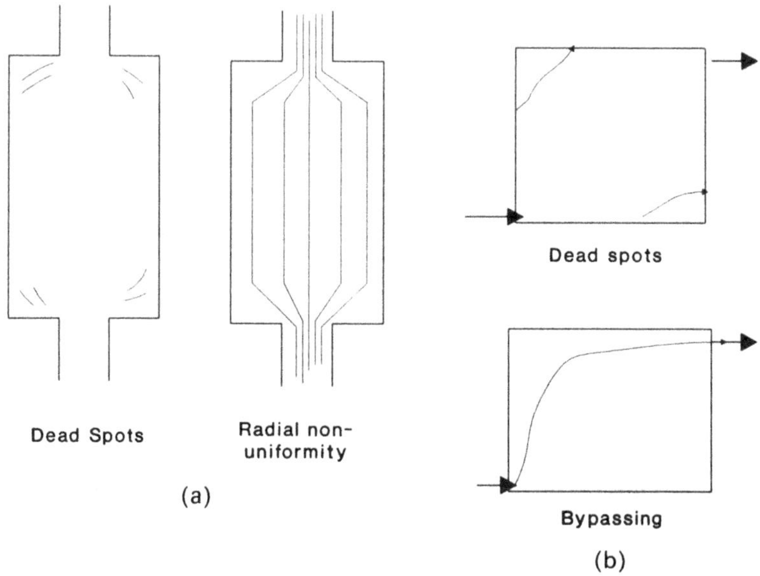

FIGURE 5.7. Causes for nonideal fluid flow in (a) ideal plug flow and (b) ideal stirred tank reactors. Nonideality is to be avoided in reactor design.

a. Efficiency of the reactor is lowered.
b. Often stagnant layers result in homogeneous reactions and create particulate problems.
c. Increased concentrations of products that are not pumped away might result in competition for adsorption sites on the substrate surface, lowering the deposition rate.
d. Local differences in concentration can lead to nonuniformity in deposition rate on the substrate surface.
e. Incorporation of the product gases in the growing film can have profound effects on its properties (see Chapter 2).

Given the importance of understanding flow patterns in the reactor, measurement of residence time in a reactor assumes paramount importance in its design. Formally, the time a molecule takes to pass through a reactor is called its residence time θ. θ is made up of two components: the time elapsed since the molecule entered the reactor, i.e., its age, and the remaining time it will spend in the reactor before exiting, i.e., its residual lifetime.[7]. It is usually impossible to measure either of these quantities inside the reactor.

At steady state, the mean residence time of the fluid elements is the same as the residence time in an ideal reactor.

$$\bar{\theta} = \frac{V}{Q} \tag{5.12}$$

where Q is the volumetric flow rate and V is the reactor volume. The residence time distribution function $J(\theta)$ is defined as the fraction of fluid leaving the reactor that has residence time less than θ. Hence the mean residence time can be written as

$$\bar{\theta} = \frac{\int_0^1 \theta dJ(\theta)}{\int_0^1 dJ(\theta)} \tag{5.13}$$

In order to normalize the function, the denominator is set to 1, so we can write

$$\bar{\theta} = \int_0^1 \theta dJ(\theta) = \frac{V}{Q} \tag{5.14}$$

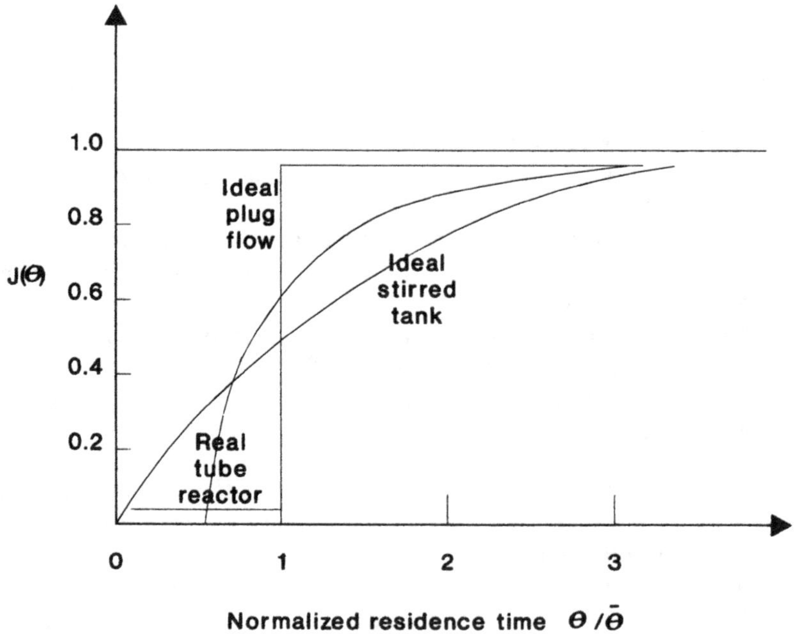

FIGURE 5.8. Residence time distributions in actual reactors. Ideal reactors are shown for comparison.

Figure 5.8 illustrates the residence time distribution for ideal and nonideal reactor configurations. Measurements of residence time distribution are made by introducing a tracer gas, either as a step function or as a pulse then measuring time versus tracer concentration distribution of the exit stream. Often in CVD reactors, residual gas analyzers are used to detect the effluent gas stream. One has to ensure in such measurements that the measurement time is small compared to the residence times being measured.

5.4. GRADIENTS IN REACTORS

Deviation from ideality produces flows with a distribution in the residence times of fluid elements within the reactor. Difference in the ages of reactants produces concentration differences between fluid elements. Flow patterns also produce gradients in concentration, velocity, and temperature throughout the reactor. As we saw in reaction kinetics, local concentration gradients can produce local variations in reaction rates. The temperature gradient normal to the substrate surface determines the extent or possibility of homogeneous nucleation. Velocity patterns and the extent of free and forced convection within the reactor determine the degree of mixing of the incoming reactants within the reactor. Even though some laboratory reactors have been built to be essentially gradient-free, almost all commercial reactors exhibit gradients in gas phase properties. Let us examine some of these gradients and their influence on film properties.

5.4.1. Concentration Gradients and Conversion

Conversion can be defined as the mole fraction of the inlet flow that is converted into product. When a single reaction is occurring within a reactor, the concentration of one of the reactants is likely to limit the reaction rate, and conversion can be expressed in terms of that reactant. The concentration of all other reactants can then be expressed in terms of the extent of the reaction and the conversion of the limiting reactant. For example, in the tube flow reactor discussed in the previous section, conversion X at the outlet of the reactor can be written as

$$X = \frac{F_i - F_o}{F_i} \tag{5.15}$$

where F is the molar flow rate and the subscripts i and o refer to inlet and outlet conditions.[8] The heterogeneous reaction rate at a given temperature can then be expressed in terms of the conversion as

$$r_o = F_i \frac{dX_o}{dA} \tag{5.16}$$

where dA is a differential reaction area for the heterogeneous reaction. Consider the following reaction occurring inside a reactor with an exit total pressure p:

$$\text{A(g)} + 2\text{B(g)} \rightarrow \text{C(s)} + 2\text{D(g)} \tag{5.17}$$

If the controlling reactant is A, and the inlet molar flow of B is twice that of A, the flow at the exit consists of $(1 - X)$ of A, $(2 - 2X)$ of B and $2X$ of C. The total flow is then $(3 - X)$. The exit partial pressure of A is $p(1 - X)/(3 - X)$ and so on for the other components.

That is why the reactor has to be operated at maximum conversion for reasons of economy. But this leads to concentration gradients in the reactor. Ideally, a uniform deposition rate requires the conversion to be as low as possible with the resultant low concentration gradients across the wafer surface.

Exercise 5.2

The elementary reaction A + B → C is taking place in the gas phase of a square pipe. The gas feed rate is pure A, and pure B evolves as vapor from liquid present at the bottom of the tube. The reactor is in a plug flow mode and B maintains its equilibrium vapor pressure. Express the reaction rate as a function of conversion. What is r_A when the conversion is 50%? Ignore the volume of the liquid B. The equilibrium vapor pressure of B is 0.25 atm, the temperature is maintained at 273 K, and the pressure is maintained at 1 atm. The rate constant $k = 10^6\ \text{l}^{-1}\ \text{kmol}^{-1}\ \text{s}^{-1}$, and the flow rate of A is 1 mol/s.

A differential reactor is one in which the conversion is very small ($X < 0.05$). Hence an ideal differential reactor exhibits only a very small gradient in concentration. Such a reactor is commonly used for the determination of reaction kinetics. For instance, consider the Arrhenius plot shown in Figure 4.6. We determined that the steep portion indicated control by the reaction rate and the gentler portion indicated control by the arrival of the reactants—mass transfer control.

If the reactor is maintained under a differential conversion regime, the conversion of the limiting reactant will be small, so we can be sure that the reactor is not *running starved*. And when we use an Arrhenius plot to determine the mass transfer kinetics, we need to ensure the conversions of all the reactants are small. To distinguish between a starved reactor and a truly mass transfer controlled reactor, we can define two dimensionless numbers, called the Damkohler numbers.[9]

$$D_1 = \frac{V/Q}{LC/r(X = 0)} \tag{5.18}$$

and

$$D_2 = \frac{L/D}{C/r(X = 0)} \tag{5.19}$$

where V is the reactor volume, Q is the volumetric flow rate, L is a characteristic dimension of the boundary layer, D is the gas phase diffusion coefficient, C is the concentration of the limiting reactant, and r is the reaction rate at differential conversion. The first Damkohler number is the ratio of the residence time to the typical reaction time. The second is a ratio of the time for diffusion through the boundary layer to the typical reaction time. Small D_1 denotes differential conversion and large D_2 denotes mass transfer control.

5.4.2. Temperature Gradients

We described hot and cold wall reactors in Section 5.1. Hot wall reactors are isothermal reactors, where the gas phase is at the same temperature as the reactor walls and the substrate. In a cold wall reactor, the substrate is maintained hot and a steep temperature gradient exists in the direction normal to the substrate surface. The walls of the reactor are cooled to maintain constant temperature. The logistics of maintaining such temperature gradients in commercial reactors are dealt with later.

The choice of hot versus cold wall reactors is based only on the chemistry of the reactants.

a. Since the reaction occurs on all available hot surfaces, if adhesion of products to the chamber walls is a problem, hot wall reactors should be avoided.
b. If the occurrence of a homogeneous reaction is possible at elevated gas temperatures, a cold wall reactor is preferable.
c. If the boiling point of any of the reactants is low enough such that condensation on the walls could occur, the walls should be heated.

In general hot wall reactors are best suited to batch processing. Intermediate wall temperatures have also been used for reactions such as

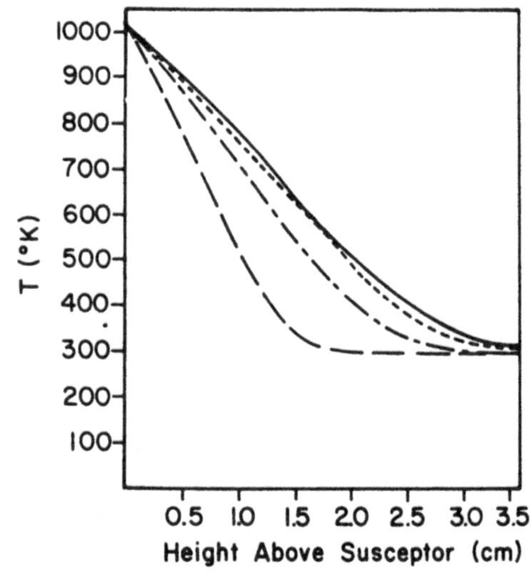

FIGURE 5.9. Carrier gases and diluent gases can have a profound influence on the temperature of the gas phase: (top) He, (upper middle) H_2, (lower middle) high flow He, (bottom) N_2. Reprinted from Ref. 11 with permission from the Electrochemical Society.

the formation of TiN from $TiCl_4$ and NH_3, where there is an alternate reaction path at lower temperatures.[10]

The temperature of the gas phase in a cold wall reactor can be critically affected by the thermal conductivity of the gas phase. Hence the choice of an inert carrier gas has to take into account its thermal properties along with its role in the reaction. Similarly, gas phase velocity can determine the temperature gradient over the hot susceptor. Figure 5.9 illustrates the effect of the carrier gases on the thermal gradient normal to the substrate surface.[11]

5.4.3. Velocity Gradients

In addition to the external flow imposed on the reactor, the presence of temperature gradients results in a convection component to the overall flow pattern in a reactor. A dimensionless number called the Grashoff number is a ratio of buoyancy forces to viscous forces in the reactor and is given by

$$Gr = \frac{g\beta_c L^3 \Delta T}{\eta^2} \tag{5.20}$$

where g is the acceleration due to gravity and L is the characteristic reactor dimension.[12] The ratio Gr/R_e^2 acts as a diagnostic for the nature of the flow.

When the ratio is large ($Gr/R_e^2 > 20$), the flow is dominated by buoyancy forces, i.e., free convection dominates the flow pattern. Between 20 and 0.5, the behavior is mixed. At ratios below 0.4, the externally imposed flow dominates.

From a purely thin film growth viewpoint, mixed flow patterns and free thermal flows should be avoided, since they lead to nonuniformities in growth. There have been no structural studies performed on the thin films at different flow regimes.

5.5. COMMERCIAL PRODUCTION CVD REACTORS

We have so far concentrated on the design of the reaction chamber itself, and what occurs within the reaction vessel. In a commercial reactor, in order to make the reaction chamber deposit thin films in a reproducible, manufacturable manner, other subsystems need to be added. Figure 5.10 illustrates the inputs required to turn a reaction vessel into a production reactor. Safety is of prime importance in any reactor and particularly in CVD reactors for microelectronics, where pyrophoric and toxic gases are commonly used. Control systems, levels of redundancy in containment of leaks, and the ability to safely power down in an emergency are essential features of a safe reactor.

The reactor inputs are controlled by personnel through software and hardware interfaces. In a laboratory reactor, manual valves and flowmeters

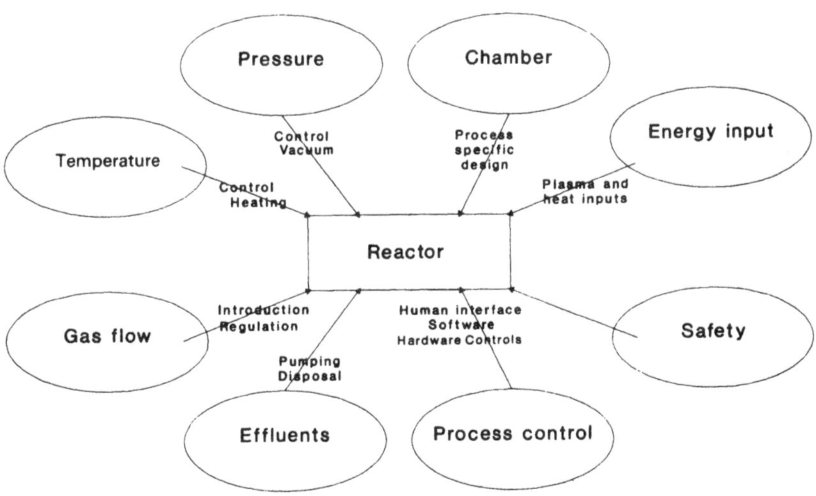

FIGURE 5.10. Inputs needed to design a reactor for a VLSI manufacturing line.

are common, but automatic controls are essential, to ensure reproducibility for all input variables such as pressure, flowrate, and temperature.

The reactor has to be environmentally safe. Since many of the effluent gases can be toxic, safe collection and disposal of the reactor effluent has to be planned before powering up. Controlled combustion and oxidation of effluents to safe and stable oxides has been used to treat CVD effluents. Toxic gases are collected in adsorptive resins or scrubbed through water columns for safe disposal.

In Figure 5.10 the electrical power supply to the reactor is shown as "energy input". The electrical power is used to heat the reactor or to produce a plasma. Plasmas are often used in thermal CVD reactors to clean unwanted deposits on the walls and the substrate holders. We will discuss coupling of plasmas to reactors in Chapter 7. Reactor heating is accomplished commonly through inductive heating or resistive heating. Inductive heating of a conductive susceptor, such as graphite, inside a reactor is shown in Figure 5.11a. Such a system uses cooled RF coils outside the reactor to induce eddy currents in the susceptor. The eddy currents cause the temperature increase. Resistive heating in a hot wall reactor, shown in Figure 5.11b, employs coils of resistive elements wrapped around the reactor tube. A feedback temperature control system using multiple thermocouples is used to maintain the reactor temperature. Often, due to process sensitivity to temperature, the control is better than 5°C at 1 000°C. Resistive heating of the substrate holder in a cold wall reactor is achieved through high current feedthroughs, which provide electrical contact from heating elements in the heater block, beneath the substrate holder, to external power supplies. High intensity infrared lamps have also been used in place of resistance coils to heat the substrate. Heat transfer between the block and the substrate may be conductive or radiative.

Pressure control within the chamber, especially in LPCVD reactors involves the design and incorporation of vacuum systems. A detailed discussion on vacuum system design, pressure, and flow monitoring and control can be found in the appendix.

5.5.1. Atmospheric Pressure Continuous Reactor

In concept the simplest CVD reactors are atmospheric pressure, continuous reactors, since there are no vacuum systems to add complexity; they cannot in general sustain plasmas. Heating of the reactor is resistive and cleaning of the reactor is not performed with plasmas. In a continuous reactor, the reaction chamber itself is at steady state, with reactant gases and wafers being continuously introduced and removed from the reaction area.

A typical example of an atmospheric pressure reactor is the Watkins Johnson Model WJ 999 APCVD system.[13,14] It is mainly used to deposit

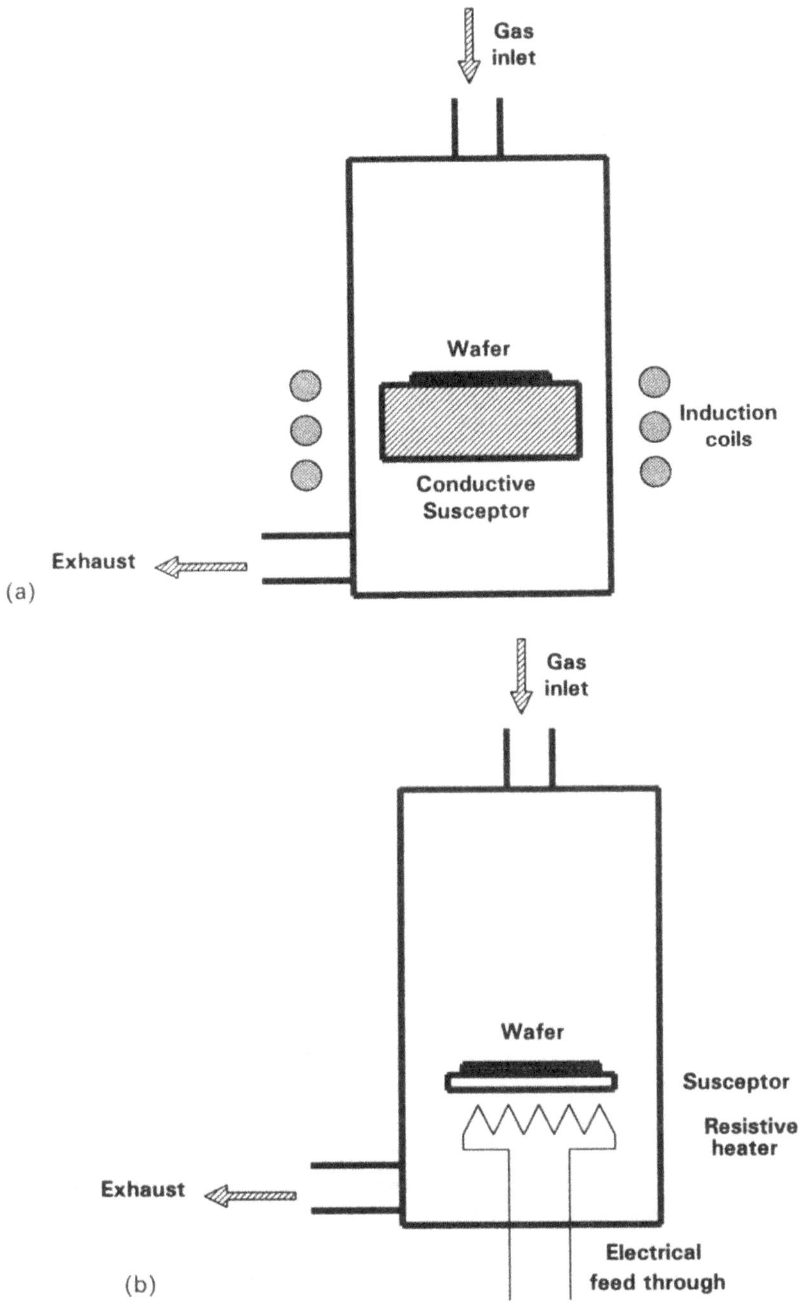

FIGURE 5.11. Heating inside production reactors: (a) inductive and (b) resistive.

SiO_2 films onto silicon substrates. Many variations of SiO_2 can be deposited from silicon-bearing hydride sources (see Chapter 9).

Instead of placing the wafers in a chamber, which is in turn sealed and pumped to remove atmospheric gases (as with LPCVD systems), in the Watkins Johnson APCVD system, the wafers simply pass through the deposition area at atmospheric pressure. A steady flow of high purity nitrogen keeps the deposition area free of contaminants. A high exhaust flow keeps gas phase reaction from reaching the substrate.

Wafers are transported to three deposition areas in sequence by a loadbelt and conveyed out of the hot area by the unload belt. The belt is continuously cleaned of the SiO_2 deposited on its surface in a belt cleaner. Central to the deposition is the delivery of the reactant gases. This is accomplished by a unique injector design. Oxidizers and hydrides, which react to form the film, are kept separate till the gases exit the injector onto the surface of the wafer.

The injector creates a chemical vapor curtain under which the wafers are transported by the belt. The curtain contains a trilinear flow of oxygen in N_2, N_2, and hydride in N_2. Deposition occurs in a small zone where the residence time of the reactants is minimized by the high gas flow volumes. The wafer is heated through a resistive heater block situated beneath the reaction surface.

Films are produced by the following general reactions:

$$SiH_4 + 2O_2 \rightarrow SiO_2 + 2H_2O$$

$$2PH_3 + 4O_2 \rightarrow P_2O_5 + 3H_2O$$

and

$$B_2H_6 + 3O_2 \rightarrow B_2O_3 + 3H_2O$$

Depending on the presence of the last two dopant hydrides, the film deposited is undoped oxide, phosphorus doped glass (PSG), or borophosphosilicate glass (BPSG). The maximum deposition temperature is under 550°C. Substitutes for the hydrides in the three reactions include tetraethylorthosilicate (TEOS, $Si(OC_2H_5)_4$), trimethylborate (TMB $B(OCH_3)_3$), and tetramethyl phosphite (TMPi, $P(OCH_3)_3$).

The silane reaction tends to have a silane conversion of about 10%, approaching that of a differential reactor. Flows in general are high (on the order of 5 standard liters per minute) to maintain the chemical curtain and to maintain desired flow patterns. Excessive hydride concentrations can lead to gas phase nucleation, producing white quartz powder.

Other features of the WJ 999 reactor, unrelated to the reactant chamber, include safe scrubbing of the pyrophoric and toxic gases through a properly designed exhaust and scrubber system. Economics of the reactor are realized through the high wafer throughput obtained by the atmospheric pressure reactor. For a film thickness of 5 000 Å the wafer throughput is higher than 50 wafers per hour for both TEOS and silane-based films. Continuous cleaning of the belt after it passes through the deposition chamber assists in maintaining low system downtime and low particulates.

Exercise 5.3

In the reactor described above, calculate the conversion efficiency of SiH_4 for the following conditions: growth rate 5 000 Å/min, belt speed 15 cm/min, injector width 15 cm, and silane gas flow rate 50 sccm. Assume the growth rate does not change for a 1% B_2H_6 gas phase composition. If the segregation coefficient of B_2H_6 is 10%, calculate the weight and molar percentages of boron in the film.

5.5.2. Low Pressure, Single-Wafer Reactors

In contrast to the atmospheric pressure reactor discussed above, a low pressure single-wafer reactor offers increased gas phase diffusion, lower gas flows, and often better wafer-to-wafer process control. The chamber can sustain a plasma, and often chamber cleaning is accomplished using reactive plasmas, either after each deposition or after a set of depositions. Low pressure, single-wafer reactors suffer the disadvantage of reduced wafer throughput, often compensated by having multiple reaction chambers on a single chassis which provides the wafer handling and control capabilities. Such a *cluster* tool is illustrated in Figure 5.12.[15]

Typical of LPCVD cluster reactors is the Applied Materials Precision 5000 series, used in various modifications to produce SiO_2, Si_xN_y, and tungsten.[16,17] We will follow the construction and film properties pertaining to the Applied Materials 5000 blanket tungsten reactor. A schematic of the reactor is shown in Figure 5.13. Wafers are loaded onto the susceptor face up through an automated loading mechanism. The susceptor is radiatively heated from below using an array of quartz lamps. Gas introduction is through a showerhead, small holes on the lower wall of a mixing chamber into which reactant gases are fed. Vacuum manifolds, situated around the circumference of the reactor, lead to mechanical pumps and on to the exhaust. The reactor volume is small, dictated by the size of the wafer and the spacing between the wafer and the shower head.

An RF discharge (see Chapter 7) can be sustained between the susceptor and the chamber walls. This feature, along with the ability to purge the space below and around the susceptor, helps prevent/remove backside deposition of tungsten. Frequent cleans of the chamber and the wafer rear help to

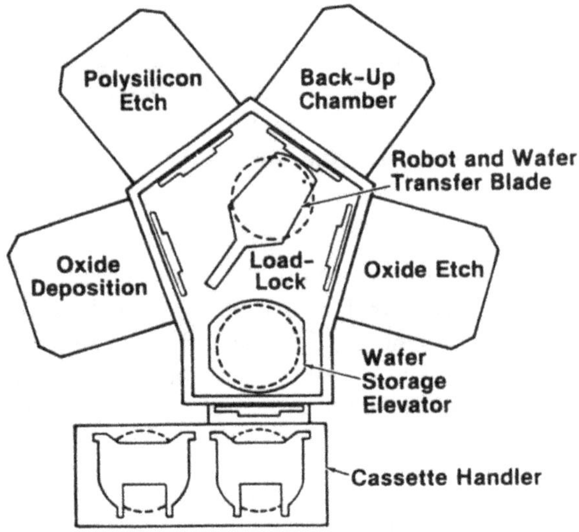

FIGURE 5.12. A schematic cluster production tool. In one housing and handler assembly the tool contains deposition, etching, and plasma cleaning stations. Reprinted with permission from Applied Materials, Inc.

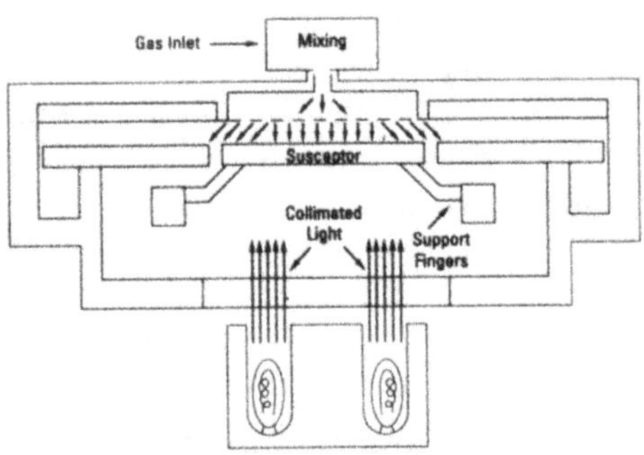

FIGURE 5.13. Applied Materials Precision 5000 series reactors can be configured for dielectric or conductive deposition processes. Reprinted with permission from Applied Materials, Inc.

minimize particle contamination arising out of tungsten deposition and subsequent peeling of other surfaces.

Tungsten deposition chemistry (Section 8.3) consists of the following reactions:

$$2WF_6 + 3SiH_4 \rightarrow 2W + 3SiF_4 + 6H_2$$

for tungsten nucleation and

$$WF_6 + 3H_2 \rightarrow W + 6HF$$

for the bulk growth. Partial pressure of the reactant gases and the total pressure in the chamber, along with the temperature of deposition, affect film properties such as resistivity, resistance uniformity, step coverage, and stress. The other important reactor-related factor is the spacing between the showerhead and the substrate. This spacing changes the reactor volume and affects both the deposition rate and the WF_6 conversion efficiency.

Film growth rate in this reactor varies between 2 500 and 5 000 Å/min, with W resistivity at 9 $\mu\Omega$ cm. Step coverage can be optimized to better than 90%.

Exercise 5.4

For a processing pressure of 40 torr, at a gap between the susceptor and the wafer of 600 mils, calculate the Reynolds number of the flow in the chamber. The flows are respectively 200 sccm of H_2, 16 sccm of WF_6, and 110 sccm of Ar. What is the thickness of the boundary layer over the wafer? The wafer size is 200 mm. If the growth rate is 4 000 Å/min at 420°C, and the growth is diffusion-controlled, calculate the gas phase and surface concentrations of WF_2 and H_2. State any assumptions made.

References

1. M. L. Hammond, *Solid State Technol.* **21**(11), 69 (1978).
2. S. Dushman and J. M. Lafferty, *Scientific Foundations of Vacuum Technique*, John Wiley, New York, 1962; G. N. Patterson, *Introduction to Kinetic Theory of Gas Flows*, University of Toronto Press, Toronto, 1971.
3. J. B. Hudson, private communication.
4. A. Sherman, *Chemical Vapor Deposition for Microelectronics*, p. 16, Noyes, Park Ridge, N.J., 1987
5. J. M. Smith, *Chemical Engineering Kinetics*, p. 104, McGraw-Hill, New York, 1981.
6. O. Livenspiel, *Chemical Reaction Engineering*, John Wiley, New York, 1982.
7. A. S. Inamdar, and C. M. McConica, in *Tungsten and Other Refractory Metals for VLSI/ULSI Applications V* (Wong and Furukawa Eds.), p. 93, Materials Research Society, Pittsburgh, 1990.
8. G. P. Raupp, in *Tungsten and Other Refractory Metals for VLSI Applications III*, (Wells Ed.), p. 15, Materials Research Society, Pittsburgh, 1988.
9. J. Bloem, and L. J. Giling, *Current Topics in Materials Science* **1**, 147 (1978).
10. A. Sherman, *J. Electrochem. Soc.* **137**(6), 1892 (1990).
11. M. E. Coltrin, R. J. Kee, and J. A. Miller, *J. Electrochem. Soc.* **131**, 425 (1984).
12. K. Jensen, and W. Kern, in *Thin Films Processes II* (Vossen and Kern, eds.), p. 284, Academic Press, New York, 1991.

13. Watkins Johnson Ltd., private communications.
14. L. Bartholomew, and J. Sisson, in *Proc. 3rd Ann. Dielectrics and Metallization Symp.*, J. C. Schumacher, San Diego, Calif. 1981.
15. E. A. Matsuzak, C. M. Hill, and D. V. Horak, *SPIE Proc.* 1188 (1989).
16. Applied Materials Ltd., private communications.
17. T. E. Clark, and A. P. Constant, *Microelectronic Manuf. and Testing* May/June, 8 (1990).

Chapter 6

Fundamentals of Plasma Chemistry

Up to now we have considered CVD processes in which the source of energy for the forward process of endothermic reactions was purely thermal. However, interactions involving charged particles produced in a plasma have been effectively utilized in various CVD processes to reduce reaction temperatures. Figure 6.1 illustrates the familiar energy diagram for a reaction: reaction pathway X is the one we have previously considered in thermal CVD, where the forward reaction between reactants A and B has to overcome the potential hill, corresponding to an activation energy. However, the presence of charged particles opens up new reaction pathways such as Y, with a lower activation energy.[1] The lowering of the activation energy through the formation of excited species A* and B* allows the forward reaction to proceed at lower substrate temperatures or at increased rates for the same temperature when compared with thermal CVD.

The need for lowering the temperature in IC processing arises from many requirements. For example, in an integrated circuit, once aluminum-based metallization has been deposited and patterned, the highest allowable processing temperature for the wafer is constrained below the melting point of the aluminum alloy (about 650°C). In most cases, due to other diffusive processes at temperatures close to the melting point of aluminum, operating temperatures have to be maintained below 450°C.

The use of a plasma also provides other advantages. For instance, energetic charged particle collisions can produce metastable species (such as C* in Figure 6.1), new reaction paths and produce species not available through thermal CVD; control of film microstructure and mechanical properties can be enhanced through ion bombardment; and directionality in film deposition

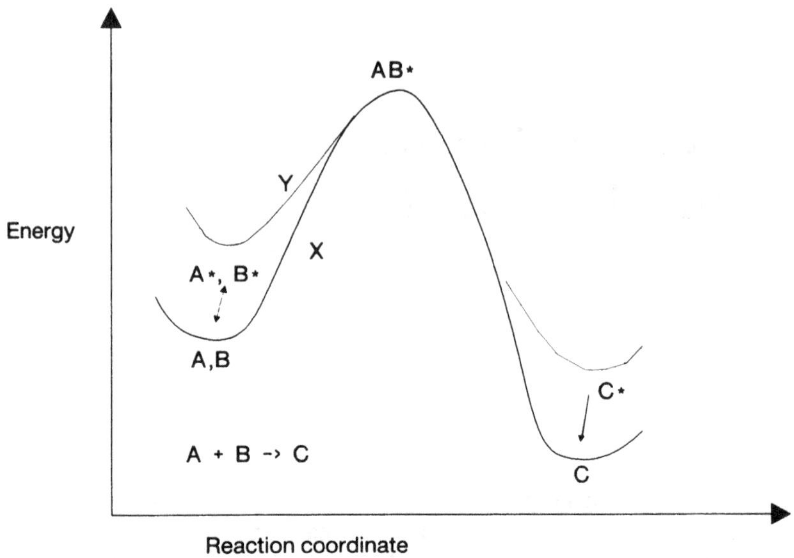

FIGURE 6.1. Energy change and species formed (X) with a thermal CVD path and (Y) with a plasma CVD path. The plasma reduces the activation energy.

and etching can be achieved through the acceleration of charged species in a directional potential gradient.

In this chapter we consider the fundamentals of plasma chemistry by focusing on the interactions between electrons, ions and neutrals in a discharge. We do this from a physical viewpoint and from the viewpoint of chemical reactions between them. Chapter 7 will focus on the production and sustenance of processing plasmas in commercial CVD reactors optimized for thin film deposition.

6.1. BASICS OF PLASMAS

Let us begin with a definition:

> A plasma is a quasi-neutral gas of charged and neutral particles which exhibits collective behavior.

What we mean by *quasi-neutral* and *collective behavior* will become clear shortly. In this section we study the behavior of a plasma in the following sequence:

a. Understanding the constituents of a plasma, their relative concentrations and energies.

b. Since the plasma contains charged species, we develop expressions for plasma potential with reference to the confining walls and the substrate.
c. Understanding the collective behavior of the plasma under external perturbations.

6.1.1. Physical Characteristics of Plasmas

Any body of gas typically contains three species: neutral atoms or molecules, ions and electrons. Their relative concentrations (i.e., the degree of ionization) are considerably different in a plasma. The concentration of ions in the atmosphere is negligibly small. In a typical glow discharge plasma used in CVD processes, ionic concentrations are of the order of 10^{10} per cm^3. The electron concentration is the same as the ion concentration, so the overall plasma is electrically neutral.

From fundamental kinetic theory, the kinetic energy of a particle is given by

$$\varepsilon = \tfrac{1}{2}mv^2 \tag{6.1}$$

If we substitute the Maxwellian velocity distribution into the right-hand side of equation (6.1) after some algebraic manipulation, we can derive the average energy of the gas phase as

$$\bar{\varepsilon} = \frac{\int \varepsilon n(\varepsilon) d\varepsilon}{\int n(\varepsilon) d\varepsilon} = \frac{3}{2}kT \tag{6.2}$$

If we extended equation (6.2) which was derived for neutral atoms and molecules, to include electrons and ions, we can see that the energies of electrons and ions can be represented in terms of a temperature. And in a plasma the electrons and ions can have different velocity distributions (or energy distributions) so there can be many different temperatures in a plasma. Typically, in glow discharge plasmas the electron temperatures are on the order of 1–10 eV (10 000–100 000 K), and the ion temperatures are on the order of 0.1 eV (1 000 K). The temperature of neutrals is similar to the ion temperature and is indicated by more routine temperature measurements.

Figure 6.2 illustrates some characteristics of plasmas, including those that occur naturally (terrestrial and celestial), and some that are artificial.[2] The portion relevant to CVD processing occurs in a region known as *glow discharge*. The term glow discharge arises because atoms excited by electron impact release energy in the form of photons when they relax; the photons produce the glow. The degree of ionization in each of the different plasmas

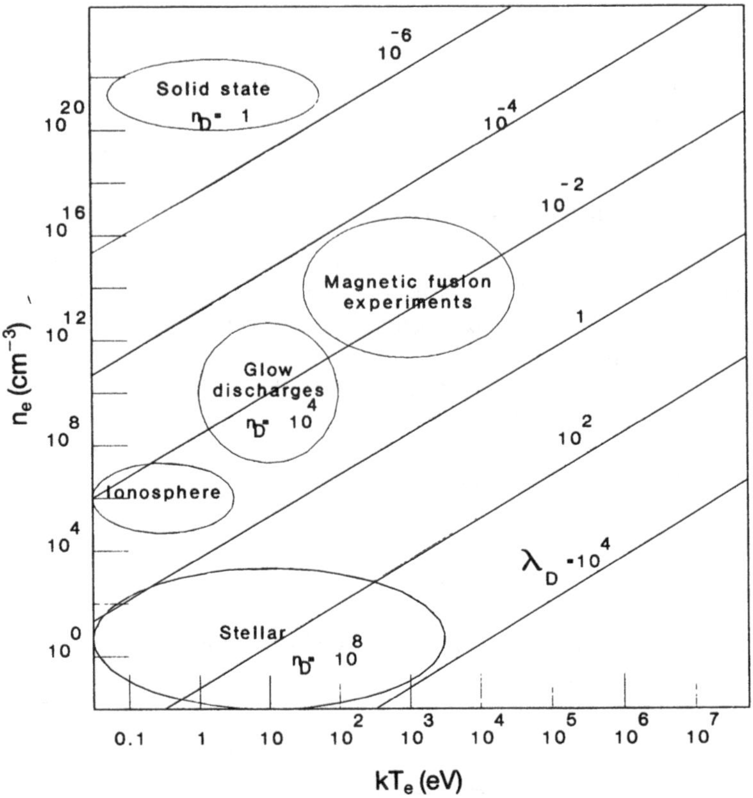

FIGURE 6.2. General characteristics of some natural and man-made plasmas. The illustrated parameters are discussed in the text. λ_D = Debye length (cm) and n_D = the number of electrons in a Debye sphere (order of magnitude). Adapted from Ref. 2 with permission from John Wiley & Sons.

is also shown. Table 6.1 summarizes properties of a typical glow discharge used in CVD reactors. Throughout the rest of this book, the terms plasma, glow discharge and discharge are used loosely and interchangeably. In all cases we refer to the region in Figure 6.2 marked as glow discharge plasma.

An understanding of the interactions between the three types of species and the energetics involved in these processes is essential for us to make use of plasmas in producing thin films. Energy transfer into a plasma and between the species, diffusional phenomena within the plasma and chemical reactions in the presence of charge species are some of the processes that depend on collisions between the three species.

TABLE 6.1 Properties of a Typical Glow Discharge

Electron concentration	10^{10} to 10^{11} per cm^3
Ion concentration	10^{10} to 10^{11} per cm^3
Electron temperature	3–5 eV
Ion temperature	0.05 eV
Debye length	0.1 cm
Electron velocity	10^7 cm/s
Ion velocity	10^4 cm/s

6.1.2. Collisional Processes in a Plasma

In a field-free space, charged particles behave the same way as neutrals, and their behavior can be treated similarly using the kinetic theory of gases. Collisions between particles can be written in terms of a mean free path λ given by

$$\lambda = \frac{1}{\sqrt{2}n\sigma} \tag{6.3}$$

where n is the number density of the particles and σ is the collision cross section. The collision cross section is a measure of the probability that a collision process between the particles in question will occur. Even though the process might be more complicated than the collision of two hard spheres, similar to billiard balls, the collision cross section retains the unit of area. For electrons, the collision cross section is a small number; ions, which are large and more massive, have larger collision cross sections. Electron–electron collisions dominate when the degree of ionization exceeds 10^{-4} or 10^{-3}, i.e., $n_e = n_i > 10^{15}$ per cm^3, seldom seen in processing plasmas.

The mass difference between the electron and the ion is manifested as a difference in their respective accelerations in the presence of a field. The acceleration of a charged particle in an electric field is given by

$$m\frac{dv}{dt} = -ZeE \tag{6.4}$$

where dv/dt is the acceleration of a particle with mass m. Z is the number of charges on the particle and e is the electronic charge. Note that the electric field E is always *as felt* by the charged particle, irrespective of the applied field. For instance, in the parallel-plate diode shown in Figure 6.3, the electron paths from the cathode to the anode are a function of the current density. At a high current density, existence of other electrons in the proximity of a

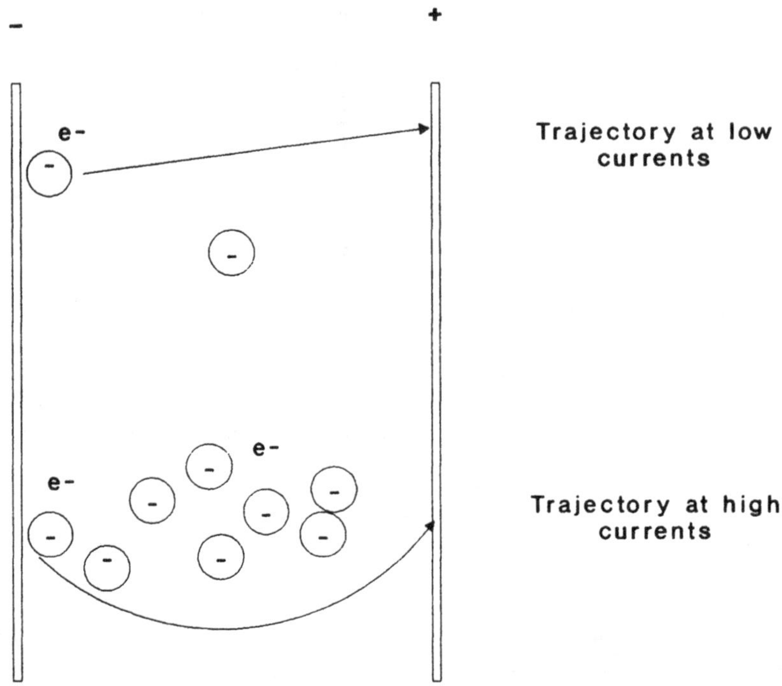

FIGURE 6.3. The electron path from the cathode to the anode depends on the spare charge, the charge density that screens some of the potential.

particular electron can completely shield it from the anode, resulting in a curved trajectory. The shielding electrons are called space charge and we will be studying its effect when we deal with many plasma properties.

Due to their small mass electrons acquire kinetic energy from the electric field much more rapidly than ions, hence energy from an electric field is coupled to a gas entirely through the kinetic energy of electrons. Electrons in turn lose their energy by collisional processes with ions and neutral species present in the gas phase. However, from the classical theory of elastic collisions, the elastic energy transfer per two-body collision (as in a collision between an electron and a neutral/ion) is given by $2m/M$ where m and M are the mass of an electron and neutral/ion respectively. Since m can be more than three orders of magnitude smaller than M, the fractional energy loss per elastic collision between an electron and a massive ion or a neutral particle is negligible.

Inelastic processes, those in which kinetic energy is not conserved, are the primary means through which the externally applied field accelerates electrons and transfers energy to other species in the plasma. Such processes include impact ionization, electron attachment and chemical processes in the gas phase. We will treat them in more detail when we address issues relating to reactions in the plasma. These processes make up the fundamentals of plasma chemistry and their effective application directly contributes to the usefulness of CVD plasmas.

Exercise 6.1

What is the velocity of a 15 eV electron? What is its temperature in Kelvin? If its collision cross section with an Ar atom is $2.5 \times 10^{-15}\ \text{cm}^2$, what is the probability it will collide with an Ar atom when it travels a distance of 1 cm. The chamber pressure is 10 millitorr.

6.2. PLASMA POTENTIAL

In our earlier definition of a plasma, we used the terms quasi-neutrality and collective behavior without dwelling on their true implications. We will develop our understanding of these terms by examining the perturbation caused by a CVD substrate. If the substrate is grounded, it can be considered as similar to the grounded walls confining the plasma. If it is left floating, it assumes a potential V_f. Let us examine the relationship between the potential of the plasma V_p and the potential of an electrically isolated surface, V_f, before considering the case of the grounded substrate.

6.2.1. Electrically Isolated Surface in a Plasma

The impingement rate of any particle arriving at a surface was calculated in Chapter 2 from the kinetic theory of gases. For the arrival of charged particles at the electrically isolated surface, the flux can be modified to a current density J by multiplying the charge carried by the species.[3]

$$J = \frac{ne\bar{v}}{4} \tag{6.5}$$

where n is the density of charged species and $\bar{v}$ is the average velocity of charged species arriving in a random fashion from the bulk of the plasma. Electron velocities in a plasma are much higher than the ion velocities; this is because the electron has a smaller mass (Table 6.1). As a result the electron current arriving at the floating substrate is initially larger than the ion current. The substrate becomes negatively charged and starts to repel further electron flux.

Since the region around the substrate is depleted in electrons, there are fewer collisions between electrons and ions, hence less glow. This region is called the dark space or sheath. Even though the plasma as a whole (including the dark space) is neutral, the local densities of ions and electrons can be different, especially around a point of disturbance. This is the origin of the term quasi-neutrality.

Since V_f acts to repel electrons, by convention it is smaller than V_p. Since electron velocities are always larger than ion velocities, we can generalize this conclusion and state that "the plasma potential is more positive than any potential exposed to the plasma".[4] As there is no other reference potential, $V_p - V_f$ is the only meaningful parameter we can model.

The immediate region around the substrate gets depleted of electrons and has a higher density of positively charged ions. Only those electrons in the body of the plasma that possess energies larger than the repulsive barrier can get through to the substrate. From the Maxwell–Boltzmann distribution, the number of electrons, n_f, out of the electron density n_e, with sufficient energy to overcome the potential hill $V_p - V_f$ is given by

$$\frac{n_f}{n_e} = \exp[-e(V_p - V_f)/kT_e] \tag{6.6}$$

At steady-state conditions, the electron and ion fluxes have to be equal. So, we can individually write out the two fluxes and equate them to calculate $V_p - V_f$.

$$V_p - V_f = \frac{kT_e}{2e} \ln \frac{m_i T_e}{m_e T_i} \tag{6.7}$$

Thus electron and ion temperatures in the plasma and the mass of the ion determine the potential of a floating substrate with respect to the plasma. Conversely, measurement of this potential difference provides a means of measuring electron temperature, as we will see in our discussion of the Langmuir probe in Section 6.4.

6.2.2. Nonisolated Surface in a Plasma

We can also derive an expression for the current through the sheath if the substrate were made conducting, as in the case of a grounded wall or a conducting cathode. And we can calculate the current when conduction is limited by the space charge in the sheath.[5] Intuitively, it would be a function of sheath width d and voltage V across the sheath. Let us consider an electrode at a large negative potential V, similar to a cathode in a DC discharge. Since there is a large negative potential let us ignore the electron current and assume no electrons in the sheath. Current density J at any instant is given by

$$J = nev \tag{6.8}$$

where v is the instantaneous velocity of the ions, in contrast to the electron flux in equation (6.5), the result of random arrival of electrons at an isolated surface. The change in voltage across the sheath can be expressed in terms of Poisson's equation as

$$\frac{d^2V}{dx^2} = -\frac{ne}{\varepsilon} \tag{6.9}$$

where ε is the permittivity. Also the kinetic energy of the ions due to acceleration in the field can be written as

$$\frac{mv^2}{2} = eV \tag{6.10}$$

Eliminating v from the previous two equations, we can write

$$\frac{d^2V}{dx^2} = \frac{1}{\varepsilon}\sqrt{\frac{m}{2eV}} \tag{6.11}$$

The solution to maximum space charge limited current from equation (6.11) is given by

$$j_{\max} = \frac{4\varepsilon\sqrt{2\dfrac{e}{m}V^3}}{9d^2} \tag{6.12}$$

Equation (6.12) gives the maximum current through the sheath when the current is limited by space charge; it is called Child's law. As we expected, the current through the electrode of a discharge or a grounded wall is a function of the voltage across the sheath V and the width of the sheath d. For a grounded wall the voltage is just the plasma potential; for a non grounded conduction electrode, the voltage depends on the applied potential.

6.2.3. Debye Shielding

Collective behavior arises because the plasma reacts to oppose an externally applied charge by forming sheaths. For instance, if an external battery is used to apply a potential V_c at location x_0 in a plasma, the plasma forms a sheath around the field so the effect of the applied potential is diminished at all locations away from x_0. The radial distance from x_0 at which the potential perturbation is reduced by a factor of $1/e$ (to approximately 37%) is known as the Debye length λ_D[6] and is given by

$$\lambda_D = \sqrt{\frac{\varepsilon_0 kT}{n_e e^2}} = 743\sqrt{\frac{T_e(\text{eV})}{n_e(\text{cm}^{-3})}} \tag{6.13}$$

The net effect of the Debye shielding phenomenon is that the plasma readjusts itself to attenuate any perturbation from the plasma potential. The plasma thus attempts to maintain a constant potential. Conversely, when attempting to study the effects of individual charged species, it is not necessary to consider coulombic interaction between the particle and its neighbors more than $2\lambda_D$ or $3\lambda_D$ away. For a typical glow discharge used in CVD, λ_D is on the order of a few hundred microns and sheath thicknesses are many times the Debye length. Debye lengths corresponding to the difference kinds of plasmas are noted in Figure 6.2.

The concentration of electrons inside a Debye sphere is a good estimation of the degree of ionization.[7] The number of electrons n_D inside a Debye sphere is given by

$$n_D = \frac{4\pi}{3}\lambda_D^3 n_e \tag{6.14}$$

$n_D \gg 1$ is often treated as a criterion for the definition of a plasma. Figure 6.2 also shows the number of electrons inside the Debye sphere for various plasmas.

Another implication of the Debye shielding phenomenon is the response of the plasma to alternating fields. The time it takes for an electron to move one Debye length, t_p is given by

$$t_p = \frac{\lambda_D}{v_e} \tag{6.15}$$

This is a reasonable estimate for the time it takes for the shielding to "get in place," i.e., the response time for the plasma to react to an external charge.

The inverse of t_p is called the plasma frequency, ω_p[8] and is given by

$$\omega_p = t_p^{-1} = 5.64 \times 10^4 \sqrt{n_e(\text{cm}^{-3})} \tag{6.16}$$

Collisions result in the damping of an applied electric field frequency, so the plasma remains static. However, when the applied field frequency exceeds the plasma frequency, the plasma is no longer able to shield the applied frequency, resulting in charged particle oscillations within the plasma.

6.2.4. Charge Diffusion in Plasmas

When considering reactor design for thermal CVD, we stressed the importance of low pressure CVD in enhancing the diffusion of reactant species. Diffusion of neutral species under a concentration gradient ∇n follows Fick's law

$$\mathbf{\Gamma} = -D\nabla n \tag{6.17}$$

where $\mathbf{\Gamma}$ is the diffusional flux and D is the diffusion coefficient. However, the presence of charge on the diffusing species results in the diffusional flux being affected by space charge related fields, even if there is no external field gradient. For instance, in a plasma, since electrons in general are more mobile, diffusion of electrons along a concentration gradient leaves behind excess positive charge, resulting in a local field which acts to slow down further diffusion of electrons.[9] In discharges with charge concentrations $n_e > 10^3/\text{cm}^3$, the space charge related field $\mathbf{E}_{sc}$ becomes large enough such that the diffusional flux for electrons can be written as

$$\mathbf{\Gamma}_e = -D_e\nabla n_e - n_e\mu_e\mathbf{E}_{sc} \tag{6.18}$$

where μ_e is the drift velocity of electrons in a unit electric field and $\mathbf{E}_{sc}$ is the space charge related field. Similarly, the flux for ions can be expressed as

$$\mathbf{\Gamma}_i = -D_i\nabla n_i + n_i\mu_i\mathbf{E}_{sc} \tag{6.19}$$

where the subscript i denotes the corresponding quantities for a positive ion.

Let us examine each of the terms in equations (6.18) and (6.19) in turn. For a Maxwellian distribution of velocities and for collision frequencies that are independent of species velocity, the diffusion coefficient D can be related to temperature as

$$D_e = \frac{kT_e}{mv_e} \tag{6.20}$$

where ν_e is the collision frequency. The mobility μ can be defined as

$$\mu_e = \frac{e}{m\nu_e} \tag{6.21}$$

The relationship between mobility and the diffusion coefficient for a Maxwellian velocity distribution is defined by Einstein's equation

$$\frac{D}{\mu} = \frac{kT_e}{e} \tag{6.22}$$

At a steady state, due to charge balance requirements, the fluxes of electrons and ions equalize, allowing us to solve for the unknown quantity $\mathbf{E}_{sc}$. Since $n_i = n_e$, and taking

$$\mathbf{\Gamma} = \mathbf{\Gamma}_e = \mathbf{\Gamma}_i$$

we can arrive at an expression for $\mathbf{E}_{sc}$ by combining equations (6.18) through (6.22).

$$\mathbf{E}_{sc} = -\frac{(D_e - D_i)}{(\mu_e + \mu_i)} \frac{\nabla n}{n} \tag{6.23}$$

Substituting $\mathbf{E}_{sc}$ back into equation (6.18) results in

$$\mathbf{\Gamma} = -\frac{(D_i\mu_e + D_e\mu_i)}{(\mu_e + \mu_i)} \nabla n \tag{6.24}$$

where D_a is called the ambipolar diffusion coefficient.[10] Ambipolar diffusion occurs in plasmas when the electron density is larger than 10^8 per cm^3, characteristic of most discharges used in CVD.

In a plasma CVD reactor with gradients in the species caused by flow, temperature, and species consumption effects, ambipolar diffusion plays an important role in determining the concentrations of reactant species at the substrate surface. Since chemical reaction rate is determined by surface concentration, diffusional terms are critical in the design of the reactor as we will see in the next Chapter.

6.2.5. Summary

We introduced the idea of temperature to indicate the energies of electrons and ions in a plasma. We saw the potential distribution in a system containing a plasma confined between grounded walls and in the presence of an isolated surface. The plasma is the most positive body in the system. We also derived expressions for current through a conductive electrode connected to ground or an external potential. And we looked at the plasma's ability to shield out external perturbations at frequencies up to the plasma frequency. Finally we saw the effect of the space charge related field in equalizing the diffusional fluxes of the ions and electrons, despite differences in their mobilities.

Exercise 6.2

In a nitrogen plasma where the neutrals are at room temperature, the average ion temperature is 0.05 eV and the electron temperature is 2 eV. The electron density is 2×10^9 per cm^3. If a floating surface is exposed to the plasma, what is the average energy of electron bombardment on the surface? At what distance from the perturbation will the plasma return to the plasma potential?

6.3. PLASMA CHEMISTRY

From the relative inefficiency of energy transfer through elastic collisions between electrons and ions or neutrals, we surmised that inelastic processes were important for the coupling of an external power supply to sustain the plasma. Let us examine some of these inelastic processes and their role in reactions occurring within discharges. The chemical identities of the gas phase species assume increased importance in determining the specific inelastic processes. The energetics of the inelastic processes depend to a great extent on the energy levels intrinsic to the gas species.

Five fundamental inelastic processes account for the majority of chemical processes occurring in plasmas: (a) ionization, (b) excitation, (c) relaxation, (d) dissociation, and (e) recombination. Many other processes that occur in discharges are species dependent. All five fundamental processes can be found in a plasma of hydrogen as the energy input to the plasma is gradually increased. To illustrate the fundamental inelastic processes, let us begin by following the plasma chemistry of hydrogen in a hypothetical situation where all electrons have equal energies and this energy can be gradually increased.[11]

At low energies, electron–H_2 collisions are elastic and the electrons suffer momentum changes only as shown in Figure 6.4. The H–H bond strength is approximately 4.5 eV so we would expect atomic hydrogen to form when the electron energies are increased to this level. However, the cross section for this process is small and the mechanism requires the hydrogen molecule

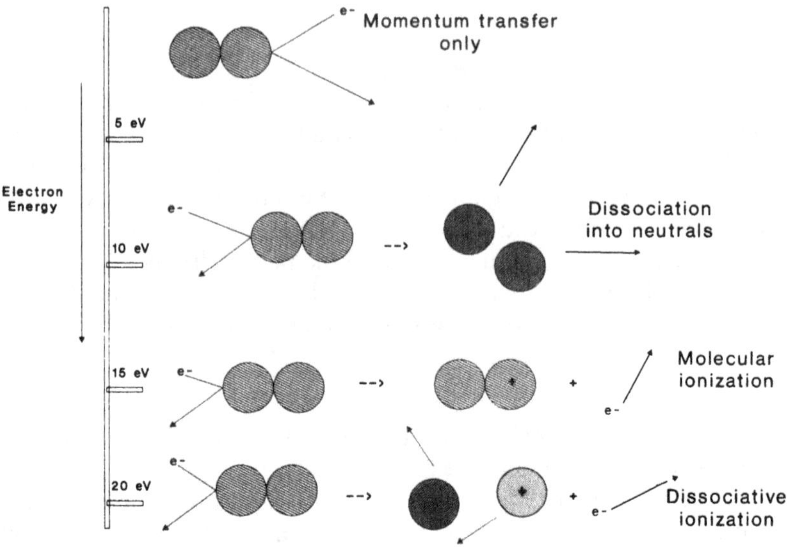

FIGURE 6.4. Plasma chemistry of the hydrogen molecule as the impinging electron energy is increased.

to be raised to a repulsive state. This excitation occurs at an energy of about 9 eV. Above this energy, electron impact results in the dissociation of the molecule to form two hydrogen atoms with each of them having significant kinetic energy. The ionization potential for H_2 molecules to form H_2^+ ions is 15.4 eV. At this energy, significant formation of singly ionized hydrogen molecules is seen. As the energy of the electrons is increased further, a repulsive state for the H_2^+ ion is reached, resulting in the formation of an H/H^+ pair. The hydrogen atoms produced have their own excited states, which can be achieved through collisions with lower-energy electrons. The excited states decay through collisions with other atoms, ions, or electrons.

In a typical glow discharge the electron energy distribution is Maxwellian, so all these processes can occur concurrently in the discharge. Bearing in mind this overview of the hydrogen molecule, let us consider each of the five processes in more detail.

6.3.1. Ionization

Ionization in the plasma can occur through many mechanisms; the simplest is electron capture. For a neutral species having high electron affinity (as in the case of halogens), the following reaction (shown here for fluorine) occurs readily:

$$F + e^- \rightarrow F^-$$

Electron capture contributes significantly to electron loss in halogen-containing plasmas.[12] An important mechanism for the maintenance of a glow discharge is the production of ions through electron impact. For instance, collision between an energetic electron and a xenon atom produces a xenon ion and another electron.

$$Xe + e^- \rightarrow Xe^+ + 2e^-$$

The two electrons are now accelerated by the potential gradient in a sheath to ionize more neutrals, starting a chain reaction. The electrons have to possess higher energy than the ionization potential of the neutral (~ 12 eV for xenon) for this process to occur.

For oxygen, the production of O^- ions occurs through a process known as resonance capture,[13] explained in the following sequence of reactions:

$$O_2 + e^- \rightarrow O_2^-$$
$$O_2^- \rightarrow O + O^-$$

The electron energy required for an ion to be produced is called its appearance potential, for oxygen it is 4.53 eV.

Photon emission from excited atoms can also lead to ionization.[14] Even though most of the photon energy in a discharge is dissipated as heat, energetic photons produce effects, such as Auger electrons, and can lead to significant ionization. Since the maximum photon energy can be as high as the direct potential difference between the electrodes, photoionization becomes important at high applied voltages.

Collisions between a neutral and a metastable species can produce ionization as well. This process, called the Penning ionization,[15] occurs as follows:

$$A^* + B \rightarrow A + B^+ + e^-$$

where the asterisk denotes an excited state. Similarly a collision between an ion and a neutral can result in a transfer of charge from the ion to the neutral.

Here is a typical ionization sequence of a molecular species in a plasma.

a. At low electron energies, resonance capture can give rise to a negative ion.

$$AB + e^- \rightarrow AB^-$$

b. This ion might dissociate.

$$AB^- \rightarrow A^- + B$$

c. At higher electron energies, the molecule can ionize during decomposition.

$$AB \rightarrow A^- + B^+$$

Notice the role of charged species collisions in producing unattached ionized species which would not otherwise be available for reaction. Reactions involving charged species would depend upon their concentration and their half-lives.

6.3.2. Excitation and Relaxation

Ionization is only an extreme case of the various excited states that an atom or a molecule can reach on electron impact. Figure 6.5 shows the ionization cross sections of different inert gas atoms. At energies smaller than the ionization potential, electrons with sufficient energy can raise a ground level species to one of the higher available states. Figure 6.6 shows the energy level diagram of a helium atom.[16] Notice the number of nonionized excited states available to the atom. Similar to ionization, there are electron energy thresholds equal to the energy of the first excited state, and they need to be exceeded for the excitation to occur.

Certain excited species are metastable (lifetimes longer than 1 ms), others have very short lives in the excited state (typically 100 ns or less). When the lifetime is short, excitation is frequently followed by relaxation processes such as radiative decay.

$$He^* \rightarrow He + h\nu$$

where $h\nu$ is the energy of a photon with frequency ν. As we saw earlier, this is the source of the glow associated with plasmas. Since the frequency of the emitted photon is specific to the species (corresponding to the energy difference between the excited state and the ground state), the emission

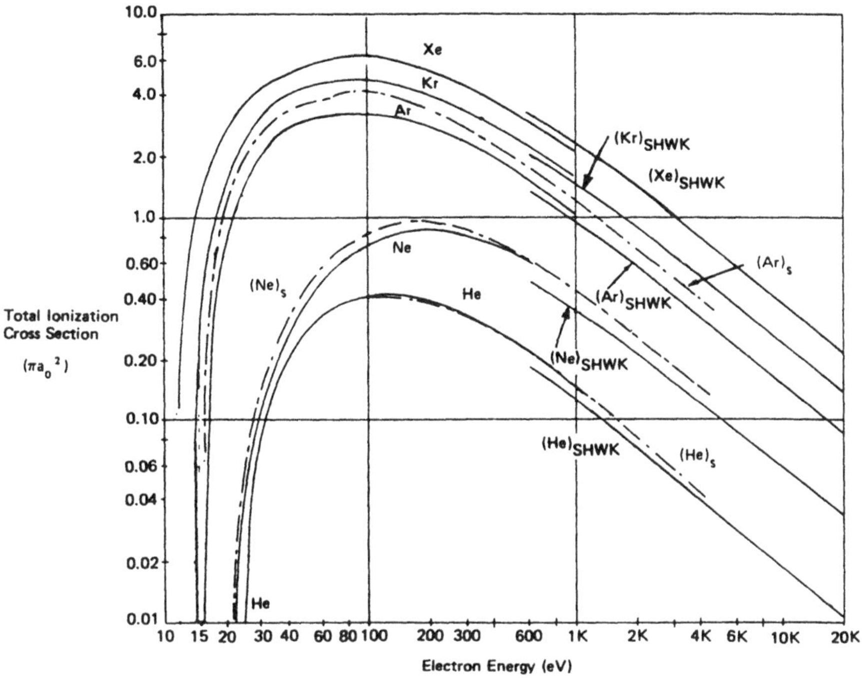

FIGURE 6.5. Ionization cross sections of different inert gases. Below a threshold energy the electrons produce little ionization. Ionization is only one of many excited states that an atom can achieve. Many curves for each gas are data from different sources. Reprinted from Ref. 3 with permission from John Wiley & Sons.

spectrum can be used in the plasma diagnostics. Optical emission spectroscopy of the plasma is often used to determine the etching endpoint in process plasmas. Figure 6.7 shows a discharge spectrum from an argon plasma. Notice the peaks corresponding to the various levels of the argon atom.

Metastable species, due to their longer lifetimes, can participate in other collisional processes, depending on their energy levels. These are called Penning processes, where collision between a metastable species and a neutral can result in the excitation or ionization of the neutral.

6.3.3. Dissociation

We have seen examples of dissociative ionization in earlier sections, when a molecule dissociates upon electron impact to either its constituent atoms or further ionized products. Of more relevance to CVD (and to etching processes in IC manufacture) is the formation of radicals in the plasma. Optical and

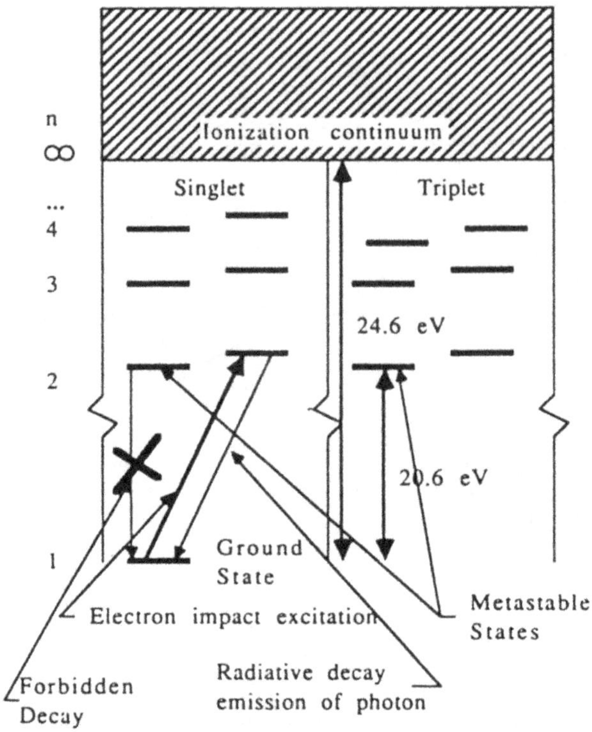

FIGURE 6.6. Energy level diagram of the helium atom. Notice the forbidden transitions. Understanding such energy diagrams is key to attributing the sources of the peaks in the glow discharge spectra. Reprinted from S. M. Rossnagel, J. J. Cuomo, and W. D. Westwood (eds.), *Handbook of plasma processing technology*, p. 34, with permission of Noyes Publications © 1989.

mass spectroscopic evidence clearly indicates the presence of various radicals such as OH, NH_2, SiH_2, WF_x, etc., in a discharge.[17] The reaction of these radicals is often the path that leads to deposition in CVD and material removal during plasma etching. For instance, the formation of silicon from SiH_4 involves the formation of the intermediate SiH_2 radical. Formation of F and CF_x radicals in a CF_4 plasma are directly responsible for the etching of SiO_2.

Bond dissociation energies for the formation of these radicals are therefore important parameters when optimizing deposition processes. Bond dissociation energies can be calculated from the ionization potential for the radical and its appearance potential in the plasma. We can consider appearance potential to be a physical analog of heat of formation.

$$e^- + AB \rightarrow A^+ + B + 2e^-$$

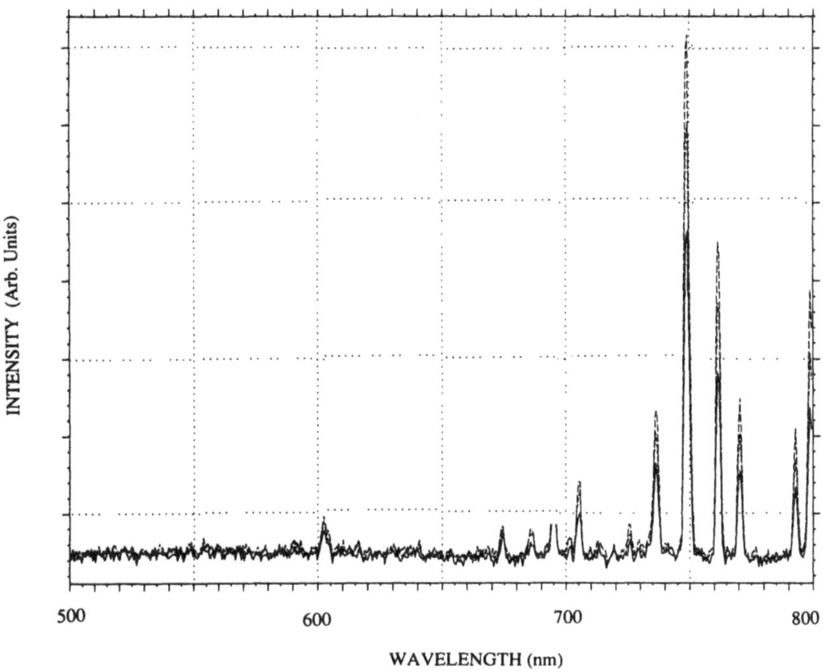

FIGURE 6.7. Glow discharge spectra can be used in qualitative and quantitative analysis of plasma reactions; this one is for argon.

and ionization potential to be equivalent to heat of reaction.

$$A \rightarrow A^+ + e^-$$

The sum of the two reactions

$$AB \rightarrow A + B$$

gives the bond energy for the formation of radical A from the compound molecule AB. Bond energies calculated by this method agree very well with those calculated by thermochemical methods. For instance, the bond dissociation energy for the following reaction

$$CH_3\text{—}Cl \rightarrow CH_3 + Cl$$

has been measured by electron impact to be 3.40 eV, and calculated from thermochemical data to be 3.47 eV.[18]

In a discharge, where the high end of the electron energy spectrum exceeds about 50 eV, a plethora of radicals is produced. In the case of tungsten hexafluoride, some of the radicals are WF_x ($x = 1$ to 5) their singly and doubly ionized ions, and if gases such as oxygen are present, various combinations of the three elements.

6.3.4. Recombination

Recombination of the atoms, ions, and radicals produced by the above processes occurs either homogeneously in the gas phase or heterogeneously at the chamber walls or on the substrate. Heterogeneous reactions at the substrate constitute the CVD and etching processes for microelectronics.

Conservation of energy and momentum implies that a two-body collisional recombination between two species with no other internal degrees of freedom is impossible.[19] Consider an electron of mass m and velocity v relative to an ion of mass M. If their combined final velocity is u and the potential energy of the ion is decreased by E due to their recombination, conservation of momentum yields

$$mv = (m + M)u \tag{6.25}$$

Conservation of energy yields

$$\tfrac{1}{2}mv^2 = \tfrac{1}{2}(m + M)u^2 - E \tag{6.26}$$

Solving for the final velocity u,

$$u^2 = -\frac{2Em}{(m + M)M} \tag{6.27}$$

However, since all the quantities on the right are positive, this yields an unreal solution for the final velocity.

Hence the only way a recombination process can occur is if there is a three-body collision (involving another ion, the wall or substrate, or a photon) or if there are other internal degrees of freedom to absorb or release energy.

The most common example of homogeneous recombination of atoms involving three-body collisions is the formation of ozone, which is commonly used in the CVD of SiO_2. The ozone-forming mechanism is as follows:

$$O + O_2 + M \rightarrow O_3 + M (M = O_2, \text{Ar, He, etc.}) \tag{6.28a}$$

The decomposition of ozone occurs as:

$$O + O_3 \rightarrow 2O_2 \tag{6.28b}$$

Since the concentration of ozone found in the discharges is relatively small, we may conclude that under the same conditions, reaction (6.28b) proceeds faster than reaction (6.28a).[20] The ozone-forming reaction (6.28a) is first-order in [O].

In general, the three-body collision reaction for the formation of the molecule of element from atoms can be written as

$$A + A + M \rightarrow A_2 + M^*$$

where M is an inert third body aiding the formation of the molecule of element A from its atom. The formation occurs in two stages.

$$A + M \rightleftarrows AM$$

$$A + AM \rightarrow A_2 + M$$

AM is an activated van der Waals' complex between the atom and the inert species. Since the rate of formation of AM is governed only by van der Waals forces between the two atoms, the rates of the reactions are fairly similar for most diatomic molecules.

Recombinations of ions by ion–ion collisions is much more probable than recombination by ion–electron collisions to form neutrals. This is due to the higher velocities of the electrons. Recombination rates for electron–ion processes are on the order of 1×10^{-9} per cm^3 per second. For ion–ion recombination, excess energy loss can be accommodated by the release of a photon, by both atoms reaching an excited state, or through three-body collisions. Three-body collisions involving the substrate are the most important processes at CVD pressures.[21]

Collisional recombination of radicals is extremely efficient. Since the large number of internal degrees of freedom of the radical allows redistribution of the energy, almost all collisions between radicals result in reactions such as disproportionation or combination.[22] The activation energy for these reactions is found to be close to zero, with rate constants as high as 10^{10} L mol^{-1} s^{-1}.

Exercise 6.3

An electron beam with an average energy of 50 eV, and a current density of 5×10^{-3} A/cm^3 is injected through a chamber containing CF_4 at a pressure of 0.1 millitorr. If the cross section for impact ionization is 1×10^{-17} cm^2, what is the rate of ion production per centimeter of beam path? The cross section of the electron beam is 1 cm^2.

6.4. PLASMA DIAGNOSTICS

Typical characteristics of the plasma, such as the presence of a glow discharge, shielding of applied potential, or the presence of ionized atoms and molecules, allow varied tools to be used in understanding the plasma. For instance, increase in pressure of a closed cell containing a diatomic gas, upon striking a plasma is an absolute measure of molecular dissociation.

Many intrusive and nonintrusive methods have been devised to measure properties of plasmas. Techniques such as mass spectrometry have been adapted to identify radicals in plasmas. Residual gas analysis of the reaction chamber in CVD is intrinsic to the qualification of most processes for manufacturing. An extensive overview is outside the realm of this book, but we will look at three techniques unique to plasma enhanced chemical processes. They are used to identify physical characteristics of the plasma, such as electron density and temperature, and the nature and concentrations of various chemical species.

6.4.1. The Langmuir Probe

The Langmuir probe uses the principles of Debye shielding and space discharge limited current to directly determine the electron density and electron temperature.[23] It is a conductive tip that can be placed anywhere in the plasma, with suitable potentials applied to it as illustrated in Figure 6.8.

Let us examine the change in the probe current as the potential is changed from negative to positive (Figure 6.8a). At very negative voltages, the probe collects only ion current. At voltages close to zero on the probe, the probe current is an exponential function of the applied voltage [$I_p \propto \exp(-V/kT_e)$]. When the voltage is made more positive, the current consists of randomly arriving electrons ($J = ne\bar{v}/4$). By plotting the log of the probe current against the applied voltage (Figure 6.8b), the slope of the line around the zero voltage gives a measure of the electron temperature. The transition point to the lower slope occurs at the plasma potential. The electron density can be found from the current at the plasma potential.

Despite its simplicity, probe results can often be difficult to interpret, especially in high frequency and electronegative discharges. High/low pass filters need to be designed to eliminate noise. Several commercially available probe packages contain appropriate software that automatically calculate

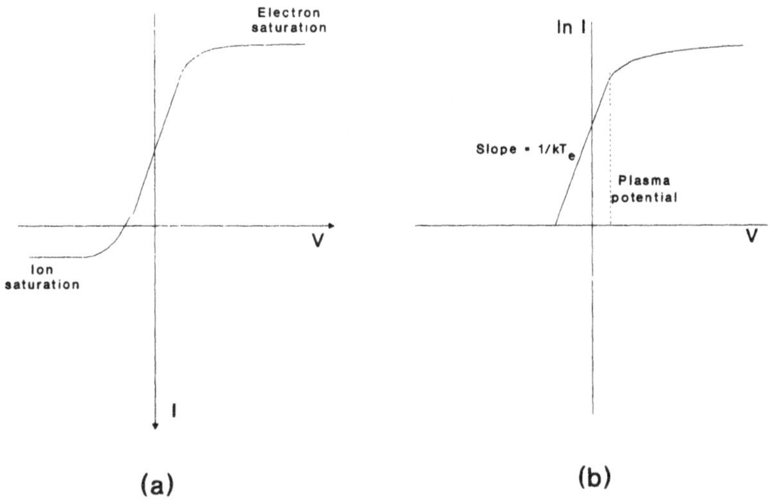

FIGURE 6.8. Voltage–current characteristics of a typical Langmuir probe. Note the ion and electron saturation currents and the calculation of plasma potential from this graph.

parameters such as plasma potential, Debye length, electron temperature, and electron density.

6.4.2. Optical Emission Spectroscopy

Recombination and/or decay of excited species results in characteristic emission patterns. Since energies in discharges are high enough to produce and excite species, optical emission spectroscopy can be used to identify the presence of a species by analyzing the intensities of the optical spectrum at different wavelengths.[24] However, quantitation of the concentration of species is much more difficult, since the intensity at any wavelength is a function of the species density and its electron impact excitation probability. However, ratios of intensities at any given wavelength under different experimental conditions can be used semiquantitatively as predictors of species concentration.

Detection systems for optical emission use many techniques for discriminating between wavelengths, such as monochromators or a narrowband filter and a photomultiplier tube. Identification of species is through comparison of the spectrum to standard signatures of the species published by NIST. Optical emission is widely used in etch systems to determine the endpoint of a process.

6.4.3. Gaseous Titration

Determination of concentrations of gaseous species can be achieved by titrating the plasma with another reactive gas.[25] The production or extinction of a certain glow discharge color can be used as the endpoint for the titration. For example, quantitative determination of oxygen concentration can be accomplished through the following reactions:

$$O + NO \rightarrow NO_2^* \tag{6.29a}$$

$$NO_2^* \rightarrow NO + h\nu \tag{6.29b}$$

The discharge has a characteristic yellowish-green glow. However, this reaction is relatively slow. Hence in order to measure oxygen concentration in a continuous flow system, measured quantities of NO_2 are added to the system. The NO_2 reacts with oxygen as

$$O + NO_2 \rightarrow O_2 + NO \tag{6.29c}$$

The NO in turn reacts with the oxygen to give the yellowish-green glow. When all the oxygen is consumed by reaction (6.29c) and none remains for reaction (6.29a), the glow is extinguished indicating the endpoint.

NO is also used for measuring active nitrogen concentration through the reaction

$$N + NO \rightarrow N_2 + O$$

Again the yellowish-green glow is used for endpointing.

6.5. SUMMARY

A review of the fundamental characteristics of plasmas was presented. We examined the relative concentrations of electrons, ions, and neutrals in the plasma, and studied their collective behaviour in modulating externally applied perturbations. We then examined the processes that occurred when these species collided with each other. The chemical nature of the atoms became important at this point, leading to the development of plasma chemistry. We studied the chemical reactions that occur in plasmas. Finally, we studied some plasma diagnostics to optimize the plasma for CVD applications.

References

1. R. Reif, in *Handbook of Plasma Processing Technology* (Rossnagel, Cuomo, and Westwood, Eds.), p. 268, Noyes, Park Ridge, N.J., 1990.
2. J. R. Hollahan and A. T. Bell, *Techniques and Applications of Plasma Chemistry,* John Wiley, New York, 1974.
3. B. Chapman, *Glow Discharge Processes,* p. 53, John Wiley, New York, 1980.
4. F. F. Chen, *Introduction to Plasma Physics and Controlled Fusion,* Plenum Press, New York, 1983.
5. S. C. Brown, *Introduction to Electrical Discharges in Gases,* John Wiley, New York, 1966.
6. J. L. Cecchi, in *Handbook of Plasma Processing Technology* (Rossnagel, Cuomo, and Westwood, Eds.), p. 35, Noyes, Park Ridge, N.J., 1990.
7. N. A. Krall, and A. W. Trivelpiece, *Principles of Plasma Physics,* McGraw-Hill, New York, 1973.
8. M. Mitchner, and C. H. Kruger, *Partially Ionized Gases,* John Wiley, New York, 1973.
9. F. F. Chen, *Introduction to Plasmas and Controlled Fusion,* Plenum Press, New York, 1983.
10. J. R. Hollahan and A. T. Bell, *Techniques and Applications of Plasma Chemistry,* p. 28, John Wiley, New York, 1974.
11. A. G. Engelhardt, and A. G. Phelps, *Phys. Rev.***131**, 2115 (1963).
12. M. V. Kurepa, and D. S. Belic, *J. Phys. B* **11**(21), 3719 (1978).
13. D. Rapp, and W. E. Francis, *J. Chem. Phys.* **37**(11), 2631 (1962).
14. G. L. Weissler, *Handbuch der Physik*, pp. 21, 304, Springer, Berlin, 1956.
15. J. W. Coburn, and E. Kay, *Appl. Phys. Lett.* **18**(10), 435 (1971).
16. P. J. Ficalora, private communication.
17. F. Kaufman, *Production of Atoms and Simple Radicals in Glow Discharges,* Advanced Chemistry Series No. 80, p. 29, 1969.
18. S. Sivaram, unpublished.
19. B. Chapman, *Glow Discharge Processes,*p. 35, John Wiley, New York, 1980.
20. F. Moghadam, Intel Corp., private communications.
21. T. S. Carlton, and B. H. Mahan, *J. Chem Phys.* **40**, 3683 (1964).
22. M. Hayashi, Report IPPJ-AM-19, Nagoya Institute of Technology, Nagoya, Japan, 1981.
23. I. Langmuir, and H. Mott-Smith, *General Electric Review* **26**, 731 (1923).
24. W. R. Harshbarger, R. A. Porter, T. A. Miller, and P. Norton, *Appl. Spectr.* **31**(3), 201 (1977).
25. R. V. Giridhar, "Thermal nitridation of silicon," Ph. D. thesis, Rensselaer Polytechnic Institute, Troy, N.Y., 1984.

Chapter 7

Processing Plasmas and Reactors

In our discussion of the fundamental characteristics of plasmas, we did not place emphasis on the means of generating and sustaining the plasma. Nor did we consider the proper confinement of the plasma and the reactants, or the optimization of the plasma in order to produce a solid film on the substrate. In this chapter we will address these issues and study the coupling of external power to the discharge for its generation and maintenance. We will use a simple DC diode plasma in order to illustrate electron production and loss mechanisms. However, since most useful CVD processes use AC power sources, we will also examine RF and microwave discharges.

Reactors for plasma CVD have several unique features compared with their thermal counterparts. We will consider different plasma reactor configurations and study two commercial plasma CVD reactors in detail.

7.1. DC BREAKDOWN AND DISCHARGE

The simplest (and historically the earliest means) of producing a laboratory plasma is by applying a DC voltage between two metal electrodes in an evacuated chamber containing an inert gas at low pressure. Figure 7.1a illustrates the apparatus required for such a setup. The details of the processes that occur when the voltage of the power supply is increased are covered by the references 1–3 at the end of this chapter. We will confine ourselves to two processes: the electrical breakdown of the gaseous column and the maintenance of the discharge after breakdown has occurred.

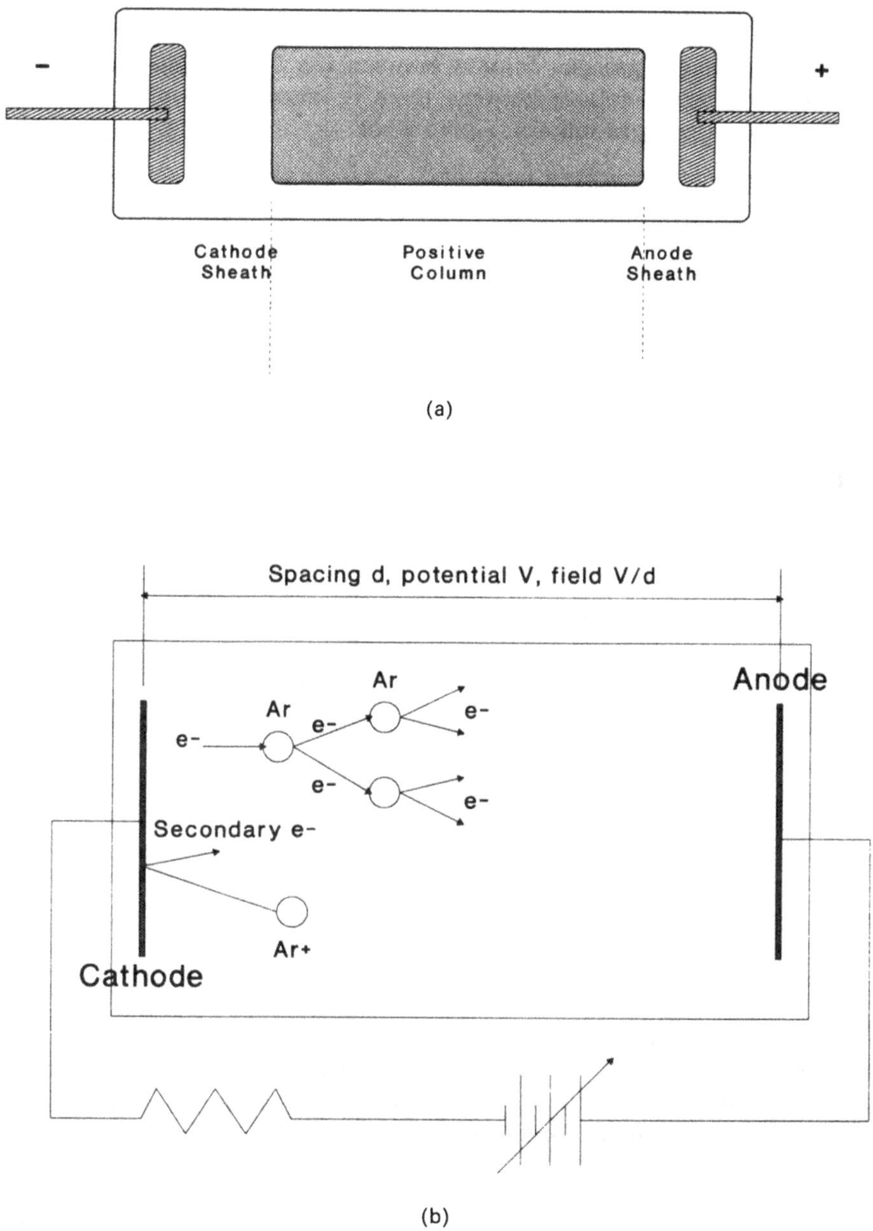

FIGURE 7.1. (a) A simple DC discharge column with a partially evacuated chamber and two electrodes. The potential distribution in a discharge is also shown. (b) Electron multiplication in a discharge column that leads to breakdown.

7.1.1. DC Breakdown

Qualitatively, the gaseous column between the two electrodes begins to conduct when the voltage between them is raised above a threshold V_B. The process occurs as follows (Figure 7.1b):

a. A stray electron near the cathode is accelerated towards the anode by the applied field.
b. Upon gaining sufficient energy, due to its acceleration in the field, the electron collides with a neutral atom in the discharge, causing impact ionization. The collision produces one more electron and an ion.
c. The two electrons are now accelerated in the field, causing more ionizations. The cascading processes of electron acceleration, ionization, and electron production results in the electrical breakdown of the gaseous column between the two electrodes.

In order for breakdown to occur, certain criteria have to be satisfied. First, the voltage between the two electrodes needs to exceed the critical breakdown voltage, V_B. Second, the distance d between the electrodes has to be sufficient for the electron to gain enough energy to cause ionization. Third, the loss of electrons by various processes should at least be counteracted by new electron production processes for a continuously increasing number of electrons. Let us try to quantify these processes and arrive at a criterion for the breakdown to occur.

We define a coefficient α (called the Townsend coefficient) which is the probability that an ionization will occur in a unit length. α is proportional to the number of collisions per unit length and the probability that a collision will cause an ionization. The probability of collision is proportional to the gas pressure in the plasma chamber. The mean free path λ is given by

$$\lambda = \frac{k_1}{P} \tag{7.1}$$

where P is the chamber pressure and k_1 is a constant. The probability that a collision will cause ionization is dependent exponentially on the ratio of an effective ionization potential V_i and the energy gained by the electron when travelling a distance equal to λ. Hence α can be written as

$$\alpha = \frac{1}{\lambda}\exp(-V_i/eE\lambda) \tag{7.2}$$

Let us consider now the total electron current reaching the anode. If the starting current from the cathode is I_c, the anode current rises from this value exponentially with α, resulting in an anode current I_a given by

$$I_a = I_c e^{\alpha d} \tag{7.3}$$

The same current due to ions will reach the cathode. However, when energetic ions impinge on the cathode, secondary electrons are emitted from the cathode with a probability of γ. The importance of the secondary electrons in maintaining the plasma will become evident shortly. However, for the present purpose of accounting for all the electrons, we can write the number of secondary electrons as

$$I_{se} = \gamma I_a(e^{\alpha d} - 1) \tag{7.4}$$

where γ is the number of secondary electrons produced for every ion impact, also called the secondary electron yield. Hence the total current arriving at the anode will have successive I_{sc}'s also being accelerated towards the anode, and I_a needs to be modified as

$$I_a = \frac{I_c e^{\alpha d}}{[1 - \gamma(e^{\alpha d} - 1)]} \approx \frac{I_c e^{\alpha d}}{[1 - \gamma e^{\alpha d}]} \tag{7.5}$$

to account for the electrons produced by impact ionization in the gas phase and the secondary electrons produced by secondary emission at the cathode. When the denominator in equation (7.5) tends to zero, the current I_a tends to rise rapidly, resulting in breakdown. Noting that the breakdown voltage $V_B = E_B d$, we can combine this with equations (7.5) to arrive at

$$E_B = \frac{AP}{C + \ln(Pd)} \tag{7.6}$$

and

$$V_B = \frac{APd}{C + \ln(Pd)} \tag{7.7}$$

where A and C are constants that depend on the gas. Notice the electrode spacing and the gas pressure are intimately related in determining the breakdown conditions. At higher pressures, since the number of collisions is high, the distance between electrodes for breakdown can be small. Conversely, at lower pressures, the electrons travel longer distances before collision and

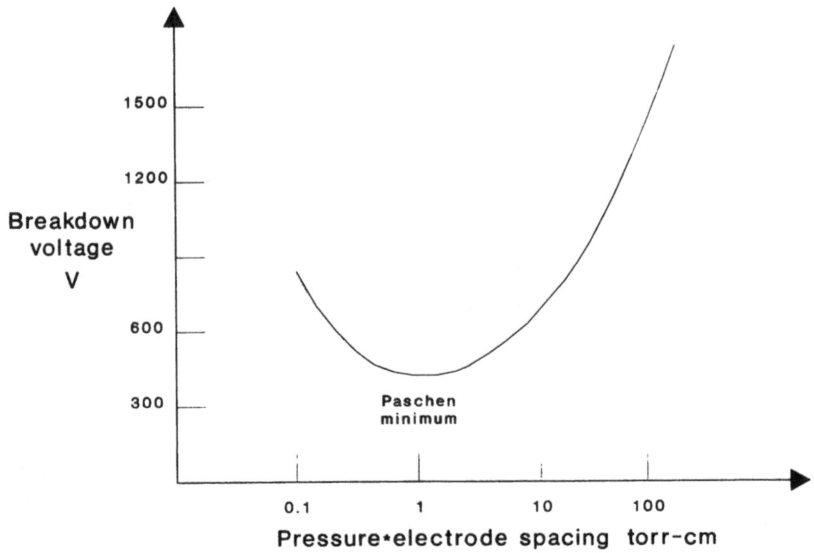

FIGURE 7.2. The Paschen curve for argon which illustrates the breakdown characteristics of the chamber as a function of pressure–electrode spacing product. The minimum in the curve is called the Paschen minimum.

a larger electrode spacing is needed for the same breakdown characteristics. Figure 7.2 plots the breakdown voltages for argon as a function of the product *Pd* of pressure and electrode spacing. The minimum in the breakdown curve is called the Paschen minimum.

7.1.2. Maintaining the DC Discharge

Once breakdown occurs, the various electron and energy loss mechanisms are simultaneously established: electron loss to the external circuit, recombination of electrons and ions at the walls, energy loss through heating of the electrodes and through optical emission, etc. To maintain the discharge at steady state, the production and loss of the species need to be balanced; energy from the external power supply needs to be coupled to the discharge to compensate for the energy lost. Before going into great detail, let us reiterate some of the fundamental characteristics of plasmas from Chapter 6.

a. The plasma is the most positive body in a discharge; it does not take a potential intermediate between the two electrode potentials.
b. The DC plasma is field-free; all fields are restricted to two sheaths, one surrounding each electrode.
c Each sheath field acts to repel electrons.

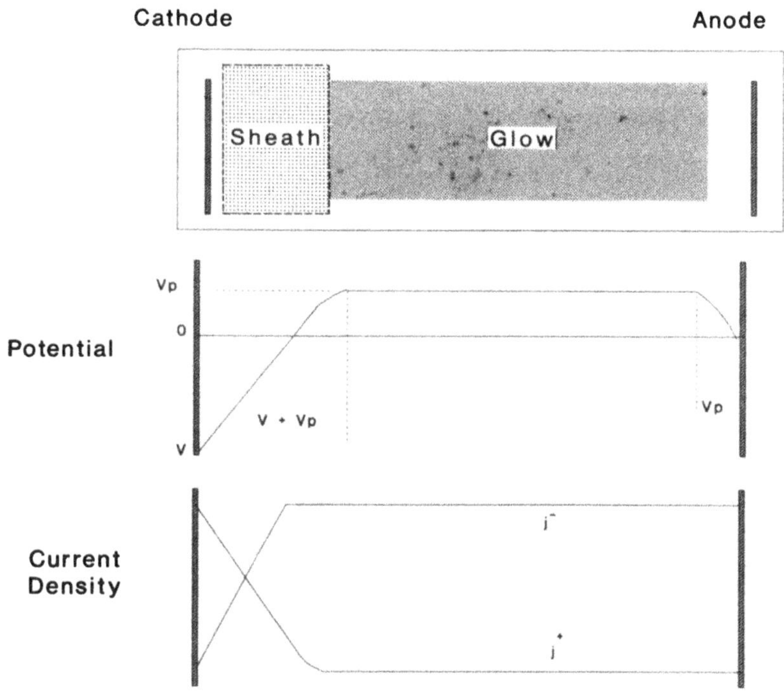

FIGURE 7.3. Potential and current density distributions in a DC discharge.

Using these ideas we can draw the potential curve of a CD discharge with a grounded anode and a cathode voltage of 1 000 V, as shown in Figure 7.3. Since there are no fields in the glow discharge itself, ion and electron accelerations occur in the sheaths only. And, most of the ionizations occur in the sheath and the sheath–bulk plasma transition, called the negative glow. This creates large gradients in current density in the sheath regions. A schematic of the ion and electron current density is also shown in Figure 7.3.

Even though ionization in the sheath and the glow produces electrons, the most important source of electrons, required for maintenance of the glow, is secondary electron emission from the electrodes and the chamber walls. Table 7.1 shows the secondary electron yield, γ, for various solid materials. Although secondary electrons can be produced from the bombardment of the solid by neutrals, electrons, and photons, the key mechanism is the ion current striking the cathode. Secondary electron production due to ion bombardment at low energies is an Auger neutralization process and is energy-independent up to several hundred electronvolts.

The DC diode plasma described so far is an inefficient process for the

TABLE 7.1 Secondary Electron Yields of Selected Materials

Metal	Ion	Yield per ion, γ	Ion energy (eV)
Si (1 0 0)	Ar	0.027	100
W	Ar	0.095	100
Al	Ar	0.12	slow
Al	N_2	0.10	slow
Ni	Ar	0.034	100
Fe	Ar	0.058	slow

production of radicals compared with the other types of discharges described below. It is infrequently used in this configuration in CVD reactors. In particular, when depositing dielectric materials, a DC discharge cannot be coupled to a nonconducting electrode. Hence other means of producing a discharge, such as RF and microwave power sources are found more commonly in CVD reactors. Inductive coupling of the plasma in place of the two electrodes in the DC diode plasma has also been a popular configuration. Since alternating current discharges are so widely used, let us examine what the effect of the input frequency is on the discharge.

7.2. FREQUENCY EFFECTS ON DISCHARGES

A DC discharge requires us to use conductive electrodes. Let us follow what happens if one of the electrodes, say the cathode, is made of an insulating material (this follows an explanation similar to Chapman[15]). If a DC voltage larger than the breakdown voltage is applied to the cathode, current flows until the insulator becomes charged up and terminates the discharge. If instead we apply a very low frequency AC voltage, during the first half of each cycle the insulator will charge up and extinguish the discharge. During the second half of each cycle, the insulator will discharge, and the current will flow in the opposite direction till the insulator charges up again. So the plasma essentially behaves like a capacitor, charging and discharging as the direction of the voltage changes. Since the capacitance C = charge Q/voltage V, and the charge Q = current $i \times$ time t, the typical charging time for a plasma using a quartz electrode is about 1 μs. So if the applied frequency does not allow sufficient time for the electrode to charge up and extinguish the discharge, i.e., it exceeds about 1 MHz, then a continuous discharge can be maintained. In reality, discharges can be maintained above about 100 kHz. Applied frequency above this level has some important effects on the discharge.

a. It can change the spatial distribution of species and electric fields in the discharge.
b. It determines whether the energy and the concentration of species are constant over time or whether they fluctuate over each period of the applied field.
c. It determines the minimum voltage required to maintain the discharge and hence the energy of ions bombarding the cathode.
d. It shapes the electron energy distribution function.

7.2.1. RF Breakdown and Discharge

Most commercial plasma reactors operate at 13.56 MHz, licensed by the Federal Communications Commission (FCC) as an industrial frequency at which commercial generators are readily available. At such high frequencies, the breakdown of a gaseous column between two electrodes is actually easier than with a DC field. Electrons undergo an oscillatory motion due to the applied frequency, so they gain energy more efficiently. And the oscillations minimize any loss mechanisms due to recombination at the walls, so the ionization efficiency increases. Breakdown occurs at lower peak-to-peak voltages. In effect, the impedance of the discharge decreases with increasing frequency, hence we can drive a higher current through the discharge for the same applied voltage. Secondary electrons, which are so important in maintaining the discharge in the DC case, are less important in the RF discharge. Furthermore, the efficiency of ionization in RF discharges is increased by phenomena such as the *surf riding effect*, where electrons gain energy by oscillating in and out of the sheath boundary.

At RF frequencies, the more massive ions have too much inertia to respond to the instantaneous electric field, unlike the electrons. This creates a time-averaged negative DC bias on one electrode, if that electrode is geometrically smaller than the other. This negative bias is often called self-bias, since it is not externally applied and develops due to electrode asymmetry. Such a configuration, where the effective area of one electrode is different from the other is called an asymmetric system. Asymmetric systems can be produced from electrodes with different geometrical areas or by confinement of the plasma. Confinement results in different *wetting* surface areas on the two electrodes; for instance, grounded walls effectively increase the total grounded electrode area. The self-bias is related to the effective area as

$$\frac{V_1}{V_2} = \left(\frac{A_2}{A_1}\right)^4 \tag{7.8}$$

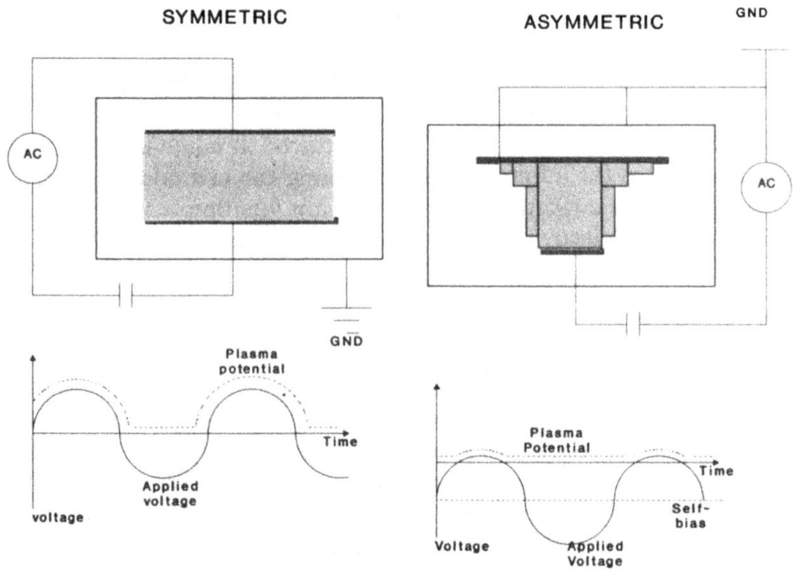

FIGURE 7.4. Symmetric and asymmetric discharges. Development of DC self-bias is also illustrated.

Either electrode may be grounded, but it is usually the larger electrode, including the chamber walls. The smaller electrode is often called the cathode. Since the plasma potential is the most positive body in the discharge, the potential follows the instantaneous fluctuations and the magnitude of the plasma potential is very small for the nonsymmetric system, as illustrated in Figure 7.4. This is important because the difference between the plasma potential and the negative DC bias on the cathode determines the energy of ions bombarding the cathode surface.

On a similar note, the DC bias developed in an asymmetric RF discharge is also inversely proportional to the gas pressure in the system. At high neutral densities, the discharge is able to put energy directly into the glow electrons and does not need as a high a sheath field to sustain itself.

Finally, notice the blocking capacitor and the matching network in Figure 7.5. In order to efficiently couple the RF energy to the plasma, an inductance/capacitance network is included between the plasma load and the generator.

RF generators are designed to have purely resistive internal impedance which as a standard is set to be 50 Ω. It can be shown that the maximum dissipation of power in a load occurs when the impedance of the load is the conjugate of the impedance of the generator. Since the generator is resistive

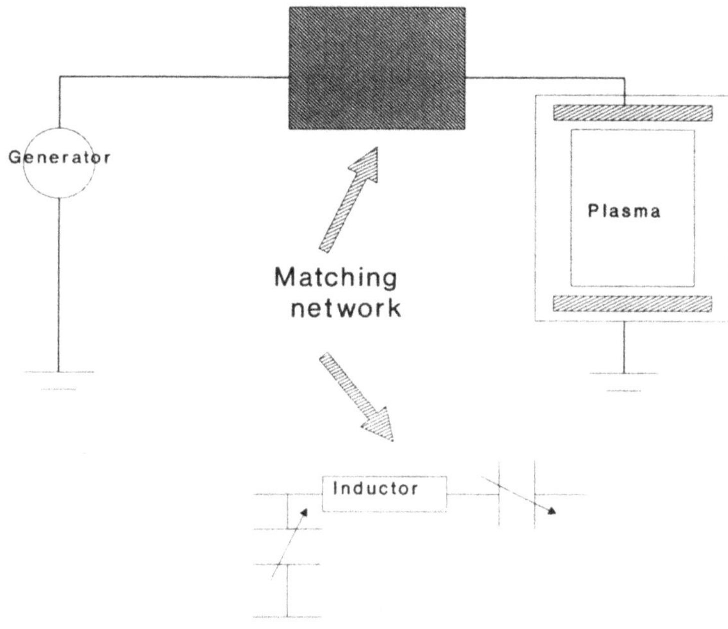

FIGURE 7.5. Matching network for an RF plasma. The capacitor/inductor network matches the plasma load to the generator output impedance of 50 ohms.

at 50 Ω, the changing impedance of the discharge needs to be tuned to this value by the matching circuitry. To get good power coupling, the matching circuitry often contains a fixed inductance coil and a variable capacitor whose value is adjusted with changes in the plasma. Obviously we would like to have zero power losses in the matching network, but that is seldom the case with most RF discharges.

7.2.2. Commercial RF Plasma Reactors

RF plasma CVD reactors have been in use since the mid-1970s for the deposition of dielectric films. Figure 7.6 shows four commonly used chamber and RF coupling configurations of PECVD reactors. In general, other than RF coupling, isolation of electrically hot surfaces, and proper grounding, gas flow and pressure control systems for PECVD reactors are similar to the LPCVD situation. Both single-wafer and batch type tools are available in the market, with sophisticated handlers and controlling software. Use of PECVD for the deposition of silicon nitrides, oxynitrides and oxides in

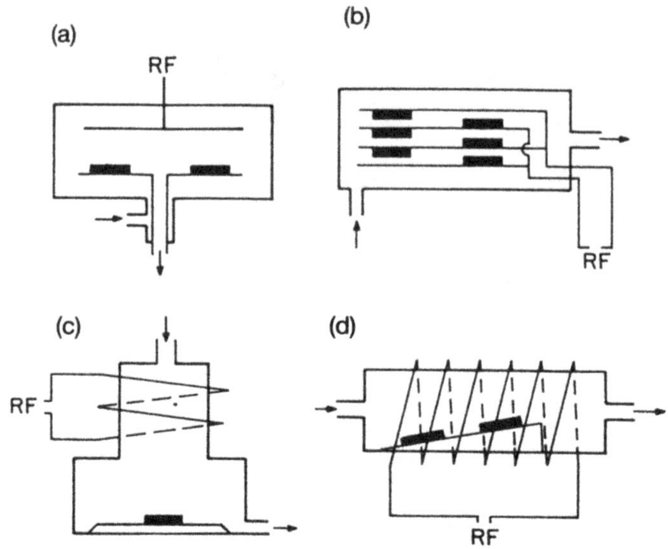

FIGURE 7.6. Four common RF coupling configurations in commercial reactors: (a) and (b) are capacitatively coupled; (c) and (d) are inductively coupled. Adapted from Ref. 9 with permission.

microelectronics is common. Amorphous and epitaxial silicon, metallic films including tungsten and TiN are in development.

Let us explore a typical commercial PECVD system made by Novellus Incorporated. The Novellus Concept One PECVD system is designed for the manufacture of SiN_x (using SiH_4, NH_3, and N_2), SiON (by adding N_2O), and low-SiH/UV-transparent nitride films along with doped and undoped SiO_2 (using SiH_4, N_2O or TEOS, O_2). Applications include passivation films, isolation films and spacer technology (see Chapter 9). It is a cold wall CVD system, capable of simultaneously applying two different frequencies to the substrate. Figure 7.7 shows the physical and electrical schematic of the reactor.

Wafers loaded by the cassette-to-cassette transfer arm, sequentially pass through multiple deposition stations before achieving their full deposition thickness. This helps in averaging the effects of the individual stations. Each of the deposition stations is a complete diode RF plasma reactor, with gas injection and wafer heating. The system uses no quartzware, and is also capable of plasma-based chamber cleans after a specified number of depositions. Both factors are instrumental in maintaining low particulate levels in the reactor. The wafers spend minimal time at elevated temperatures, resulting in reduced hillocking of the underlying aluminum metal. In the case of tetraethylorthosilicate (TEOS), a special liquid reactant delivery pump

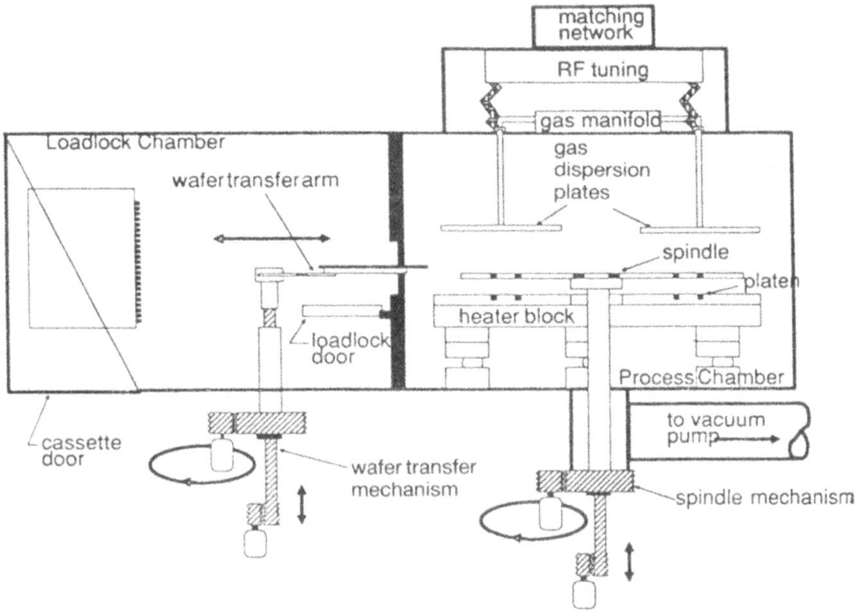

FIGURE 7.7. Schematic of the Novellus Concept One PECVD system with dual frequency for the deposition of dielectric films. Reprinted with permission from Novellus, Inc.

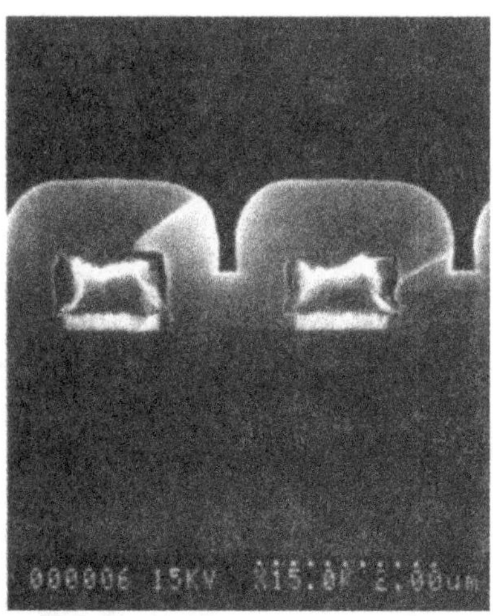

FIGURE 7.8. Film step coverage for a SiN film obtained with a Concept 1 system from Novellus. Reprinted with permission from Novellus, Inc.

precisely meters the incoming gas, which is flash evaporated inside the chamber.

The system is capable of delivering SiN films with a tunable SiH content. By changing the ratio of the high frequency to low frequency power, the ion bombardment on the substrate can be modulated, resulting in different hydrogen incorporation (see Chapter 9). The Si/H ratio is critical in determining many of the optical properties of the dielectric; films can be optimized to increase UV transparency, a requirement for passivation films in devices such as erasable programmable read-only memories (EPROMs). Similarly, film stress, controlled by ion bombardment, can also be optimized using the two dual frequency power inputs. Figure 7.8 shows the step coverage obtainable from using SiN and SiON films using the Concept One system.

7.3. MICROWAVE DISCHARGES

Frequency by itself, when increased further than 13.56 MHz does not contribute to substantial increases in electron temperature or density in the plasma. But by increasing the applied frequency to the gigahertz region, to create microwave discharges, it is possible to port the power through waveguides and resonance cavities such that there is no direct contact between the plasma and an electrode. It is also possible to use a magnetic field to interact with the electric field in order to produce high density plasmas. The ability to create a remote plasma without the presence of an electrode near the deposition region can be significant, since it allows the substrate to be separately powered, without influence from the plasma. Through proper

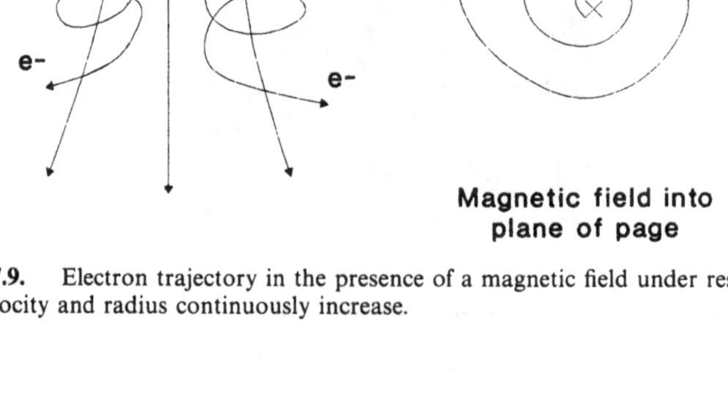

FIGURE 7.9. Electron trajectory in the presence of a magnetic field under resonance. The electron velocity and radius continuously increase.

choice of microwave coupling, microwave discharges can be maintained at pressures varying between a few millitorr and one atmosphere. Moreover, with the presence of an additional magnetic field to interact with the electric field, a variety of high density, low pressure plasmas can be created. Combining the two features, a high density plasma can be created using a microwave/ magnetic field configuration in a remote location, and active species can be extracted from the plasma to react on a substrate. This phenomenon is used in electron cyclotron resonance (ECR) plasma reactors to produce high step coverage, high deposition rate CVD processes, without causing radiation damage to the substrates, particularly for the growth of dielectrics.

7.3.1. Electron Cyclotron Resonance

In the presence of a magnetic field, electrons circle around the magnetic field lines with a frequency f where

$$f = \frac{qB}{2\pi m} \tag{7.9}$$

q is the electron charge and B is the magnet field in tesla (10^4 gauss). If an applied electric field with a frequency equal to the electron frequency is applied to the system, the electrons oscillate at resonance. Figure 7.9 illustrates the motion of an electron in a static magnetic field without a perpendicular electric field and when the condition for resonance is met. Notice the rapid rise in electron velocity due to the ECR heating of the electron.

Most ECR reactors operate at the S-band (2.45 GHz) so that moderate size magnets (875 gauss) can be used to achieve resonance. At the resonance condition transfer of electromagnetic energy to the discharge occurs simultaneously through elastic and inelastic collisions (Joule heating) and electron cyclotron heating of the electron gas. The electron gas then heats the neutral and ion species through elastic and inelastic collisions. Figure 7.10 schematically illustrates the energy transfer mechanisms in a differential volume in the reactor.

Reactor configurations vary according to the means by which the microwave energy is coupled to the reactor. In particular, waveguides and cavity applicators have become popular, their magnetic fields produced by coils or rare earth magnets. At the substrate end, often a separate RF bias is applied to extract ions and charged radicals from the microwave discharge. The substrate could have one of three configurations: (a) where the RF biased substrate is in the microwave discharge, (b) the substrate is separate but close

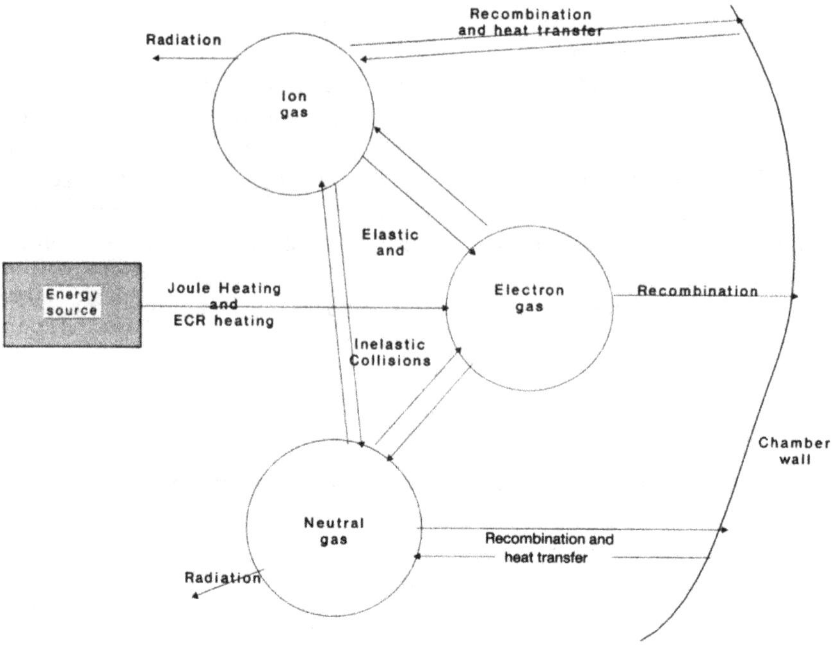

FIGURE 7.10. Energy transfer mechanisms in an ECR reactor. In the differential volume shown, ECR coupling to the electron transfers energy to the chamber. The other species are heated by collision with energetic electrons. Reprinted by permission, from Jes Asmussen, *J. Vac. Sci Technol. A.* 7(3), 883 (1989).

to the microwave discharge, or (c) the substrate is in a separate RF discharge, downstream from the microwave plasma. Each has advantages and disadvantages. In practice, the downstream plasma is more popular for commercial microelectronic CVD reactors as it protects the wafer from radiation damage that might be caused by the microwave discharge.

The main application of ECR reactors continues to be in plasma etching. However, commercial reactors for the CVD of SiO_2 and TiN are beginning to be available. We will study the configuration of the Lam EPIC ECR CVD reactor, used to deposit undoped SiO_2 for interlevel dielectric applications.

Figure 7.11 shows a schematic of the LAM EPIC reactor. 2.45 GHz microwave power is applied through a waveguide to the top of the plasma chamber. The ECR plasma chamber contains N_2O or O_2, and diluent gases such as argon. Silane for SiO_2 is introduced downstream in the wafer chamber to avoid unwanted silicon deposits. A separate 13.56 MHz RF power is applied to the substrate. This field along with the field shaping magnets

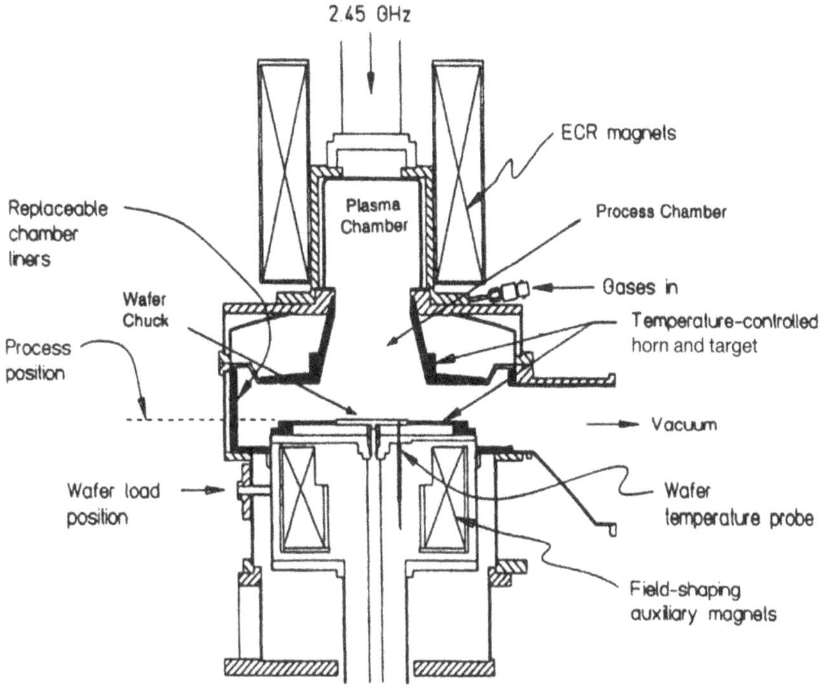

FIGURE 7.11. Lam Research Corporations EPIC ECR plasma CVD reactor. Reprint from the 1992 Proceedings of the VLSI Multilevel Interconnection Conference, p. 144.

below the substrate, help in extracting and directing the oxygen ions to the surface of the wafer, where they react with silane.

Higher RF bias on the substrate leads to higher energy ion bombardment and consequent sputter etching of the deposited film. Hence simultaneous etching and deposition processes occur, resulting in a characteristic film surface profile. Sputter etching produces enhanced etching on surfaces at 45° to the surface normal, so deposition results in increased step coverage with sharp cornered mesas, as shown in Figure 7.12. Figure 7.13 shows the surface profile as the etch/deposition ratio is varied through the bias. The higher the etch/deposition ratio, the better the step coverage. However, this results in reduced deposition rate and reduced throughput. The throughput needs to be optimized with the gap-filling requirement.

The diluent gas plays an important role in determining the sputter etch rate due to the mass of the etching ion. Hence care must be exercised in the choice of diluent gases. Film stress, silanol content, hence the refractive index and dielectric constant of the film, can also be modified by changing the O_2/SiH_4 ratio. In general better film properties are obtained at lower silane

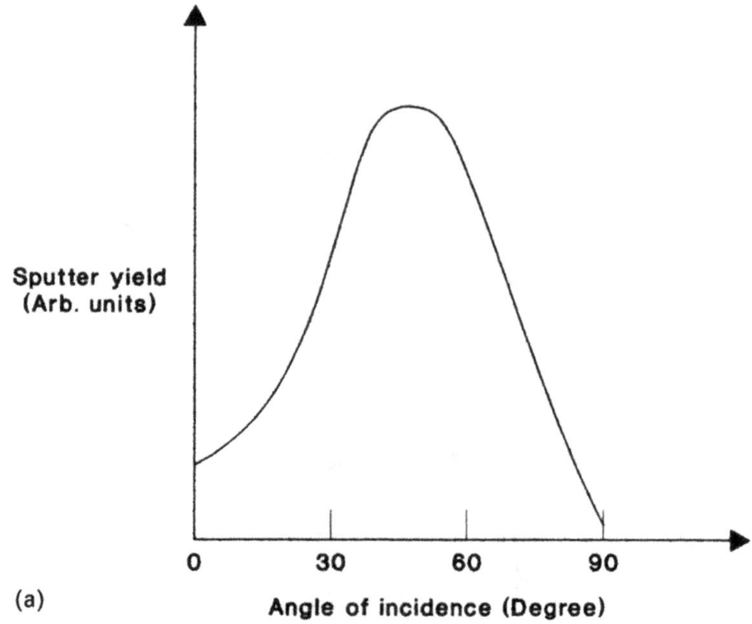

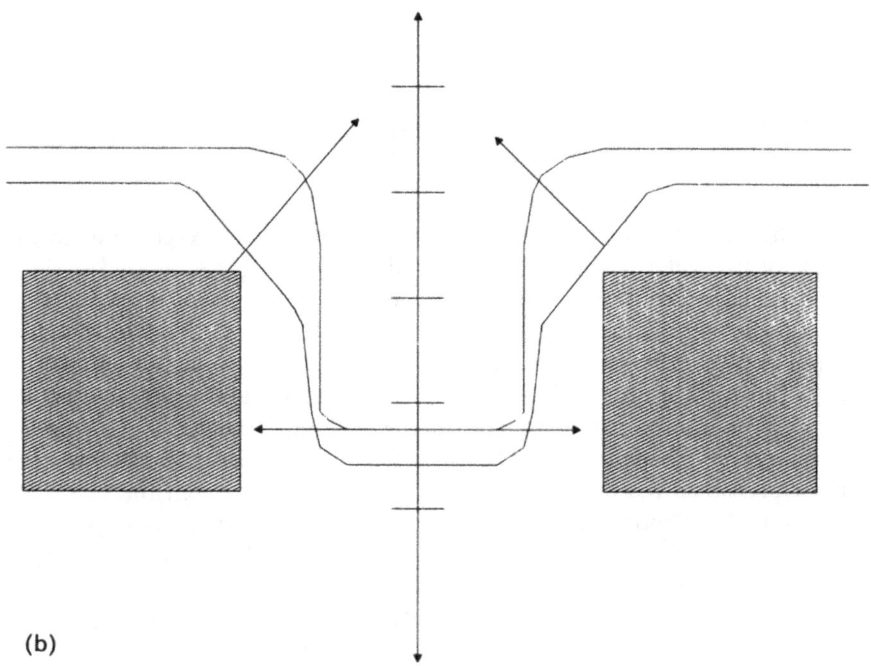

FIGURE 7.12. (a) Angular distribution of sputter etch rate. (b) The maximum at 45° leads to preferential etch and a consequent change in shape.

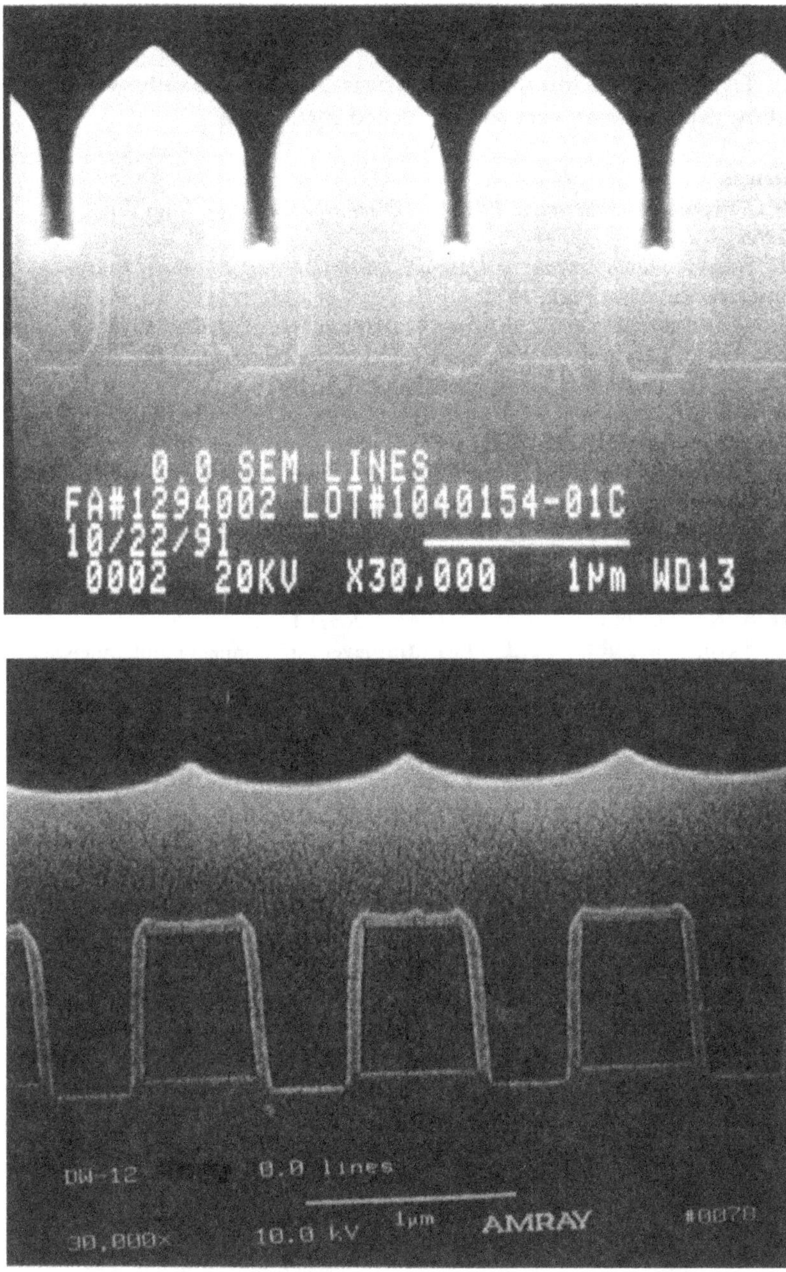

(a) (b)

FIGURE 7.13. Surface profile as the etch/deposition ratio is changed. The sharp angular features in (a) are continuously smoothed to produce the more even surface in (b), which has a higher etch component. Reprint from the 1992 Proceedings of the VLSI Multilevel Interconnection Conference, p. 145.

flows. The ECR CVD interlevel dielectric is currently the only system capable of filling gaps between very finely pitched metal lines.

References

1. S. C. Brown, *Introduction to Electrical Discharges in Gases*, John Wiley, New York, 1966.
2. E. Nasser, *Fundamentals of Gaseous Ionization and Plasma Electronics*, Wiley Interscience, New York, 1971.
3. J. S. Townsend, *Electricity in Gases*, Clarendon Press, London, 1915; J. J. Thomson and G. P. Thomson, *Conduction of Electricity Through Gases*, Vol. II, p. 512, Cambridge University Press, Cambridge, UK, 1928.
4. J. Slepian, *Conduction of Electricity in Gases*, p. 78, Westinghouse Electric & Manufacturing Co., Educational Dept., East Pittsburgh, 1933.
5. R. Paschen, *Weidemann Annalen*, **37**, 69 (1889); S.M. Rossnagel, in *Thin Film Processes II* (Vossen and Kern, eds.), p. 11, Academic Press, New York, 1991.
6. F.F. Chen, *Introduction to Plasma Physics and Controlled Fusion*, Plenum Press, New York, 1983.
7. D. L. Flamm, *J. Vac. Sci. Technol.* **A4**(3), 729 (1986).
8. H. R. Koenig, and L. I. Maissel, *IBM J. R&D* **14**, 168 (1970).
9. W. Kern, course notes, IEEE VLSI Multilevel Interconnect Conference Seminars, 1988; R. Reif and W. Kern, in *Thin Film Processes II* (Vossen and Kern, eds.), p. 525, Academic Press, New York, 1991.
10. R. S. Rosler, and G. M. Engle, *Solid State Technol.* **22**(12), 88 (1979); **24**(4), 172 (1981); J. R. Hollahan, and R. S. Rosler, in *Thin Film Processes* (Vossen and Kern, eds.), p. 335, Academic Press, New York, 1978.
11. J. Asmussen, *J. Vac. Sci. Technol.* **A7**(3), 883 (1989).
12. P. Kidd, *J. Vac. Sci. Technol.* **A9**(3), 466 (1991).
13. R. Chebi, and S. Mittal, in *Proc. IEEE VMIC*, 1991, p. 61.
14. D. Webb, and S. Sivaram, in *Proc. IEEE VMIC*, 1992, p. 141.
15. B. Chapman, *Glow Discharge Processes*, John Wiley, New York, 1980.

Chapter 8

CVD of Conductors

Conductor systems in very large scale integrated VLSI circuits act as the conduits for signals to be transported to and away from electrical devices. As device dimensions and film thicknesses scale down, thin film properties of conductors (see Chapter 2) begin to dominate and special processing conditions become necessary. For instance, it becomes essential to lower the processing temperature so as to minimize undesirable thermally activated processes, such as hillock formation in aluminum-based metallization, or interdiffusion and reaction between adjacent films. Similarly, incorporation of impurities in the film matrix can result in significant degradation in properties, such as resistivity, film roughness, and stress, as the films get thinner. In this chapter, we will first review the requirements for conductor systems in VLSI circuits, without regard to the mode of deposition of the conductors. We will then consider the CVD of individual films, keeping in mind the device requirements. Finally we will examine more recent trends in CVD of conductors, with respect to new materials and with respect to newer CVD processes for established films.

8.1. GENERAL REQUIREMENTS FOR CONDUCTORS IN MICROELECTRONICS

Of all the properties of conducting films, electrical conductance is paramount in all interconnect applications. The choice of the material for the interconnect is based on the bulk value of the material's resistivity, and most of the work during the development of the deposition process is geared toward achieving a thin film resistivity close to the bulk value. The importance of electrical conductivity can be illustrated by the following analysis.[1]

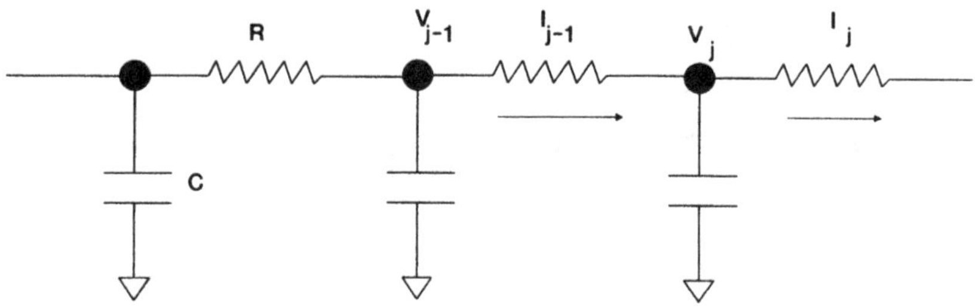

FIGURE 8.1. Analysis of delay times in a distributed network: a network of line resistances and capacitances that lead to delay in signal propagation.

Consider a long metal line on a VLSI circuit represented in Figure 8.1 in terms of distributed resistor–capacitor sections. Each of these sections represents the line resistance; the parasitic capacitances and the nodes might represent a signal input to a gate, etc. The circuit delay in the transmission of a signal from one end of the line to the other is caused by the time it takes to charge and discharge the capacitors. Over time, the voltage V_j at any node j is given by the simple capacitor charge-up equation.

$$C\frac{dV_j}{dt} = I_{j-1} - I_j = \frac{(V_{j-1} - V_j)}{R} - \frac{(V_j - V_{j+1})}{R} \tag{8.1}$$

As the number of sections in the network becomes large and the sections become small, this reduces in the differential form to

$$rc\frac{dV}{dt} = \frac{d^2V}{dx^2} \tag{8.2}$$

where r and c are resistance and capacitance per unit length and x is distance from the input. The signal propagation time t_d for a large number of such sections can then be approximated to

$$t_d = rc\frac{l^2}{2} \tag{8.3}$$

where l is the length of the wire. Hence the terms that are important for the delay are the resistance of the wire segments, determined by the metallization, and the capacitance of the segments, determined by the dielectric that separates adjacent conductors (see Chapter 9). The signal delay scales with the square of the overall length of the line. The length of the line is fixed by the circuit requirements. Hence the resistivity of the conductors and the dielectric constant of the dielectric are crucial in determining circuit performance.

8.2. OTHER PROPERTY REQUIREMENTS

Conductive layer deposition in microelectronic processing can be classified as occurring at three distinct processing stages (Figure 8.2).

a. Gate level: conductors such as transition metal silicides are used directly atop the transistor gate; often called polycides, since they are deposited on a layer of polysilicon.
b. Contact level: conductors such as Ti and TiN are used in direct contact to the underlying semiconductor in a source or drain contact, so as to provide reproducible metal/semiconductor contacts. Metals such as tungsten are deposited over these layers to smooth out harsh topographies created by the contact hole.
c. Interconnect level: interconnection between devices is accomplished by very low resistivity conductors such as aluminum alloys. This level may be repeated in second-, third- or fourth-level metallization over the first layer.

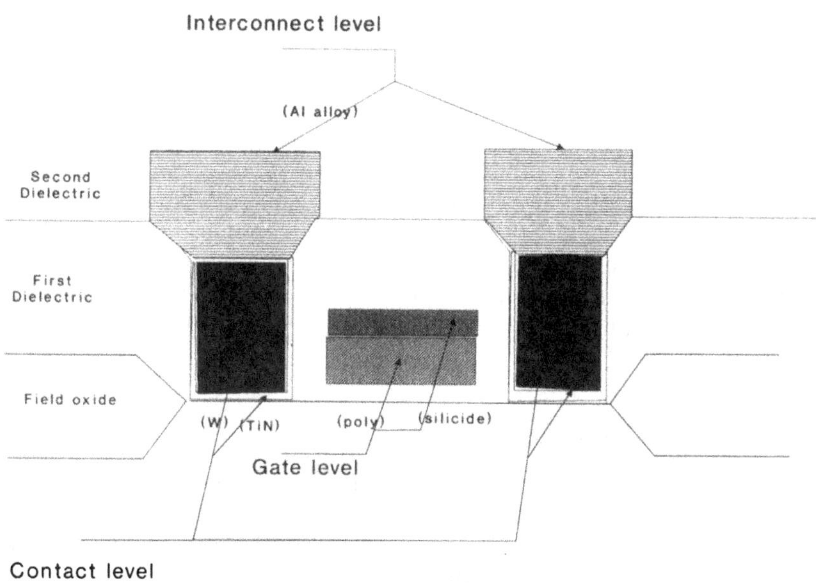

FIGURE 8.2. Classifications of conductors in VLSI devices based on their application: device level, and interconnect level conductors.

We will examine the general property requirements for each of these levels and then examine film structure requirements to satisfy them.

8.2.1. Gate Level Conductors

The gate oxide dielectric is at the heart of a MOS transistor shown in Figure 8.2. The integrity of the gate oxide determines such properties as the threshold voltage V_t, at which the transistor switches on or off, and the properties of the MOS capacitor formed at the gate. Hence gate oxide integrity critically determines the performance of the circuit.[2] To provide a reproducible interface to the gate oxide, doped polysilicon is often the first conductive layer deposited on the dielectric. Polysilicon is considered along with other semiconductors in Chapter 10, even though it acts as a conductor in this application. The resistance of the structure that acts as the gate electrode over the dielectric determines the gate delay of the device. Heavily doped polysilicon has a resistivity of about 500 $\mu\Omega$ cm. In order to reduce the gate resistance, a parallel layer of a more conductive film is added on top of the polysilicon. Commonly used gate level conductors include the silicides of titanium, tungsten, and molybdenum. Furthermore, the conductive layer may also be used directly on the diffusion area so as to reduce diffusion resistance.

These conductive films are deposited relatively early in the processing scheme, so they need to withstand high processing temperatures that occur later on. Since processing temperatures could be as high as 1 000°C, metallic elements should be bonded well enough not to diffuse into the gate dielectric. Similarly, residuals from their deposition chemistries should not damage the underlying films at elevated temperatures. They should be oxidizable in order to form stable insulators to encapsulate them, and suitable chemistries for their etching should exist. Stress levels should be low in order to eliminate film cracking, delamination, and other subtle stress-related phenomena such as band gap change and enhanced diffusion. They need to be fine-grained and isotropic so that property variations on certain axes are prevented.

8.2.2. Contact Level Conductors

The dimension of the drain contact is often the smallest patterned feature on a device.[3] Coupled with the thickness of the first-level dielectric, this results in the contact being a narrow and deep right circular cylinder, with a ratio of height to diameter, called the aspect ratio, often greater than 3 (see Figure 8.2). Hence as a first requirement we should be able to deposit the film uniformly at the bottom of the contact. Once the film is in contact with the substrate, it has to satisfy the following requirements:

a. It should not itself be detrimental to the functioning of the transistor; for instance, impurities in the film cannot contribute to traps in the band gap of the semiconductor that can destroy carrier lifetimes, or cause electrical shorting of the underlying junction.
b. It should provide a low resistance contact for the metal–semiconductor interface between the metallization and the silicon. The formation of native oxide on silicon could be a serious impediment.
c. It should protect the contact from the metallurgies and chemistries of subsequent layers.
d. If the intent is to fill the contact cylinder, the deposition step coverage needs to be good, with void-free filling of the contact.

Other requirements in this level include good thermal stability up to 500°C, little reaction with silicon or aluminum, and a smooth surface morphology. Thermal expansion coefficients compatible with silicon are beneficial.

8.2.3. Interconnect Level Conductors

At the interconnect level, lines of very low resistivity conductors, often narrow and sometimes many centimeters long, carry high current densities across the chip. Logically, integrity of the line is of paramount importance. Of the three levels of conductors, low electrical resistivity is crucial for the interconnect level, since higher electrical resistance results in larger signal delay. Resistive heating of the line results in enhancement of unwanted thermally activated processes, and a higher resistivity material requires a larger film thickness to maintain the same line resistance. This causes problems for further processing due to more severe topography.

Other impediments to maintaining line integrity are two diffusive processes which result in voiding and necking in the lines: electromigration and stress migration (or creep). Low melting conductors such as aluminum are particularly susceptible to both these phenomena. Alloying of the aluminum with metals, such as copper, titanium, and palladium, has been shown to alleviate these problems.[4] Other requirements for the interconnect level include control of grain size distribution, prevention of hillock growth, thermal stability, and a smooth surface.

8.3. CVD OF TUNGSTEN

Metallic films such as tungsten, aluminum, titanium, and copper find applications in the contact and interconnect levels. In studying their deposition characteristics, we will follow a format that consists of a review of the properties of the metal from a metallurgical standpoint, examination of possible CVD precursors, and the kinetics of deposition from some of the chemistries. Where available, a commercial reactor for the deposition of the film will be presented.

Significant research and development activity has occurred on the development of CVD tungsten for VLSI metallization in the last 10 years. The reader is strongly recommended to refer to a series of publications by the Materials Research Society from 1987 to 1992,[2] on the proceedings of the workshops on tungsten and other refractory metals.[5]

A group VIB metal, tungsten has the highest melting point of all metals (about five times the melting temperature of aluminum). It also possesses high electrical conductance (about twice the resistivity of aluminum).[6] This unique combination of high thermal stability and low electrical resistivity has made it a very attractive candidate for VLSI metallization. Table 8.1 lists relevant bulk physical, mechanical, and chemical properties of tungsten.

Since device processing and operating temperatures are relatively low compared with the melting point of tungsten, diffusivity of tungsten metal into either aluminum or silicon is low. Hence in the 1970s attempts were made to incorporate tungsten as part of the gate conductor without using an intermediate polysilicon layer. However, these efforts were not altogether successful due to nonreproducible interface properties between tungsten and the gate dielectric.

Current applications of tungsten in VLSI metallization are (a) to fill contact holes, so as to reduce the severe topography created by high aspect ratio

TABLE 8.1 Properties of tungsten

Property	Value	Units
Isotopes	180, 182, 183, 184, 186	AMU
Heat capacity at 300 K, c_p	5.1	cal K^{-1} mol^{-1}
Free energy of formation at 300 K	190	kcal/mol
Crystal structure	BCC, $a = 3.165$	Å
Theoretical density	19.35	g/cc
Activation for self-diffusion	120.5	kcal/mol
Melting point	3410	°C
Coefficient of expansion at 20°C	4.44	ppm
Thermal conductivity at 50°C	0.425	cal cm^{-1} $°C^{-1}$
Electrical resistivity (single crystal at 20°C)	5.3–5.5	$\mu\Omega$ cm
Young's modulus	38.7×10^{11}	dyne/cm^2
Chemical etchants	10% NaOH and 30% $K_3Fe(CN)_6$, 10% H_2O_2 boiling, 1:3 HF and HNO_3	

contacts, by taking advantage of CVD processes; and (b) as interconnects. Since the resistivity of tungsten is twice as high as aluminum, interconnect applications are limited to very short line segments and at the first level of a multilevel metallization, where we can take advantage of its ability to withstand higher processing temperature.

Chemical vapor deposition of tungsten by the reduction of tungsten hexafluoride has been used for the production of tungsten filaments since the 1950s.[7] The following chemistries have been examined as possible sources for the production of tungsten thin films.

8.3.1. Pyrolysis of Tungsten Hexacarbonyl

Pyrolysis of tungsten hexacarbonyl[8] at over 250°C proceeds by the following reaction:

$$W(CO)_6 \rightarrow W + 6CO$$

The tungsten film produced contains carbon and needs to be annealed at 700°C or higher to reduce resistivity.

8.3.2. Reduction of Tungsten Hexachloride

Reduction of tungsten hexachloride[9] using hydrogen proceeds according to the following reaction:

$$WCl_6 + H_2 \rightarrow W + 6HCl$$

WCl_6 is a solid at room temperature and is hence inconvenient as a CVD source. Early work on deposition of tungsten from WCl_6 suggests non-reproducible growth in the presence of oxygen-bearing impurities. The reaction preferentially occurs on silicon surfaces compared with adjacent SiO_2 surfaces.

8.3.3. Reduction of Tungsten Hexafluoride

Reduction of tungsten hexafluoride is the most popular method of tungsten film growth for integrated circuits. WF_6 is a clear liquid that boils near 25°C. The reduction of WF_6 is selective, i.e., it occurs preferentially on conductive surfaces or, more importantly, if the conditions are right, it does not deposit on SiO_2 while depositing on adjacent silicon. The reducing agent need not be hydrogen. Silicon, SiH_4, GeH_4, SiH_2Cl_2, $TiSi_2$, etc. have been used successfully as reducing agents.[10]

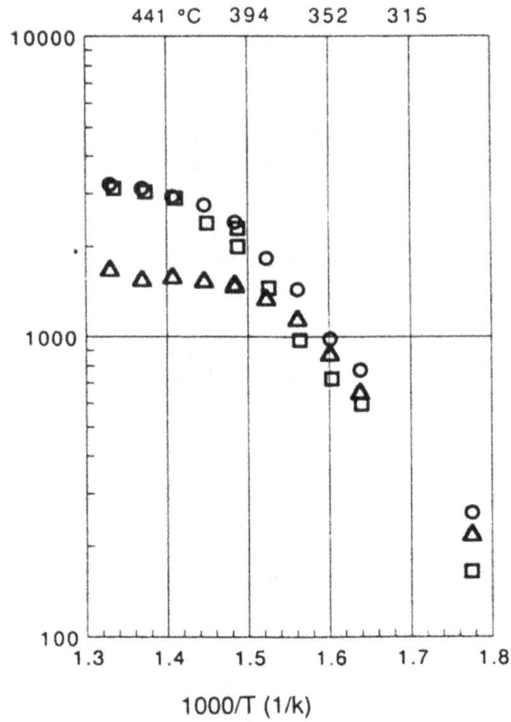

Symbol	O	□	Δ
[H_2] (torr)	*32*	*16*	*32*
[Ar] (torr)	*47*	*23*	*47*
[WF_6] (torr)	*0.8*	*0.4*	*0.4*
WF_6 (sccm)	*25*	*25*	*12*

FIGURE 8.3. Arrhenius plot for the deposition of tungsten from WF_6 and H_2. The data show different pressures and flow conditions. Reprinted, by permission, from E. J. McInerney, B. Chin, and E. K. Broadbent, in *Proceedings of the Workshop on Tungsten and Other Refractory Metals for ULSI Applications VII*, Materials Research Society, Pittsburgh, 1991.

Hydrogen Reduction of WF_6

The Gibbs free energy change for the reaction

$$WF_6 + 3H_2 \rightarrow W + 6HF$$

is -278 kcal/mol at 600°C.[11] The Arrhenius plot for the reaction in the temperature regime 300–500°C is shown in Figure 8.3.[12] Notice the reaction is strongly thermally activated up to a temperature of about 400°C, at which

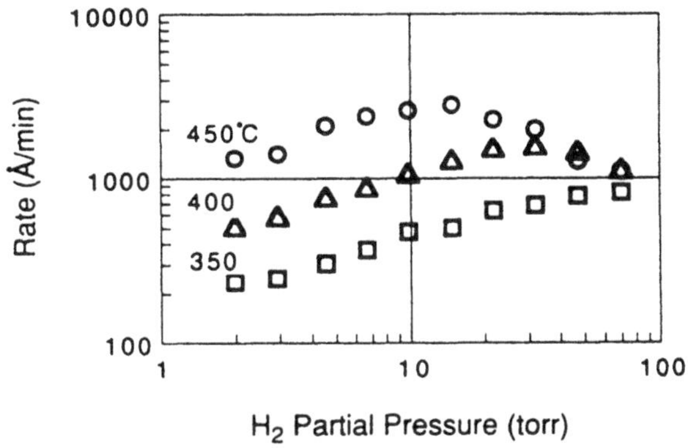

FIGURE 8.4. Dependence of tungsten growth rate on the partial pressure of hydrogen. The one-half order is important in explaining the rate-controlling step. Reprinted, by permission, from E. J. McInerney, B. Chin, and E. K. Broadbent, in *Proceedings of the Workshop on Tungsten and Other Refractory Metals for ULSI Applications VII*, Materials Research Society, Pittsburgh, 1991.

point the rate is no longer strongly temperature dependent. The activation energy for the thermally activated region has been measured by various researchers to be 67–73 kJ/mol.[13] The reaction exhibits no measurable dependence of rate on the partial pressure of tungsten hexafluoride and shows one-half order dependence on the partial pressure of hydrogen below 400°C (Figure 8.4).[14]

The mechanism for growth below 400°C is widely assumed to be the desorption of HF from the surface. There has been some controversy in earlier literature regarding dissociative chemisorption of hydrogen on the surface being rate-controlling and also being a possible explanation for the observed selectivity in growth between conductive and nonconductive surfaces. This was apparently supported by the one-half order of the reaction on the partial pressure of hydrogen. However, if the rate-controlling reaction is written as

$$1/2H_2 + * \rightarrow H^*$$

where * represents a surface site. To account for the one-half order dependence of rate on hydrogen partial pressure, we assume a gas phase dissociation of the molecular hydrogen to individual atoms. If, however, we correctly write this expression as

$$H_2 + 2^* \rightarrow 2H^*$$

or as

$$H_2 + *_2 \rightarrow H_2*_2$$

both reactions result in a first-order dependence of rate on hydrogen partial pressure, in contrast to the data.[10]

If we write out all possible sequences of reactions in the WF_6/H_2 system as

$$WF_x + * \rightarrow WF_x*$$
$$WF_x* + H* \rightarrow WF_{x-1}* + HF*$$
$$HF* \rightarrow HF$$

where x is an integer between 1 and 6. Assuming W* = *, we can solve for the overall rate expression. If we further assume the HF desorption reaction is rate-controlling, we can write the rate expression as

$$r = k[WF_6]^{1/6}[H_2]^{1/2}$$

This shows the weak dependence of rate on the tungsten hexafluoride partial pressure and the one-half order dependence on hydrogen partial pressure.[15]

Since the rate is dependent on the partial pressure of hydrogen, the total pressure is very effectively used in rate control, while simultaneously optimizing other properties. Commercially viable reactions have been reported from pressures as low as 200 millitorr to atmospheric pressure.

Reduction of WF_6 by Silicon-Bearing Species

The free energy change for the reaction

$$2WF_6 + 3Si \rightarrow 2W + 3SiF_4$$

is strongly negative in the normal operating pressure regime for CVD. In the absence of other reducing agents, the deposited tungsten forms an impervious layer with a self-limited thickness, which prevents further diffusion of the gas species to the interface. The self-limiting thickness seems to be a function of the cleanness of the surface prior to growth. Native oxide on the surface results in a discontinuous first layer containing channels which allow further gas diffusion. This results in a thicker and porous tungsten deposit.[16] The silicon reduction reaction is extremely rapid and is often the starting reaction, even when a WF_6/H_2 mixture is flowed into a reaction chamber. This is a mixed blessing. The substrate reaction results in better film adhesion and often better contact resistance. However, the consumption of silicon results in encroachment of the tungsten deposit, both laterally towards the gate and towards the shallow junction (see Figure 8.9). The preceding discussion applies equally for other silicon-bearing substrate materials, such as refractory metal silicides. As a rule, precautions need to be taken to minimize the substrate reduction of tungsten hexafluoride.

Reduction of tungsten hexafluoride by silane has found wide application as the nucleating layer for hydrogen-reduced tungsten. The reaction proceeds as follows:

$$2WF_6 + 3SiH_4 \rightarrow 2W + 3SiF_4 + 6H_2$$

The reaction produces tungsten films when the gas phase mixture has excess tungsten hexafluoride and silicides of tungsten with a silane excess (see Section 8.6). The free energy change for this reaction is more negative than for the direct hydrogen reduction of WF_6. The kinetics of this reaction are not completely understood, even though there is indication the rate might be first-order dependent on the partial pressure of silane.[17] Rate also seems to be a function of silane flow. There seems to be a small negative dependence of the rate on the partial pressure of WF_6, suggesting a competition for surface adsorption sites. The activation energy for the reaction is very small (on the order of 0.1 eV) and seems to be a function of the deposition pressure.

The difficulty in measuring the reaction kinetics comes from the fact that a tungsten film is produced only when the silane concentration is lower than the WF_6 concentration This results in a reactor condition where the conversion of silane is very high, and the condition of a differential reactor for the measurement of kinetics is hard to achieve.

It is, however, speculated that the sequence of events during deposition involves decomposition of SiH_4, releasing molecular hydrogen and the reduction of WF_6 by either Si or SiH_2. The reaction product is molecular H_2, not HF, even though the formation of HF is thermodynamically preferred.[18]

Silane reduction of WF_6 results in higher growth rates and better nucleation on most surfaces. Under relatively low temperatures (250–350°C) and under controlled flow and pressure, the silane reduction of WF_6 can be selective. The regime of selectivity is bound at the higher temperatures by loss of selectivity, and at lower temperatures by poor adhesion and high stresses. The precise mechanism that controls nucleation on different surfaces, and hence selectivity, is not well understood. In general silane reduction results in more uniform nucleation that hydrogen reduction on most substrate surfaces, even if the same cannot be said of film adhesion.

8.3.4. Properties and Applications

Application of CVD tungsten to device processing has been through two parallel approaches: using a nonselective, or blanket, deposition of tungsten, which is subsequently etched to result in the desired patterns; or using the selectivity of the deposition process to obtain tungsten films only at desired locations, without further etching. Figure 8.5 illustrates the two techniques and some applications to actual devices.

Blanket CVD of Tungsten

Even though blanket tungsten seems to be more complicated than selective tungsten, it has found wider acceptance. Some of the basic properties required of the blanket tungsten process are (a) void-free filling of contact holes and good step coverage over severe topographies, (b) economy of operation (WF_6 gas is expensive), and (c) low particle levels, which prevent delamination of the deposit from either the substrate or the chamber walls. We will address void-free filling in this section and examine the other two in the context of a commercial reactor later in the chapter.

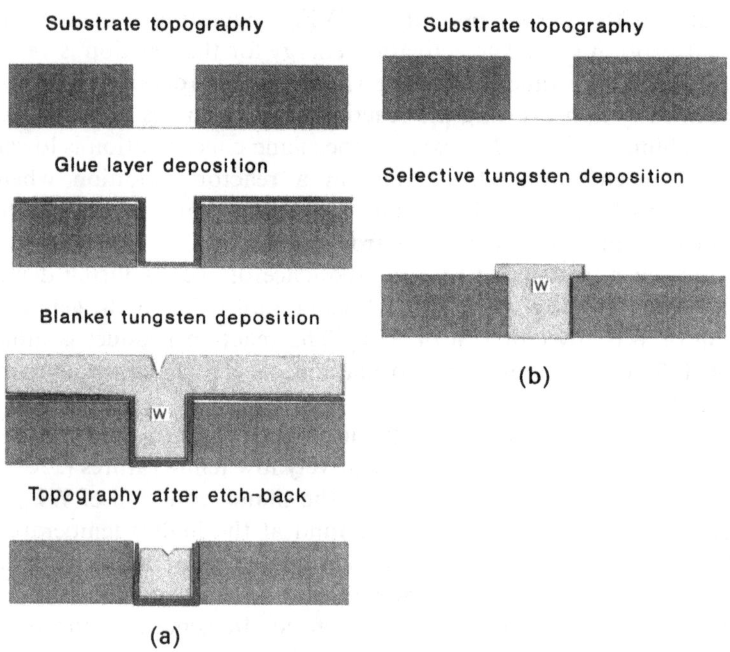

FIGURE 8.5. Two methods of incorporating tungsten plugs in an actual process flow: (a) blanket tungsten and (b) selective tungsten.

FIGURE 8.6. Step coverage of tungsten films grown by high pressure hydrogen reduction of WF_6. Reprinted from J. E. J. Schmitz, *Chemical Vapor Deposition of Tungsten and Tungsten Silicides*, p. 58, with permission of Noyes Publications © 1992.

Void-free filling and good step coverage of the tungsten films in submicron contacts, whose aspect ratios might be greater than 2–3, has been reported widely in the literature (Figure 8.6). Let us examine the problem from first principles for both silane- and hydrogen-reduced tungsten films. At a pressure of 1 torr, the mean free path for an H_2 molecule is approximately 200 μm, and for WF_6 is close to 100 μm. The mean free path for silane lies between the two. Since the contact opening is smaller than 1 μm, the Knudsen number for this situation is much larger than unity (see Section 5.2). Hence we need to consider the diffusions in a molecular flow mode, even if the flow in the reactor is in a viscous flow mode. The conductance of the contact to the various gas species can be treated as that of a short tube in molecular flow and can be written as

$$C(\mathrm{L/s}) = 3.81\left(\frac{T}{M}\right)^{1/2}\frac{D^3}{x} \tag{8.3}$$

where M is the molecular weight of the gas species, D the diameter of the tube, and x is the length of the tube (both in centimeters). The net arrival of the species can then be written as a product of the pressure difference and the conductance as

$$Q(\mathrm{torr\ L/s}) = \Delta P_x(\mathrm{torr})C \tag{8.4}$$

where P_x is the pressure difference between the entrance to the tube and the axial location, which along with the tube conductance, determines the arrival rate at that location.

We can now see that diffusivity (in g/mol) is a function of the molecular weight of the species, in our case 2 for H_2, 32 for SiH_4, and 298 for WF_6. Thus the concentrations of the different species at the bottom of the contact can be quite different and a gradient exists along the axis of the tube.

The rate equation for the hydrogen reaction was expressed earlier as

$$r_H \sim [H_2]^{1/2}[WF_6]^0 \tag{8.5}$$

and that of the silane reaction as

$$r_{SiH_4} \sim [SiH_4]^1[WF_6]^0 \tag{8.6}$$

Despite the zeroth-order dependence of both rates on $[WF_6]$, this assumes that the reaction will stop if the concentration drops to zero. At very low partial pressures of WF_6 it has been proposed that the reaction has one-sixth order dependence on WF_6. Thus we can see that the concentration gradient created along the axis of the contact results in a gradual lowering of the growth rate. The higher the order of the reaction with respect to any reactant, the more strongly is the growth affected by the concentration gradient. In our case, silane reduction results in poorer step coverage precisely because of the higher order of that reaction. Whereas it is possible in the hydrogen reduction case to increase the overall pressure outside the contact to ensure that the concentration of WF_6 does not go to zero at the bottom of the contact, the same technique would result in worsening of the gradient in the silane case. Hence most of the commercial reactors have developed a two-step, two-pressure process, where the silane reaction is used for nucleation purposes only at relatively low pressures and the hydrogen reaction is then used to fill contacts at higher pressures, often close to one atmosphere.

Precise modeling of the step coverage problem has been treated in a relatively detailed fashion in Refs 19 and 20. In general, the following guidelines help to ensure good step coverage:

a. Operate in the kinetically controlled regime (less than ($<450°C$ for the hydrogen reaction) so arrival of the reactants to the surface is not rate-controlling.
b. Ensure the reactor is not running starved of any reactant, particularly silane for silane reduction.
c. Rapidly pump product gases away from the substrate.
d. Make a judicious choice on the reaction chemistry by using multistep depositions, instead of using a silane-based reaction to fill narrow contacts, despite the high rates obtained through silane reaction.

The microstructure of deposited tungsten films holds the key to many film properties. In order to prevent film delamination, blanket tungsten films are deposited on top of glue layers, such as sputtered TiN or a Ti/W alloy. The underlayers determine the nucleation characteristics of the film, resulting in finer or coarser grain sizes.[21] Furthermore, the kinetics of growth, i.e., whether reaction- or diffusion-controlled (along with an independent role by the temperature) modifies the morphology, the stress, and the texture in the film.

The transmission electron micrographs (TEMs) shown in Figure 8.7 compare the morphologies of tungsten films produced from hydrogen and silane reductions. BCC tungsten with a [1 0 0] texture can be identified from the planar TEM and the corresponding electron diffraction pattern shown in the figure. In contrast, the TEM micrograph for the silane-reduced film of the same thickness shows a finer grain structure, due to its ease of nucleation, and no detectable texture. Similarly, hydrogen-reduced tungsten on Ti/W alloy shows finer grains than on TiN. Nucleation of hydrogen-reduced film on a TiN underlayer is often erratic.

Table 8.2 summarizes properties of both hydrogen- and silane-reduced tungsten films.[22] Resistivity of the hydrogen-reduced films can be as low as 6 $\mu\Omega$ cm (which is very close to the bulk resistivity of 5.6 $\mu\Omega$ cm) compared

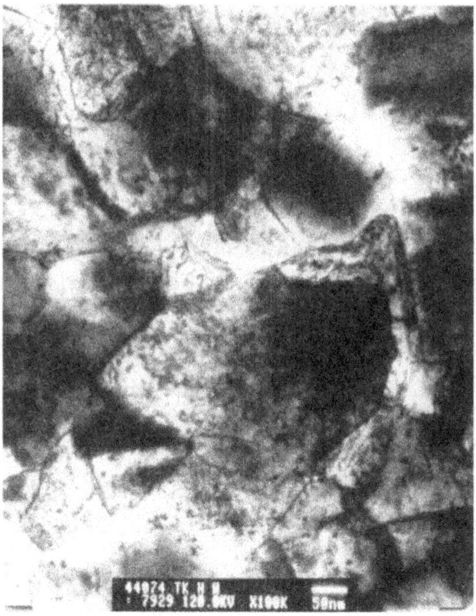

(a)

(b)

FIGURE 8.7. Grain size and grain morphology of tungsten films produced by (a) hydrogen reduction and (b) silane reduction. Silane reduction produces a finer, more faulted structure. the corresponding electron diffraction patterns are also shown. Picture courtesy L. Dass, Intel Corp.

TABLE 8.2 Properties of Tungsten Films Grown by Hydrogen and Silane Reduction of WF_6

Property	H_2 reduction	SiH_4 reduction
Density (g/cc)	15–16	15–16
Hardness (HK)	1 250	900
Stress (MPa)	600–700	830–950
Young's Modulus (MPa)	250	423
Microstructure	normal	faulted
Texture	random	[1 1 0]

with the higher resistance of the silane films. This has been attributed to the incorporation of small quantities of silicon in the silane-reduced films and to their finer grain size. Slower-growing hydrogen-reduced films are often harder than the silane films, probably due to the incorporation of oxygen-bearing species from the ambient and their subsequent precipitation. The TEM micrograph in Figure 8.7 actually shows tungsten oxide precipitates in hydrogen-reduced films. The surface morphology of the films, and hence the reflectivity, is a direct function of the grain size.

Silane-reduced films show extremely high tensile stresses, many times the bulk tensile strength. Stress is again a strong function of the mode of growth, the vacuum level in the system, and any subsequent anneals done on the films. Thus we can see that application of relatively simple thin film growth principles can assist one in tailoring film properties to suit one's needs.

Selective Tungsten

Eliminating the need to pattern and etch in VLSI processing has been a strong incentive to pursue selective tungsten. Furthermore, the consumption of WF_6 when using a selective process is greatly reduced compared with the blanket tungsten process, providing an economic motive. Manufacturing technology, however, has proven difficult to master due to (a) a lack of reproducible and quantifiable selectivity, (b) an inability to protect the junction and the semiconductor surface from the reaction chemistry, (c) the simultaneous presence of n^+ and p^+ silicon, polysilicon, and/or transition metal silicides, requiring selectivity on all of them, (d) the existence of contacts with different hole depths, each of which requires a different deposition thickness, and (e) a very narrow process window for a commercial reactor to provide robust margins. Despite these difficulties, researchers have been pursuing selective tungsten technology for VLSI over the last 15 years.

Both silane and hydrogen reduction of WF_6 show selectivity in growth. Obtaining selectivity is easier with the hydrogen chemistry than with the silane chemistry, given its nucleation characteristics. The origin of selectivity during hydrogen reduction on different surfaces is related to the ease of dissociative chemisorption of the H_2 molecules. For silane, the selectivity mechanism is not well understood.

Selectivity loss has been attributed to product gases such as SiF_x which adsorb on the dielectric and reduce the tungsten fluorides.[23] Selectivity is

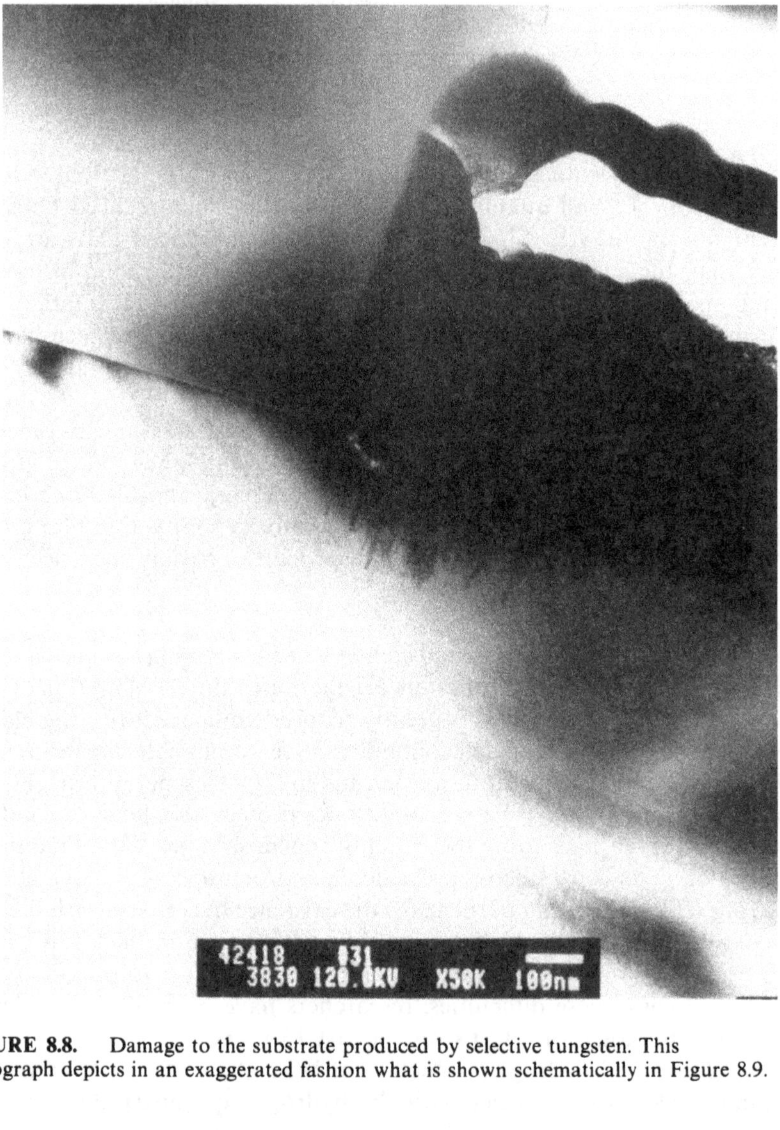

FIGURE 8.8. Damage to the substrate produced by selective tungsten. This micrograph depicts in an exaggerated fashion what is shown schematically in Figure 8.9.

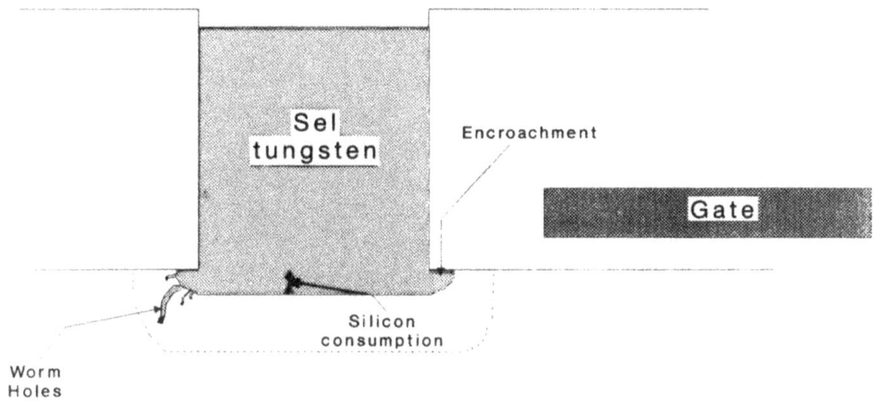

FIGURE 8.9. Schematic representation of the substrate damage produced by selective tungsten. Notice silicon consumption, encroachment, and wormhole formation.

often enhanced by (a) lower processing temperature, (b) rapid pumping, (c) minimizing actual growth surfaces and hence product species, and (d) using large concentrations of diluents such as Ar or H_2.

Wormholes and encroachment present dramatic examples of undesirable interaction between the growth chemistry and the substrate.[24] Figure 8.8 shows TEM micrographs at the bottom of a hydrogen-reduced selective tungsten plug. Reduction of WF_6 by the substrate silicon has resulted in the consumption of the silicon and the formation of a void. Tungsten grows in the lateral direction, encroaching on the gate–drain spacing. Filaments are often observed in thin film growth. The opposite of filaments, tiny wormholes, whose formation is catalyzed by tungsten particles, grow into the silicon and can short the junction. Figure 8.9 schematically summarizes the potential hurdles of the selective tungsten technology.

Even though selective tungsten technology has advanced enough to minimize substrate interactions, the technology itself has not found wide commercial acceptance.

Commercial CVD Tungsten Reactors

Both single-wafer and batch reactors are now commercially available for the deposition of tungsten films. Almost all are low pressure, thermal CVD reactors, utilizing a two-step deposition. The first step is silane reduction of WF_6 to provide consistent nucleation of tungsten on different underlayers. The bulk of the deposition is carried out at much higher pressures (around 25 torr) through hydrogen reduction. This results in superior step coverage and growth rate.

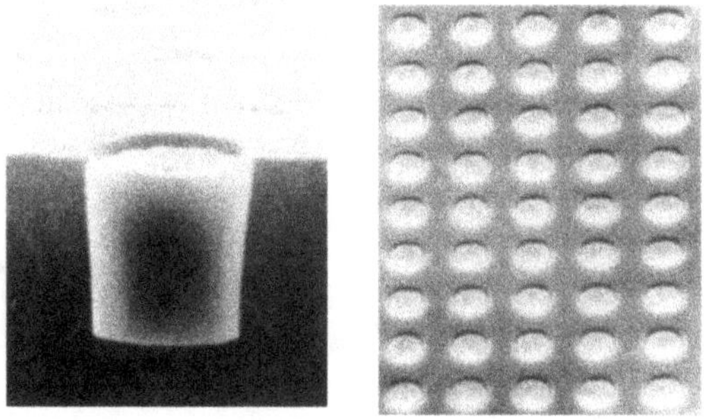

FIGURE 8.10. An individual plug and a tungsten plug array produced by the Applied Materials 5000 cluster system (Courtesy, Applied Materials, Inc.).

The reactors are cold wall or they control the wall temperature at levels significantly lower than the substrate. They introduce a premixed flow of WF_6 and the reducing agent. Control of wall temperature is crucial for particulate control. Hot surface area is in general minimized. However, deposition around the wafer on the chuck and holding clamps is periodically cleaned using a plasma with reactive gases such as CHF_3 or NF_3. The residual fluorine from these plasma treatments can affect the growth rate after a clean and the chamber often needs to be *seasoned* with a thin coat of tungsten.[25]

Deposition on the edge of the wafer and on the wafer backside can lead to film delamination either in the chamber or outside. A contacting or non-contacting clamp around the wafer periphery is often used to prevent deposition on the edges and wafer backside.

Figure 8.10 shows an electron micrograph of tungsten plugs produced in a deposition-etch cluster produced by Applied Materials, Inc. (see reactor in Figure 5.13). Figure 8.11 shows the Novellus Concept One tungsten system, which accomplishes the deposition over a series of deposition stations. A modified version of the Genus 8710 reactor shown in Figure 8.12 is also popular as a tungsten tool.

8.4. CVD OF COPPER

Metallic copper has a bulk resistivity of 1.6 $\mu\Omega$ cm, almost 50% better than aluminum. Its atoms are also heavier than aluminum so they are less prone to diffusional problems that afflict aluminum-base metallizations.

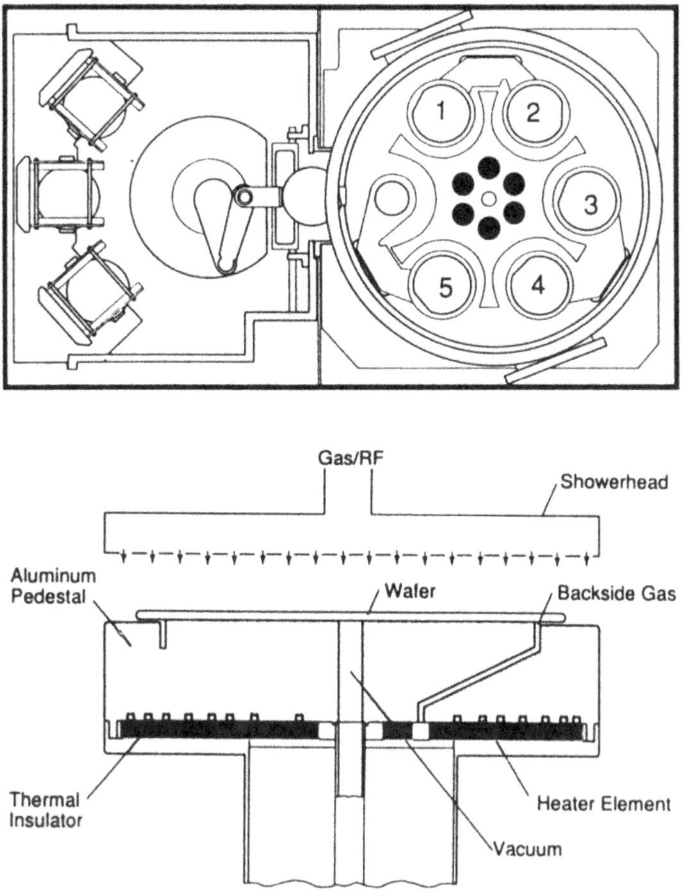

FIGURE 8.11. Wafer transport sequence and the individual cell in the sequential tungsten deposition from Novellus. Reprinted with permission from Novellus, Inc.

Copper-based processing at the time this book is being written is in the early development phase. It still has the following shortcomings:

a. Lack of volatile copper fluorides and copper chlorides creates plasma etching problems.
b. Copper readily diffuses through SiO_2; a suitable barrier is yet to be developed.
c. Copper produces mid-band-gap traps in silicon hence diffusion of copper to the transistor region could be detrimental to device performance.

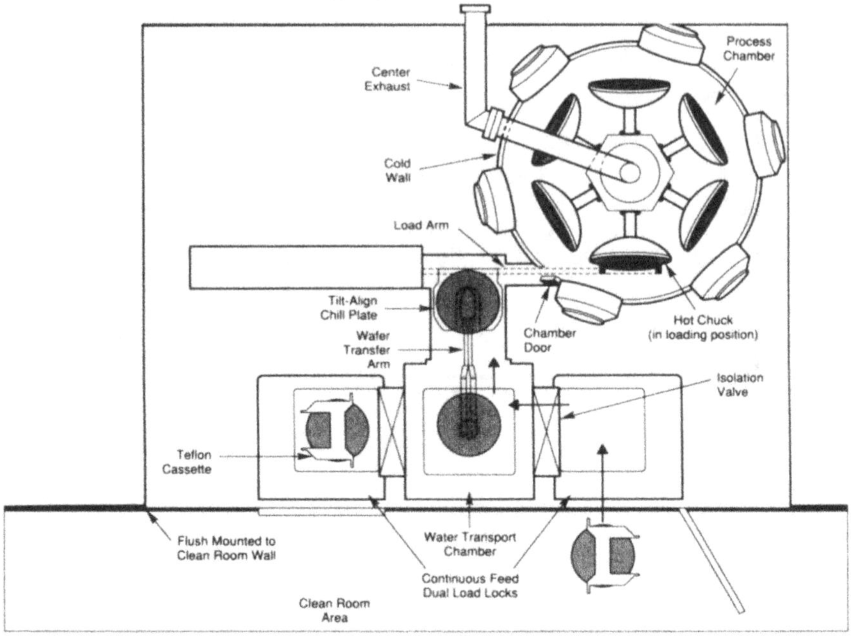

FIGURE 8.12. The Genus 8710 LPCVD system can be used for tungsten silicide and tungsten deposition with suitable modifications. Reprinted with permission from Genus, Inc.

d. Manufacturable CVD processes for the deposition of copper have not succeeded in achieving bulk resistivity values alongside other required properties. Hence the overwhelming advantage of copper metallization is lost when trying to obtain good step coverage.

8.4.1. Growth Chemistry

Let us begin the discussion on CVD copper with a general note on choice of precursors for CVD processes. An obvious requirement for CVD precursors is volatility. A gaseous source, such as silane, is better than a liquid such as WF_6, which, in turn, is better than a solid such as CuCl. Safety is another requirement. Many precursors, such as phosphene, are very toxic and some, such as silane, are pyrophoric. Stability is also important. For microelectronic applications, some of the following additional requirements are essential: (a) zeroth-order dependence of rate on precursor concentration to satisfy step coverage requirements, (b) control of surface decomposition characteristics, i.e., selectivity control and (c) ease of decomposition, where by-products desorb cleanly without leaving residues.[26]

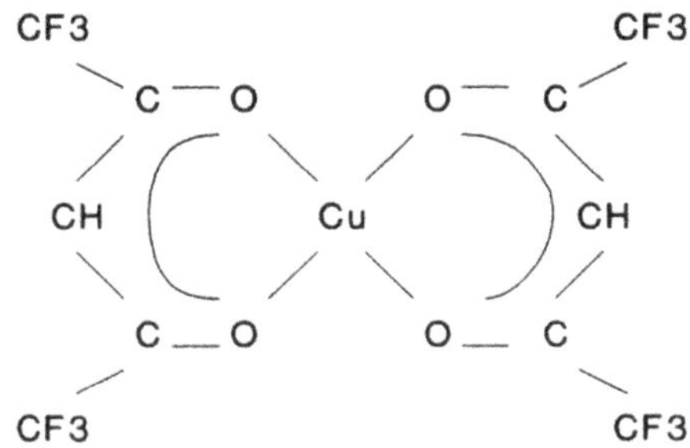

FIGURE 8.13. Structure of the $Cu(hfac)_2$, organic precursor that has been successfully used for the deposition of copper.

Organometallic sources are currently the most popular precursors of copper CVD. However, some inorganic sources, such as CuCl, have also been attempted in the literature. Reduction of CuCl by hydrogen on silicon surfaces produces a mixture of Cu and $CuSi_x$. CuCl is volatile and can use Ar or Ar + H_2 mixtures as carrier gases for the vapor. However, CuCl is hydroscopic and on the surface forms hydroxides which lower the partial pressure. Even though this reaction is reversible, it makes for an inconvenient copper source. The rest of the discussion will concentrate on organic sources.

Among organometallic compounds, Copper (II) hexafluoroacetylacetonate or $Cu(hfa)_2$ has received a lot of attention. The compound has the chemical formula $(CF_3—CO—CH=CO—CF_3)_2Cu$ and its structure is illustrated in Figure 8.13.[27] This compound has high volatility, high solubility in organic solvents, good thermal stability, and low toxicity. The enhanced volatility is attributed to the extent of fluorination of the hfa ligand. The synthesis of Cu(hfa) involves the addition of a neutralized solution of hexafluoroacetylacetone into a stirred aqueous solution of copper (II) chloride. The yellowish-green precipitate that is obtained is then dehydrated to obtain the monohydrate.[28]

The kinetics of the reaction involving $Cu(hfa)_2$ and H_2 in an atmospheric pressure, hot wall reactor have been studied by Lai et al.[29] The reaction proceeds as follows:

$$Cu(hfa)_2(g) + H_2(g) \rightarrow Cu(s) + 2H(hfa)(g)$$

The growth rate is nearly independent of the concentration of $Cu(hfa)_2$, suggesting the reaction proceeds through an adsorbed intermediate, which follows Langmuir type kinetics. The growth rate is inhibited when H(hfa) is added to the inlet, indicating that H(hfa) is adsorbed on the surface in competition with $Cu(hfa)_2$. The rate also increases with the concentration of hydrogen. The proposed rate equation for the overall reaction is as follows:

$$r_{\text{overall}} = \frac{k_{\text{ads}}k_{\text{des}}C_{\text{Cu(hfa)}_2}P_{\text{H}_2}}{[2k_{\text{ads}}C_{\text{Cu(hfa)}_2} + k_{\text{ads}}C_{\text{H(hfa)}}] + k_{\text{des}}P_{\text{H}_2}} \tag{8.7}$$

At high $Cu(hfa)_2$ concentrations, the rate drops to zeroth order with respect to $Cu(hfa)_2$ and first order with respect to hydrogen. The estimated parameters are

$$k_{\text{des}} = 2.2 \times 10^{23} \exp(-80\ \text{kJ}/RT)\text{molecule cm}^{-2}\ \text{s}^{-1}\ \text{atm}(\text{H}_2)^{-1}$$

and

$$k_{\text{ads}} = 0.1\ \text{cm/s}$$

The deposited films show a strong dependence of composition on temperature. In the absence of hydrogen, only copper is detected below 340°C and no copper is seen at 650°C. In the presence of hydrogen and argon, even though the deposition rate is enhanced, the film shows large concentrations of carbon, fluorine, and oxygen. With only hydrogen, film purity is greatly enhanced and the film is 99% copper. The film is polycrystalline and shows good adhesion to TiN glue layers. Film resistivity for 5000 Å films is stable near 1.9 $\mu\Omega$ cm for deposition temperatures above 350°C.

Nucleation of the copper is not uniform on all surfaces. Selective copper deposition processes based on the difference in nucleation rate between conductive and dielectric surfaces have been proposed. Selectivity is modulated by the presence of water and some organic solvents. Other precursors, such as cyclooctadiene copper, (I) hexafluoroacetylacetonate (or COD-Cu-hfac), and acetylacetanatocopper show better nucleation properties. Researchers have also attempted to overcome the nucleation problems by using plasma processes, such as diode RF reactors and microwave plasma enhanced CVD. In both cases the step coverage and resistivity are compromised.

8.4.2. Film Properties and Applications

Copper films deposited by CVD, especially from $Cu(hfac)_2$, are still not manufacturing worthy. The paramount problem has been obtaining consistently close to bulk resistivity. Figure 8.14, for instance, shows the variation in resistivity obtained during the CVD of copper from a Cu(I) hfac tmvs (Cu(I) hexafluoroacetylacetonate trimethylvinylsilane) source using hydrogen reduction (this helps remove water-based reduction of the precursor). Notice

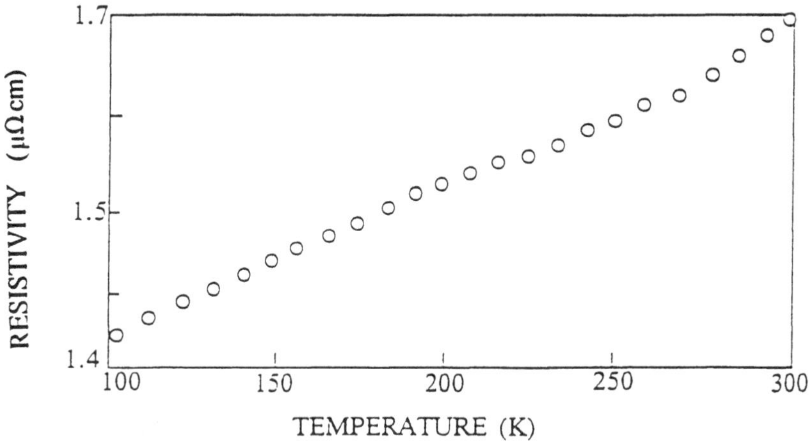

FIGURE 8.14. Resistivity of copper films from $Cu(hfac)_2$ as a function of the deposition temperature. Reprinted, by permission, from A. E. Kaloyeros, *J. Electron. Mater.* **19**(3), 271 (1990). © IEEE.

the sensitivity of the resistivity to process parameters.[30] Often the film resistivity is a function of its impurity content, and carbon residues from the organic sources contribute to the problem. No commercial reactor is currently available and most experiments have been restricted to hot wall tube reactors. Hence optimization of film properties through modifications in reactor configuration has not been attempted.

Most of the current work is being done on the major integration problems of incorporating copper into an IC line: how to pattern copper films to fine line interconnects, how to prevent corrosion of copper once deposited, and how to prevent migration of the copper through SiO_2. More subtle issues regarding the material, its microstructure, and the influence of microstructrure on interconnect reliability performance have yet to be dealt with.

8.5. CVD OF ALUMINUM

Metallic aluminum and its alloys are universally used in metallization in VLSI circuits. However, the primary mode of deposition of aluminum alloys is DC magnetron sputtering from a powder compressed alloy target. There are several inherent advantages to aluminum-based metallization. After the group IB metals, silver, copper and gold, aluminum has the lowest resistivity of any metal. However, aluminum exhibits better chemical stability than silver and copper. Gold has adverse effects on carrier lifetimes in semiconductors and is seldom used. Table 8.3 compares the properties of aluminum with other possible interconnect choices.

TABLE 8.3 Comparison of Metals for Interconnects

Property	Al	W	Cu	Au
Resistivity ($\mu\Omega$ cm)	2.66	5.6	1.67	2.35
Thermal conductivity (W/cm)	2.38	1.74	3.98	3.15
Thermal expansion (10^{-6} K^{-1})	23.5	4.5	17	14.2
Corrosion resistance in air	good	good	poor	good
Adhesion to oxide	good	poor	poor	poor
CVD status	difficult	advanced	early	?
Dry etching	yes	yes	?	?

Reprinted with permission from S.P. Murarka.

Also, several aluminum halides are volatile, providing suitable etching agents for aluminum. As we saw earlier, etching of copper has been difficult. Sputter deposition of aluminum is well characterized and exhibits excellent growth control and manufacturability. Improvement of diffusion characteristics through alloying is easy to achieve through sputtering.

The one major drawback of sputtered aluminum is its poor step coverage. This results in poor thickness control on vertical surfaces, and incomplete coverage inside contacts and vias. Localized thinning results in enhanced I^2R heating, poor electromigration, and difficulty in processing ensuing layers due to poor planarity. All attempts at developing a CVD process for aluminum center around improving the step coverage of the sputtered layer.

Even though step coverage is indeed improved in a CVD aluminum film compared with the sputtered film, the CVD film comes with its own set of problems. Alloying the aluminum with elements such as copper or silicon to improve film properties is not easy to accomplish through CVD. Residues from the precursor, which are often organic, result in carbon incorporation in the film, degrading properties such as resistivity.

Grain size and roughness are not as well controlled in a CVD film as in sputtered aluminum films. Nucleation of aluminum CVD on many surfaces requires activation, such as exposure to $TiCl_4$.

Even though there are commercial tools available for CVD aluminum, at the time this book is being written, CVD aluminum has not gained full acceptance as a manufacturing worthy process. Development is ongoing at several industrial and academic sites.

8.5.1. Growth Chemistry

The most common mode of CVD growth for aluminum is through the pyrolysis of triisobutyl aluminum (TIBAL).[32] Other solid sources such as $(CH_3)_3Al$ and $Al—H_3—(NR_3)_x$, where R is an organic group, have also been reported.[33]. Triethylamine alane is a liquid with decomposition properties similar to the solid precursors and the ease of handling of TIBAL; it has started receiving attention as a possible precursor.[34] TIBAL, the smallest branched-chain alkyl, is expected to provide the lowest carbon content in the deposited film. It also has a high vapor pressure, which is helpful in precursor transport to the reactor, and requires a moderate deposition temperature. At room temperature, TIBAL is a clear liquid with a density of 0.8 g/cc; it melts at $-6°C$ and boils at 40°C. It is pyrophoric and mildly toxic. The pyrolysis reaction occurs as follows:

$$[(CH_3)_2CH{-}CH_2]_3Al(TIBAL) \rightleftarrows [(CH_3)_2CH{-}CH_2]_2AlH(DIBAH) + (CH_3)_2C{=}CH_2$$

Diisobutyl aluminum hydride (DIBAH) and isobutylene are produced in the first stage of the reaction. DIBAH in turn decomposes to form aluminum hydride.

$$[(CH_3)_2CH{-}CH_2]_2AlH \rightarrow AlH_3 + 2(CH_3)_2C{=}CH_2$$

Aluminum hydride then decomposes to form aluminum, releasing hydrogen.

$$AlH_3 \rightarrow Al + 3/2H_2$$

The first stage occurs at temperatures as low as 50°C. Complete pyrolysis takes place at over 120°C, with appreciable rates at over 200°C.[35]

At the conditions discussed above, the growth of aluminum films is selective. Freshly grown aluminum catalyzes the decomposition and results in further deposition. Hence the nucleation of the first aluminum layer is critical. Uniform nucleation is promoted by the exposure of the substrates to $TiCl_4$ vapor.[36] This results in the displacement of the surface hydroxyl groups by titanium chlorides. The small amount of titanium introduced by this technique has beneficial effects on both the physical and electrical properties of the film.

As we saw in the decomposition reaction, two of the three isobutyl groups in TIBAL are removed easily upon adsorption onto the substrate, but the third one does not decompose easily, especially on oxide surfaces. Ultraviolet laser irradiation and X-ray irradiation are effective in removing the last isobutyl group, and result in the formation of aluminum metal particles on the surface. These particles act as nucleation sites for further film growth. This technique, known as photoenhanced CVD, has been effectively used in the prepatterned nucleation of aluminum and its subsequent growth.[37]

Though the feasibility of direct patterning of the predeposition nucleation layer and its subsequent growth has been demonstrated, the method still remains experimental.

Given the nature of the very reactive gases used in conductor CVD, plasma enhanced processes have not become very popular. Confinement of the plasma, even with magnetron type configurations, is incomplete and contamination from the electrodes and walls is still a problem. But this has not prevented researchers from attempting plasma enhanced deposition of conductor films. Results on the decomposition of trimethyl aluminum (TMA) in the presence of hydrogen proceeds in the presence of the plasma according to the reaction

$$Al_2(CH_3)_6 + 2H_2 \rightarrow C_2H_6 + 2Al + 4CH_4$$

even below the thermal decomposition temperature of TMA. At very low temperatures and with proper plasma confinement, aluminum films produced by this method show little oxygen contamination from the chamber walls.[38]

8.5.2. Structure and Properties

Sputtered aluminum films currently used in VLSI metallization often use as much as 4 wt% copper as a dopant to improve electromigration and stress migration performances. Silicon is also often added to prevent further silicon dissolution from the substrate in the absence of a diffusion barrier. The result is an equiaxed grain morphology with excess second- and third-phase precipitates of Cu or $CuAl_2$ precipitated at the grain boundaries. Silicon precipitates are often seen at the film/substrate interface. The grain size is also optimized so as to minimize triple points in fine lines. Triple points suffer a net outflow of atomic flux during electromigration, resulting in void formation at these locations. There are reports of a strong correlation between the slope of the log-normal grain size distribution in the metallization and the scatter in the mean time to failure of the interconnect line. The wider the distribution in grain sizes, the broader is the distribution in failure times. Often a bamboo-like structure is preferred, with grain boundaries orthogonal to the interconnect line and grain sizes larger than its width.[39] Noncoherent precipitates at the grain boundary slow down atomic flux due to the stress fields around them. Figure 8.15a shows a typical surface morphology and a cross-sectional TEM micrograph of the grain structure of a typical Al–Cu film; the film was sputter deposited from a planar magnetron source. Figure 8.15b illustrates schematically the ideal grain morphology desired for metallization.

This is the structure we would like to obtain from the CVD process, along

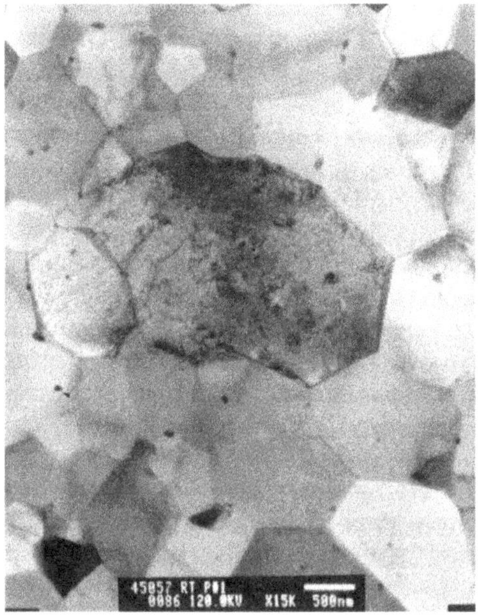

(a)

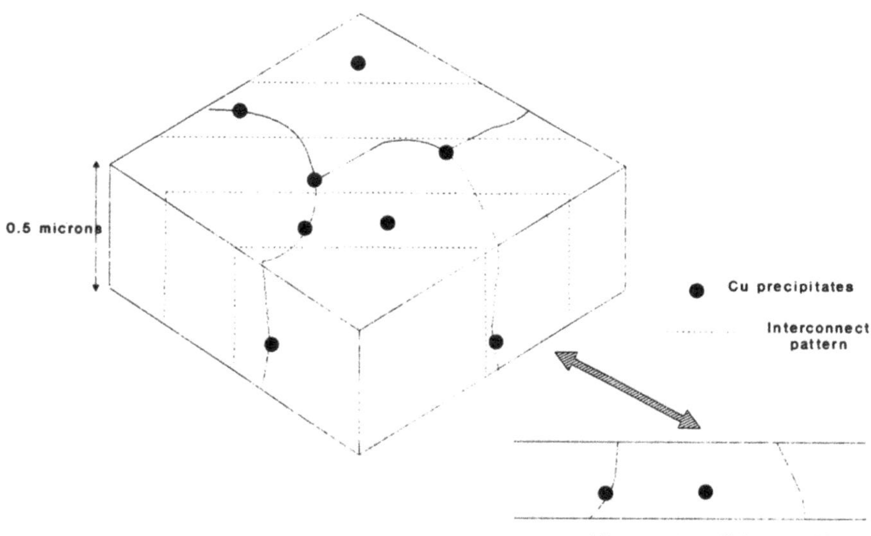

(b)

FIGURE 8.15. (a) Grain morphology obtained in a sputtered Al–Cu thin film. Notice the grain size distribution and the presence of the copper precipitates. (b) Schematic illustration of the Al–Cu microstructure suitable for interconnect applications. Photo courtesy L. Dass, Intel Corp.

with the promised step coverage improvements. Other constraints include film purity, since most impurities such as nitrogen and carbon degrade reliability performance; specularity of the surface, since subsequent imaging by lithographic systems needs to focus on the surface; close to bulk resistivity (2.8–3 $\mu\Omega$ cm) and reproducible hardness to allow wire bonding. All have been difficult to achieve with CVD aluminum films.

The overwhelming problem with CVD films has been the degree of difficulty in controlling nucleation and hence film roughness. Specularity of the film is often poor. Surface treatments alleviate the problem, but it has not been easy to convert the columnar grains into equiaxed grains, seen in sputtering. $TiCl_4$ exposure is not the preferred technique, as residual chlorine can cause corrosion of aluminum lines in the presence of moisture, so presputtered nucleation layers, akin to tungsten CVD, have been attempted. Some of these, such as a thin sputtered layer of copper also are used as dopant sources.[40] TiN layers give the best surface smoothness among the various underlayers reported. Figure 8.16 shows the surface morphology of a CVD aluminum film.[42] The specularity of the film is about 70% of the corresponding sputtered film.

However, all these techniques suffer from increased process complexity. Uniformity of the underlayer and step coverage of the underlayer over complex topography add to the challenge of process integration. None of

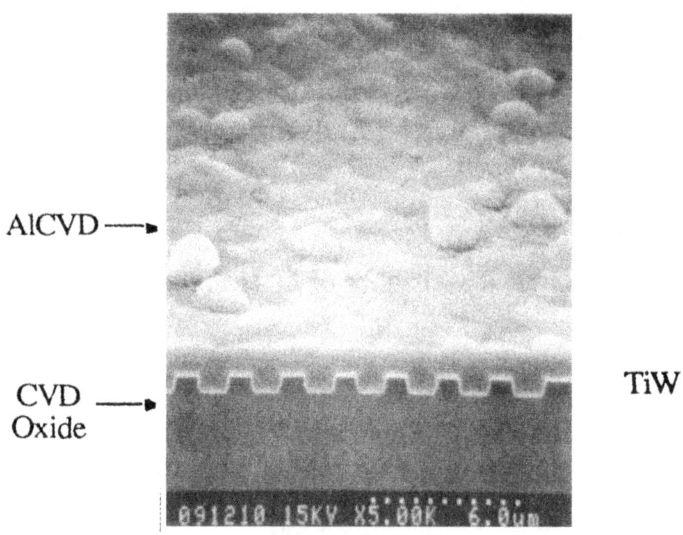

FIGURE 8.16. Surface roughness of the CVD aluminum thin film. The poor specularity of the film is indicative of the coarse grain structure. Reprint from the 1989 Proceedings of the VLSI Multilevel Interconnection Conference, p. 127.

the process parameters, such as growth pressure or flow, have significant influence on nucleation from TIBAL. To improve surface and grain structures, new methods of nucleation have to be identified.

8.6. TUNGSTEN SILICIDE

Silicides of transition metals became popular in the early 1980s as possible shunt materials to reduce gate and diffusion resistances. Success in commercially viable CVD deposition among these silicides has been restricted to tungsten silicide. Tungsten silicide is now widely used on the gate structure as a polycide, a polysilicon/silicide sandwich where the polysilicon imparts good interfacial properties with SiO_2, and the silicide provides the low resistance shunt. A process sequence for the fabrication of the polycide structure is shown in Figure 8.17. Even though tungsten silicide is widely used in the polycide application, other minor uses for it include local interconnects, buried contacting layers, and as an adhesion layer for tungsten.

Tungsten silicide is a hexagonal crystal in very thin films and is stable at normal temperatures in its tetragonal state. Other properties of the silicide are shown in Table 8.4. The phase diagram shows two line compounds W_5Si_3

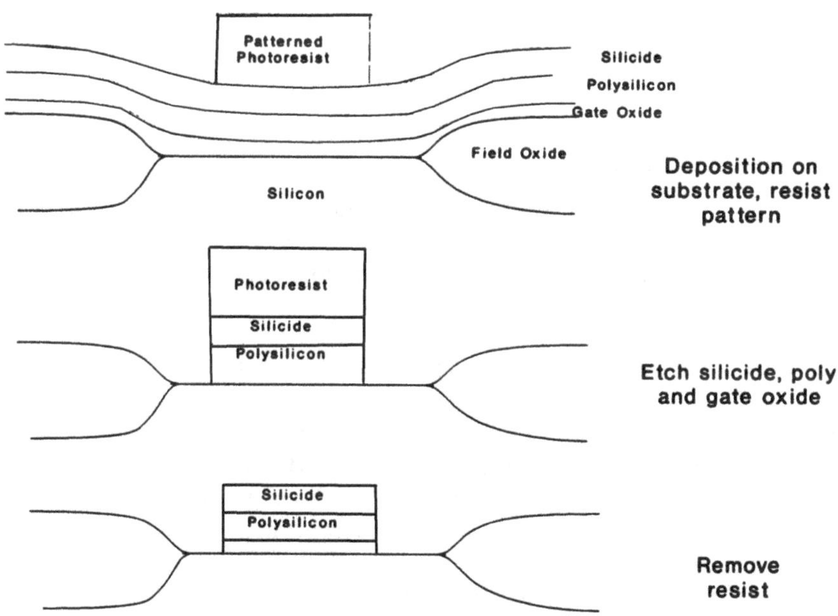

FIGURE 8.17. Process sequence for a polycide gate. CVD silicide and doped polysilicon are etched sequentially. The gate is formed below the polysilicon.

TABLE 8.4 Properties of Tungsten Silicide

Property	WSi_2	W_5Si_3
Structure	tetragonal	tetragonal
Lattice parameters (Å)	3.211, 7.868	9.605, 4.964
Density (g/cc)	9.857	14.523
Resistivity ($\mu\Omega$ cm)	30	
Temperature coefficient of resistivity (10^{-3} K^{-1})	+2.91	
Thermal expansion coefficient (10^{-6} K^{-1})	6.25–7.9	
Microhardness (kg/mm^2, 100 g)	1074	
Etchants	$HF+HNO_3$	
Room temperature heat of formation per metal atom (kcal)	22.2	9.3

and WSi_2, both stable at room temperature.[41] Normal practice obtains a continuous transition in silicon content from W to WSi_2 with excess silicon distributed in the bulk, based on growth conditions. Most of the commercially used silicide is designated WSi_x with x as high as 3. WSi_2 is a hole conductor, whereas $TiSi_2$ is an electron conductor. And upon annealing, WSi_2 achieves its bulk resistivity of approximately 50 $\mu\Omega$ cm, compared with nearly 500 $\mu\Omega$ cm for heavily doped polysilicon. As expected, the addition of excess silicon increases the resistivity.

8.6.1. Growth Chemistry

WSi_x was first grown using WF_6 and SiH_4, and this is still a popular method. The chemistry is very similar to the growth of tungsten. The ratio of SiH_4 to WF_6 dictates the composition. The reaction occurs according to the following stoichiometry:

$$2WF_6 + 7SiH_4 \rightarrow 2WSi_2 + 2SiF_4 + 14H_2$$

The CVD phase diagram in Figure 8.18 shows the final composition of the film as a function of silane mole fraction. Growth rate of the silicide is a direct function of the WF_6 flow. No predictions regarding the kinetics of growth can be deduced from this since the composition is dependent on the flows. The growth temperature is under 400°C. Silicon content and as-deposited film resistivity are both sensitive to deposition temperature.[43]

Recently, deposition of WSi_x from SiH_2Cl_2 has become more popular due to the better step coverage obtainable. The stress and the composition control

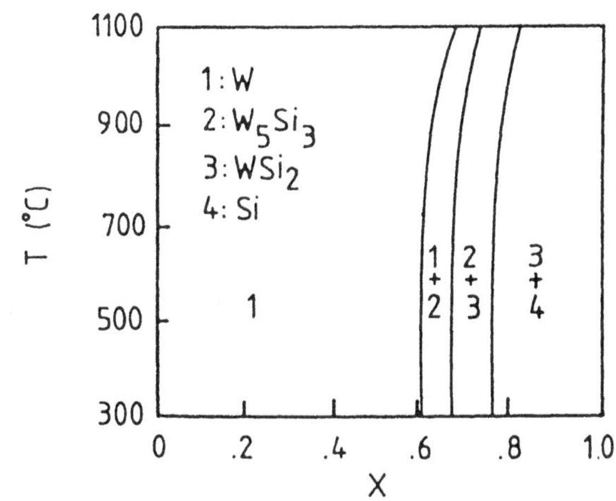

FIGURE 8.18. CVD phase diagram showing the phases formed in the solid phase as a function of the mole fraction of silane in the gas phase. S. L. Zhang, R. Buchta, Y. F. Wang, E. Niemi, J. T. Wang, and C. S. Petersson, in *Proceedings of the 10th International Conference on Chemical Vapor Deposition*, 1987, Vol. 87–8, p. 135. Reprinted by permission of the publisher, The Electrochemical Society, Inc.

with the dichlorosilane (DCS) process are also better.[44] The chemistry is as follows:

$$2WF_6 + 10SiH_2Cl \rightarrow 2WSi_x + 3SiF_4 + 3SiCl_4 + 8HCl + 6H_2$$

The operating temperature is 650–700°C. One of the advantages of using DCS is the elimination of high residual fluorine in the film, which can diffuse to the gate causing an apparent increase in the gate oxide electrical thickness! The use of DCS also enhances the amount of silicon that can be produced in the WSi_x.

8.6.2. Reactor Configurations and Film Properties

Almost exclusively all CVD WSi_x deposition is carried out in cold wall reactors. Control of gas phase temperature limits the gas phase decomposition of silane. Delamination of the film due to deposition on the chamber wall is also eliminated. Figure 8.12 shows the Genus 8710 batch reactor optimized for WSi_x deposition. The system consists of a set of heated chucks on which wafers are placed. Each chuck in this batch reactor has a flow module in front of it, from which a premixed flow of silane, WF_6, and other diluents is directed at the wafer. A central pumping port produces a laminar flow pattern with uniform flow across the surface of a 200 nm wafer.

Films produced in this reactor have a nominal resistivity of 75 $\mu\Omega$ cm after anneal, and a film stress of 10^{10} dyne/cm^2. The step coverage is better than 85% on conventional gate level topography. Adhesion of the film to the substrate is a strong function of the silicon content. At high DCS/WF_6 ratios, films with an average composition of WSi_x where $x = 3$ can be produced. Upon annealing, the excess silicon precipitates out, starting near the polysilicon interface. After film anneal, the silicide thickness shrinks while the polysilicon thickness increases. The interface is graded, resulting in good adhesion.[45]

Before annealing, the film is microcrystalline with fluorine trapped at the

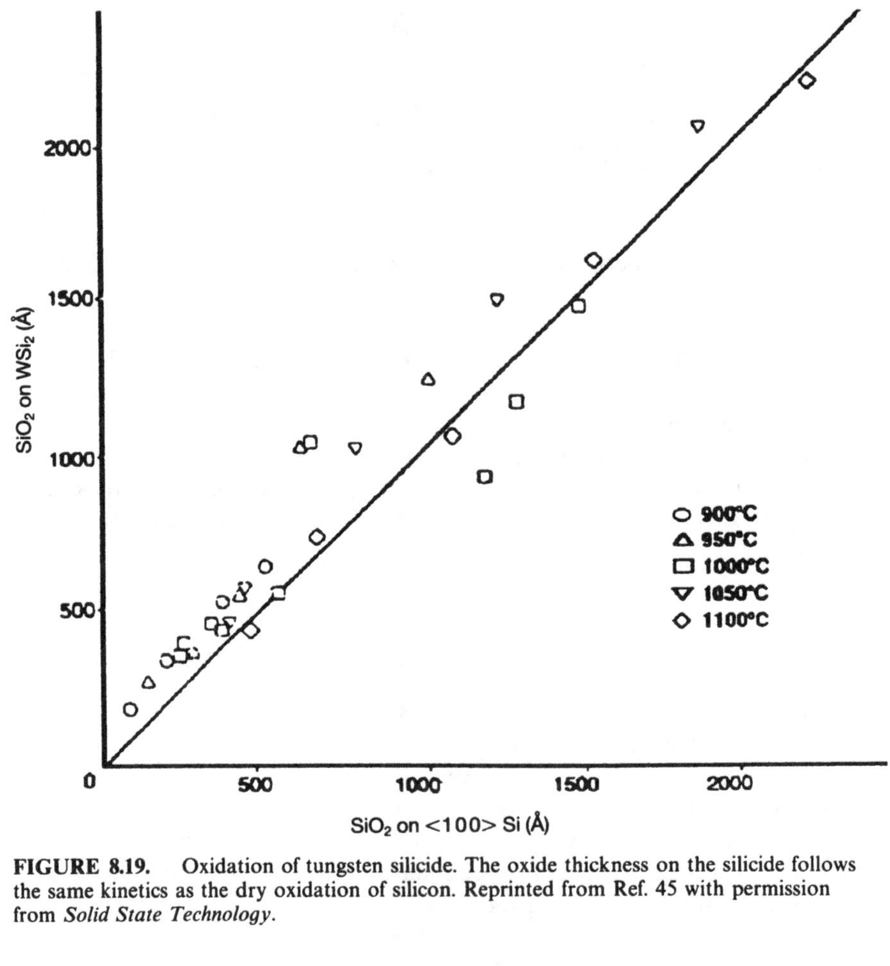

FIGURE 8.19. Oxidation of tungsten silicide. The oxide thickness on the silicide follows the same kinetics as the dry oxidation of silicon. Reprinted from Ref. 45 with permission from *Solid State Technology*.

grain boundaries and the substrate interface. Upon annealing, there is significant grain growth, the composition tends to the equilibrium disilicide state, with the silicon precipitating as the second phase. The continuous larger-grained matrix lowers the resistivity by almost one order of magnitude.

Annealing of the film during normal IC processing occurs as part of an oxidation process. Exposure of the silicide to oxygen-bearing ambients at temperatures near 900°C produces a continuous protective layer of SiO_2 on the surface. Figure 8.19 shows the thickness of oxide formed as a function of the oxidation temperature and time on WSi_x. In all cases, the oxide forms from the dissolved or second phase silicon in the silicide. Hence excess silicon during deposition can be gainfully consumed in forming the SiO_2. After the excess silicon in the film has been consumed[46] silicon diffuses from the underlying polysilicon through the silicide. Oxygen diffuses in through the growing oxide. The reaction occurs at the silicide/oxide interface. Other thermally activated processes such as grain growth occur simultaneously.

8.7 OTHER TRANSITION METAL SILICIDES

Other transition metal silicides such as $MoSi_2$, and $TiSi_2$ have all been grown by thermal CVD from their respective metallic chlorides by reacting with silane. However, stoichiometry control, resistivity, and the use of chlorine-based precursors have resulted in little commercial acceptance. Most of the experiments for these silicides have been with hot wall, tube reactors, even though the use of cold wall reactors seems to enhance the stoichiometry range obtained.

Of these silicides, only $TiSi_2$ is popular in VLSI fabrication. Since titanium-bearing species are sensitive to the presence of oxygen, rapid thermal silicidation using sputtered titanium films is the current method of choice. However, the advent of CVD TiN from organometallic sources suggests CVD $TiSi_2$ might be reconsidered.

8.8. TITANIUM NITRIDE

Titanium nitride is an interstitial titanium compound, with a resistivity lower than titanium metal. It is a refractory material with a melting point of 2 950°C. Table 8.5 lists some of the properties of TiN. Since titanium, which is more electronegative than silicon, can reduce silicon oxides, Ti/TiN sandwiches have become popular as a diffusion barrier/contact couple. The titanium reduces the native oxide and provides reproducible contact to silicon, while an upper coating of the refractory nitride prevents material diffusion into the contact.[47] Ti/TiN contacts were thus commonly used as diffusion barriers between the normally soluble Al/Si couple.

TABLE 8.5 Properties of TiN

Structure and lattice parameter (Å)	Cubic B1, 4.244
Density (g/cc)	5.43
Melting point (°C)	2 950
Room temperature heat of formation (kcal/mol)	−80.4
Young's modulus (kg/mm^2)	8 000
Hardness (Moh's scale)	8–9
Microhardness (kg/mm^2)	1 900
Resistivity of thin film ($\mu\Omega$ cm)	25–75
Homogeneity (at% N_2)	32–51

With the advent of tungsten plug technology, TiN became simultaneously a diffusion barrier for gas phase, fluorine-based reactants and products and an adhesive layer for the growing tungsten film. TiN has become the de facto standard for tungsten adhesion layers in submicron contacts. However, this TiN film was conventionally reactively sputtered from a powder sintered titanium target in the presence of nitrogen. As the contacts grew smaller, the deficiencies of the sputtered TiN, especially with respect to its step coverage, began to dominate the contact reliability. When it became essential to have a sister technology to CVD tungsten for depositing glue layers, CVD TiN started to gain widespread support.

8.8.1. Reaction Chemistries and Film Properties

Titanium nitride is a cubic compound with a bulk resistivity lower than 26 $\mu\Omega$ cm. The film color is characteristically gold, but the nitrogen content modulates it between burnt red and purple. However, color is not a reliable predictor of film quality.[48] Direct nitridation of Ti in N_2 produces films thinner than 50 nm. The properties of the film desirable for microelectronic applications include low resistivity, film purity, columnar and void-free grain morphology, conformality, and high density. Higher temperatures of deposition in all the growth techniques seem to provide better film quality. The presence of chlorine in the film can lead to substrate corrosion, and lack of adhesion. Carbon in the film leads to poor mechanical characteristics and high film resistivity.

Thermal CVD of TiN from inorganic sources involves the use of $TiCl_4$.[49] At atmospheric pressures, the following reaction produces high quality TiN films:

$$2TiCl_4 + N_2 + 4H_2 \rightarrow 2TiN + 8HCl$$

The deposition temperature required is about 1 000°C. At this temperature the reaction is limited by the surface reaction rate between hydrogen and nitrogen or the adsorption of these species on the surface. The activation energy for the process is about 101 kJ/mol.[50] The deposition rate is proportional to the square root of the partial pressures of nitrogen and hydrogen and is negatively influenced by the partial pressure of $TiCl_4$.

To reduce the temperature of deposition, the following reaction has been proposed:[51]

$$6TiCl_4 + 8NH_3 \rightarrow 6TiN + 24HCl + N_2$$

The reaction proceeds to yield good quality TiN above 700°C. Reducing the temperature of the reaction below 550°C results in the formation of a powdery complex that leads to contamination and clogs exhaust lines. The activation energy for the reaction is 61 kJ/mol. The rate is dependent slightly superlinearly on the partial pressure of ammonia and negatively on the partial pressure of $TiCl_4$.

This latter reaction is of importance to the microelectronics industry. However, the deposition temperature is still very high, especially for use over aluminum-based metallization. The film resistivity decreases with increasing temperature (Figure 8.20a) as does the film chlorine concentration (Figure 8.20b).[52] The useful range of thermal deposition using $TiCl_4$ is above 600°C.

Reduction in temperature below 600°C involves the use of plasmas or organometallic precursors. Plasma deposited TiN tool coatings have been in use since the mid-1980s. Using NH_3 and $TiCl_4$ in an RF plasma, useful films have been grown at temperatures above 300°C. Below this temperature, the formation of the solid complex between the two gases leads to poor films.[53] However, plasma deposited TiN films have not received widespread acceptance. Step coverage of the film and the need for complex equipment are the suggested reasons.

Organometallic sources such as tetrakisdiethylaminotitanium (TDEAT) and its methyl counterpart were first developed by Gordon.[54] The ethyl compound has the formula $Ti[N(CH_2CH_3)_2]_4$. A commercially available TDEAT source is made by J. C. Schumacher Corporation and has a molecular weight of 336 g. The source is a translucent orange liquid with a vapor pressure of 0.5 torr at 90°C.

Reaction of TDEAT in the presence of NH_3 above 400°C leads to acceptable TiN films.[55] The final film resistivity is still controlled by the deposition temperature. Residual carbon in the film contributes to increased resistance. Conformality of organometallic TiN films is in general inferior to those deposited using $TiCl_4$. The higher temperature of deposition for the $TiCl_4$ films and the poor surface migration characteristics of the TDEAT

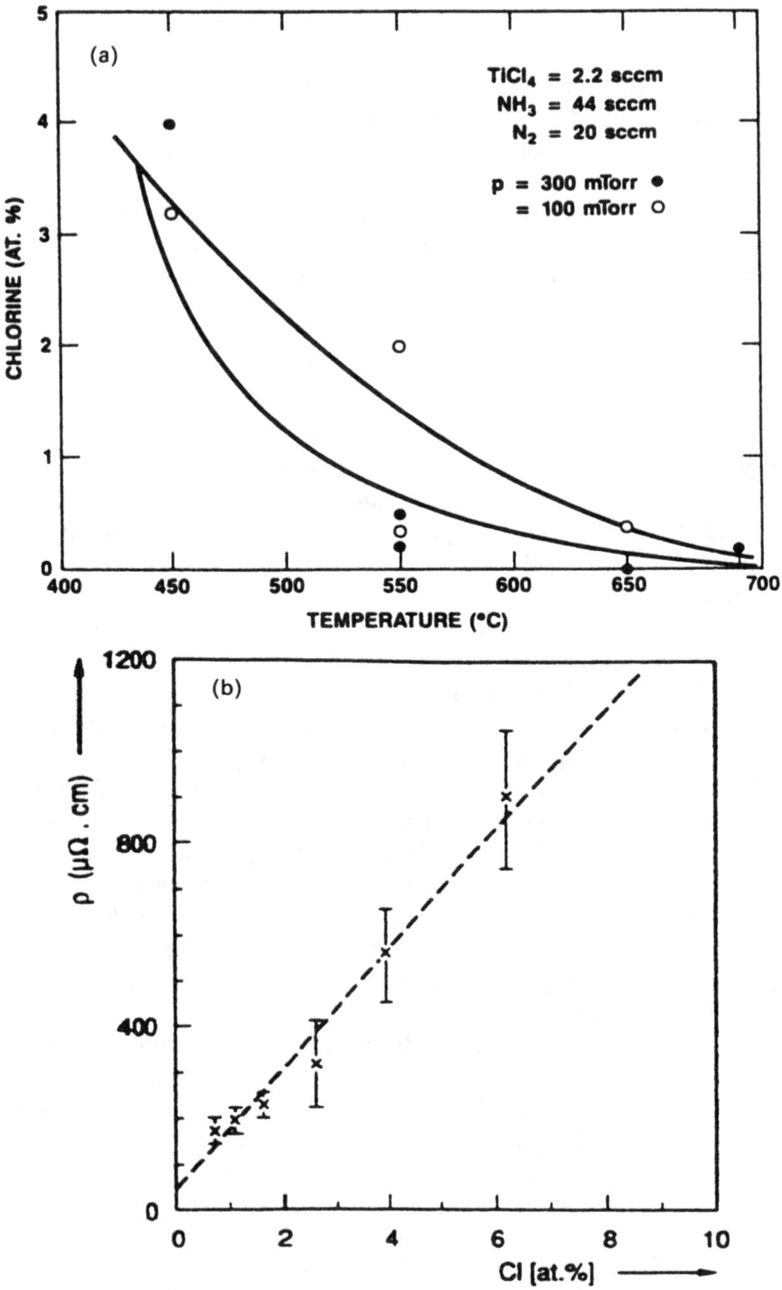

FIGURE 8.20. CVD of TiN from $TiCl_4$: (a) the concentration of chlorine decreases with increase in temperature; (b) this results in a decrease in electrical resistivity with increase in temperature. A. Sherman, *J. Electrochem. Soc.* **137**(6), 1892 (1990). Reprinted by permission of the publisher, The Electrochemical Society, Inc.

complexes have been suggested as possible sources for the poor conformality. However, all CVD films produce better step coverage when compared with their sputtered counterparts.

One final note regarding CVD of TiN is the fact that CVD titanium is undeveloped. Oxygen affinity of titanium leads to growth control. Hence the advantage that sputtered Ti/TiN couples enjoyed with respect to the reduction of native SiO_2 is yet to be realized by CVD TiN films. Use of sputtered Ti films prior to CVD TiN is economically unattractive.

References

1. N. Weste, and K. Eshraghian, *Principles of CMOS VLSI Design,* p. 141, Addison-Wesley, Reading, Mass., 1985.
2. R. S. Muller, and T.I. Kamins, *Device Electronics for Integrated Circuits*, p. 443, John Wiley, New York, 1986.
3. C. Mead, and L. Conway, *Introduction to VLSI Systems*, Addison-Wesley, Reading, Mass., 1980.
4. I. Ames, F. d'Heurle, and R. Hortsmann, *IBM J. R & D* **14**, 461 (1970).
5. *Tungsten and Other Refractory Metals for VLSI Applications*, Vols. I to V, Materials Research Society, Pittsburgh, 1986, 1987, 1988, 1989 and 1990.
6. *Handbook of Physics and Chemistry*, 65th ed. (R. C. Weast, ed.) CRC Press, 1986.
7. R. W. Haskell, and J. G. Byrne, in *Treatise on Materials Science and Technology*, Vol. I, (Herman, ed.), p. 293, Academic Press, New York, 1972; C. M. Melliar-Smith, A. C. Adams, R. H. Kaiser, and R. A. Kushner, *J. Electrochem. Soc.*, **121**, 298 (1974).
8. M. L. Green, in *Proc. 10th Int. Conf. on Chemical Vapor Deposition*, Vol. 87–8 (Cullen, ed.), p. 603, The Electrochemical Society, Pennington, N.J., 1987.
9. R.A. Levy, and M.L. Green, *J. Electrochem. Soc.*, **134**(2), 37C (1987).
10. T. Ohba, T. Suzuki, and T. Hara, in *Tungsten and Other Refractory Metals for VLSI Applications III* (Blewer and McConica, eds.), p. 17, Materials Research Society, Pittsburgh, 1989.
11. JANAF *Thermochemical Tables*, 2nd Ed. (D. R. Stull and H. Prophet, eds.) NSRD5-NBS37, 1971.
12. E. K. Broadbent, and C. L. Ramiller, *J. Electrochem. Soc.* **131**(6), 1427 (1984).
13. H. Cheung, in *Proc. 3rd Int. Conf. on CVD* (F. Glaski, ed.), p. 136, The American Nuclear Society, Hinsdale, Ill., 1972.
14. E. J. McInerney, B. L. Chin, and E. K. Broadbent, paper presented at the Workshop on Tungsten and Other Advanced Metals for ULSI Applications VII, Dallas, Texas, Oct. 22-24, 1990.
15. C. M. McConica, and K. Krishnamani, *J. Electrochem. Soc.* **133**(12), 2542 (1986).
16. K. Y. Ahn, T. Lin, and J. Angilello, in *Tungsten and Other Refractory Metals for VLSI Applications III* (Wells, ed.), p. 25, Materials Research Society, Pittsburgh, 1988.
17. E. J. McInerney, T. W. Mountsier, B. L. Chin, and E. K. Broadbent, paper presented at the Advanced Metallization for ULSI Applications Conference, Murray Hill, N.J., Oct. 8–10, 1991.

18. S. Sivaram, E. Rode, and R. Shukla, in *Tungsten and Other Advanced Metals for VLSI/ULSI Applications V* (Wong and Furukawa, eds.), p. 47, Materials Research Society, Pittsburgh, 1990.
19. J. E. J. Schmitz, R. C. Ellwanger, and A. J. M. van Dijk, in *Tungsten and Other Refractory Metals for VLSI Applications III* (Wells, ed.), p. 55, Materials Research Society, Pittsburgh, 1988.
20. R. Blumenthal and G. C. Smith, in *Tungsten and Other Refractory Metals for VLSI Applications III* (Wells, ed.), p. 47, Materials Research Society, Pittsburgh, 1988.
21. V. V. S. Rana, J. A. Taylor, L. H. Holschwandner, and N. S. Tsai, in *Tungsten and Other Refractory Metals for VLSI Applications II* (Broadbent, ed.), p. 187, Materials Research Society, Pittsburgh, 1987.
22. S. Sivaram, M. L. A. Dass, C. S. Wei, B. Tracy, and R. Shukla, *J. Vac. Sci. Technol.* **A11**(1), 87 (1993).
23. J. R. Creighton, *J. Vac. Sci. Technol.* **A5**, 1739 (1987).
24. P. van der Putte, D. K. Sadana, E. K. Broadbent, and A. E. Morgan, *Appl. Phys. Lett.* **49**(25), 1723 (1986).
25. E. G. Colgan, P. M. Fryer, and K. Y. Ahn, in *Tungsten and Other Advanced Metals for VLSI/ULSI Applications V* (Wong and Furukawa, eds.), p. 243, Materials Research Society, Pittsburgh, 1990.
26. S. P. Murarka, private communications.
27. C. Lampe-Onnerud, A. Harsta, and U. Jansson, *J. Physique* **C2**, 881, 1991.
28. D. Temple, and A. Reisman, *J. Electrochem. Soc.* **136**(11), 3525 (1989); S. K. Reynolds, C. J. Smart, E. F. Baran, T. H. Baum, C. E. Larson, and P. J. Brock, *Appl. Phys. Lett.* **59**(11), 2332 (1991).
29. W. G. Lai, Y. Xie, and G. L. Griffin, *J. Electrochem. Soc.* **138**(11), 3499 (1991).
30. C. Oehr, and H. Suhr, *Appl. Phys.* **A45**, 151 (1988).
31. A. E. Kaloyeros, et al., *J. Electronic Materials* **19**(3), 271 (1990).
32. J. A. T. Norman, B. A. Muratore, P. N. Dyer, D. A. Roberts and A. K. Hochberg, in *Proc. IEEE VMIC* 1991, p. 123; J. Pelletier, R. Pantel, J. C. Oberlin, Y. Pauleau, and P. Guoy-Pailler, *J. Appl. Phys.* **70**(1), 3862 (1991).
33. S. P. Murarka, private communications.
34. R. A. Levy, and M. L. Green, *J. Electrochem. Soc.* **134**(2), 37C (1987).
35. D. A. Mantell, *J. Vac. Sci. Technol.* **A9**(3), 1045 (1991).
36. D. B. Breach, S. E. Blum, and F. K. LeGoues, *J. Vac. Sci. Technol.* **A7**(5), 3117 (1989); M. E. Gross, K. P. Cheung, C. G. Fleming, J. Kovalchick, and L. A. Heimbrook, *J. Vac. Sci. Technol.* **A9**(1), 57 (1991).
37. G. S. Higashi, K. Raghavachari, and M. L. Steigerwald, *J. Vac. Sci. Technol.* **B8**(1), 103 (1990).
38. K. P. Cheung, C. J. Case, R.Liu, R. J. Schutz, R. S. Wagner, L. Kwakman, D. Huibregtse, H. Piekaar, and E. Granneman, in *Proc. IEEE VMIC*, p. 303 (1990).
39. J. Y. Tsao, and D. J. Ehrlich, *Appl. Phys. Lett.* **45**(6), 617 (1984).
40. T. Kato, T. Ito, and M. Maeda, *J. Electrochem. Soc.* **135**(2), 455 (1988).
41. F. M. d'Heurle, and P. S. Ho, in *Thin Film Interdiffusion and Reactions* (Tu, Poate, and Meyer, eds.), p.243, John Wiley, New York, 1978.

42. L. Kwakman, D. Huibregtse, H. Piekaar, E. Granneman, K. P. Cheung, C. J. Case, R. Liu, R. J. Schutz, and R. S. Wagner, in *Proc. IEEE VMIC*, 1990, p. 282.
43. C. Bernard, R. Madar, and Y. Pauleau, *Solid State Technol.* **2**, 79 (1989); E. J. Rode, and W. R. Harshbarger, *J. Vac. Sci. Technol.* **B8**(1), 91 (1990).
44. S. P. Murarka, *Silicides for VLSI Applications*, Academic Press, New York, 1983.
45. D. L. Brors, J. A. Fair, K. A. Monnig, and K. C. Saraswat, *Solid State Technol.* **26**(4) 183 (1983).
46. T. Hara, T. Miyamoto, and T. Yokoyama, *J. Electrochem Soc.* **136**(4), 1177 (1989).
47. C. Fuhs, private communications.
48. K. C. Saraswat, D. L. Brors, J. A. Fair, K. A. Monnig, and R. Beyers, *IEEE Trans. Electron Devices* **ED30**(11), 1497 (1983).
49. M. Wittmer, *J. Vac. Sci. Technol.* **A3**(4), 1797 (1985).
50. H. J. Goldschmidt, *Interstitial Alloys*, Plenum Press, New York, 1967.
51. A. Sherman, *J. Electrochem. Soc.* **137**(6), 1892 (1990); M. J. Buiting, A. F. Otterloo, and A. H. Montree, *J. Electrochem. Soc.* **138**(2), 500 (1991).
52. N. Yokoyama, K. Hinode, and Y. Homma, *J. Electrochem. Soc.* **138**(1), 190 (1991).
53. M. R. Hilton, L. R. Narasimham, S. Nakamura, M. Salmeron, and G. A. Somorjai, *Thin Solid Films*, **139**, 247 (1986); E. F. Gleason, Ph.D. thesis, University of California, Berkeley, 1987.
54. R. M. Fix, R. G. Gordon, and D. M. Hoffman, *Mater Res. Soc. Symp. Proc.* **168**, 357 (1990).
55. I. J. Raaijmakers in *Proc. IEEE VMIC*, 1992, p. 260.

Chapter 9

CVD of Dielectrics

Even though electrical conductivity is the main feature that separates conductors from dielectrics, the roles played by dielectrics in microelectronics are more varied when compared to the applications of conductors. Silicon-based integrated circuit technology owes its popularity in no small measure to the existence of a stable native dielectric, SiO_2. SiO_2 is used as the gate oxide in MOS devices, where it dictates their performance. Dielectrics are used to isolate electrically active components, either semiconductors or conductors. And they are used as capacitors; they provide protection for the device from ambient impurities and moisture.

In each of these roles, the requirements of the thin dielectric film are very different. For instance, resistance to moisture permeability or transparency to UV light might be important in a passivation film. Conformality and low stress levels might be more important in interlevel dielectrics that separate two metal lines. Ability to flow in order to obtain planarity is a very important characteristic of a first-level dielectric glass. A high dielectric constant is required for a capacitor and a low dielectric constant is required to reduce parasitic capacitance. It follows that both the method of deposition and the properties of the film are more application specific in the case of dielectric films than in other microelectronic thin films. Table 9.1 illustrates the growth temperature regimes of different dielectric films used in microelectronics.[1] The temperature of growth is often the deciding factor when it comes to the role played by a dielectric in some of the applications discussed in the next section.

TABLE 9.1

Film	Growth temperature (°C)
Thermal oxidation of Si	850–950
Thermal CVD silicon nitride	800–850
Thermal CVD silicon dioxide	400–700
PECVD silicon dioxide	200–400
PECVD silicon nitride	200–400
PECVD silicon oxynitride	200–400

Walter Runyan and Kenneth Bean, *Semiconductor Integrated Circuit Process Technology* (pp. 130, 341) © 1990 by Addison-Wesley Publishing Company, Inc. Reprinted by permission of the publisher.

9.1. CLASSIFICATION OF DIELECTRICS IN MICROELECTRONICS

Dielectrics for microelectronics have traditionally been formed by thermal oxidation or by CVD. We shall attempt to classify CVD dielectrics used in microelectronics based on the stage in the processing sequence at which they are deposited: isolation level dielectrics are deposited before active devices are fabricated; transistor level dielectrics may be deposited as the gate dielectrics or between two polysilicon layers in some applications; interlevel dielectrics, such as poly–metal dielectrics or metal–metal dielectrics, insulate successive levels of interconnects; and passivation dielectrics are last to be deposited during conventional front-end processing, immediately before assembly packaging activities. Figure 9.1 illustrates these different dielectric films. Let us explore their requirements before considering the CVD growth of individual dielectric films.

9.1.1. Dielectrics for Device Isolation

Device isolation in MOS devices is a complex science. Many processing activities have been devised to ensure adjacent devices do not electrically interact while maintaining high device density. Isolation through local oxidation (LOCOS) has been popular for almost a decade and uses a CVD deposited silicon nitride to prevent thermal oxidation at locations where active devices are later fabricated.[2] Figure 9.2 shows a schematic of the LOCOS process. The sacrificial silicon nitride film is typical of isolation level CVD dielectrics and needs to possess uniformity in thickness and density such that diffusion of oxygen through the film is prevented. It also needs to be low in stress so as not to damage the substrate silicon.

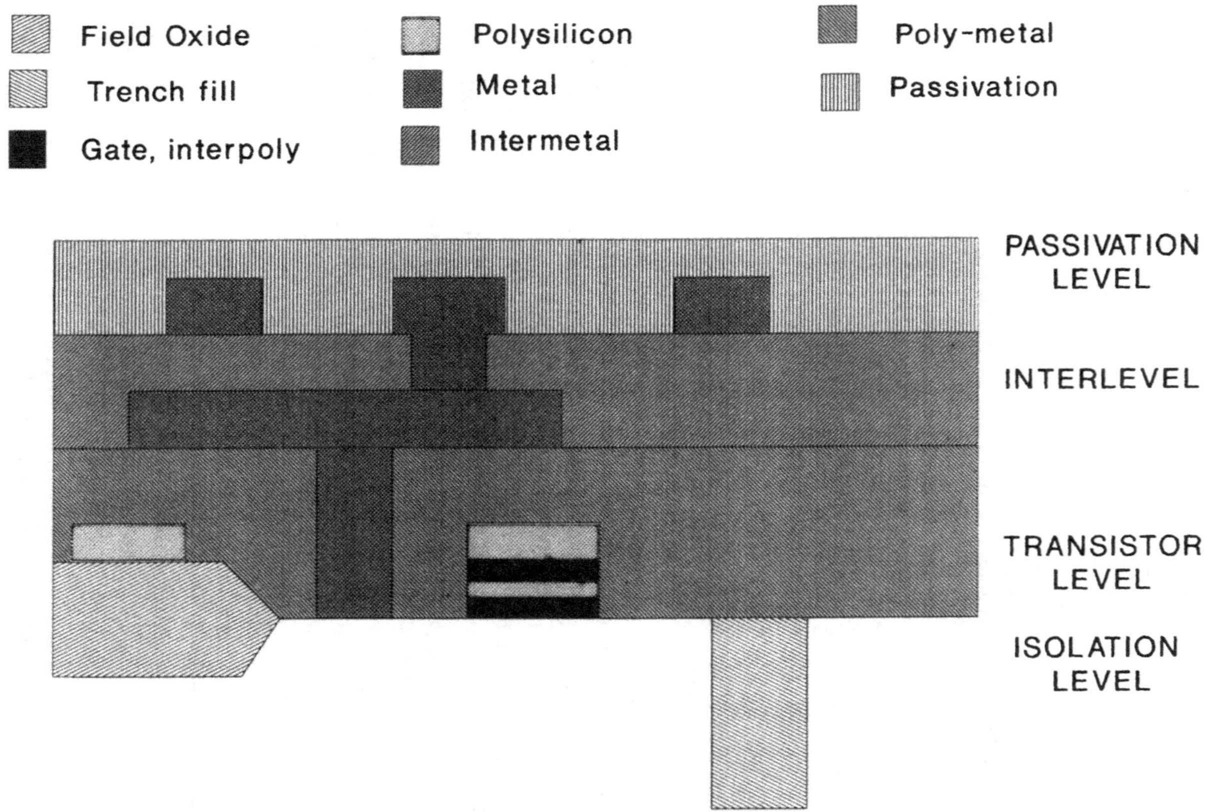

FIGURE 9.1. Classification according to application of dielectrics used in microelectronics: isolation, device level, interlevel, and passivation films.

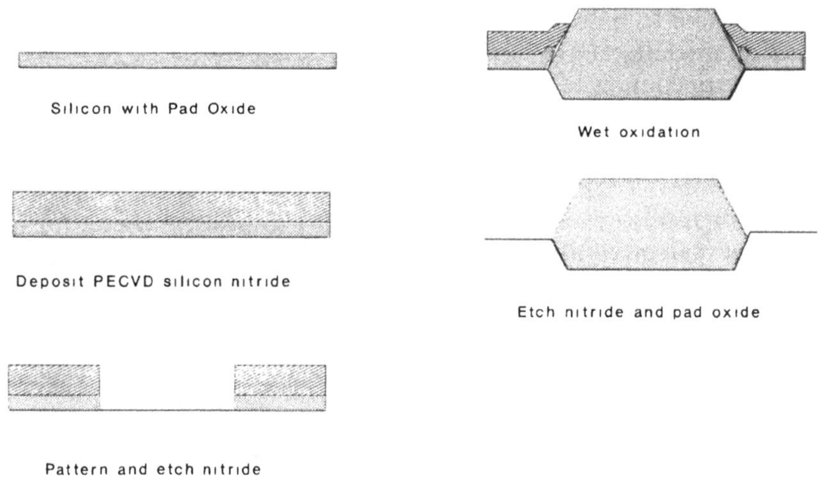

FIGURE 9.2. Process sequence for the fabrication of LOCOS isolation. The sequence involves the use of CVD silicon nitride as an oxidation suppressant.

Silicon dioxide in the isolation region is often grown by wet oxidation and is called the field oxide. In conjunction with ion implantation, CVD oxides are now being attempted to isolate adjacent devices.[3]

9.1.2. Transistor Level Dielectrics

Due to the stringent electrical requirements on the gate level dielectric, by tradition the SiO_2 layer commonly used in MOS devices has been thermally grown. However, high temperature CVD oxides and alternate dielectrics such as SiC have been tried at the gate level. These dielectrics need to be isotropic, electrically charge-free, and possess trap-free interfaces with the underlying semiconductor. These dielectrics are often very thin (less than 100 Å). Hence they require pinhole-free deposition with a highly uniform thickness. The presence of metallic impurities can be catastrophic to device performance.

Use of multiple layers of polysilicon is limited to certain special kinds of devices, such as EPROMs, DRAMs, and SRAMs. However, in all such cases, the dielectric layer which separates conductive polysilicon layers has to contain charge leakage from the polysilicon. Minimization of electrical conductivity is thus key to this type of application, especially near the edges of the polysilicon layers, where the electric field is typically higher. Use of multiple layers of dielectrics, such as thermally grown or high temperature CVD silicon dioxide, followed by an oxidized layer of CVD nitride have been developed to ensure there are no pinholes in the dielectric.[4] Again, these dielectrics need to possess excellent control in mechanical properties, so as to prevent mechanically induced defects. Also, these films need to withstand the high temperatures of the subsequent processing steps.

9.1.3. Interlevel Dielectrics

In microelectronics, dielectrics used specifically to insulate conductor lines are thicker than most other dielectrics so their conductivity is not as crucial. However, they often possess stringent temperature constraints on their growth. Often, dielectrics over polysilicon are restricted to growth temperatures under 850°C. Dielectrics over the metal are restricted to under 450°C. They are also required to maintain planarity of the top surface, even when deposited over complex topography.

To lower the glass flow temperatures, the dielectric over polysilicon often uses doped glasses such as boron and/or phosphorus in SiO_2. Viscous flowing of the glass results in a planar top surface over complex topographies. But it is important to prevent doping of the polysilicon from the glass and to control the oxidation of the underlying semiconductor during flow. Adequate control requires careful optimizing, i.e., maximizing the dopant concentration to improve flow while simultaneously preventing excessive dopant concentration from precipitating out of solution.

Intermetal dielectrics ensure planarity of the top surface while filling narrow spaces between metals through a variety of techniques, including spun-on layers, etch-back of protruding features using a sacrificial layer, and polishing the top surface. But before planarizing, gaps between narrow metal lines need to be filled without leaving voids or cracks. Techniques that utilize deposition and etching sequences and electron cyclotron resonance plasmas have widespread application at this level.

9.1.4. Passivation Dielectrics

We discussed passivation dielectrics in Chapter 3. To protect a device from the ambient, these films are deposited at the end of the process sequence, so they have severe restrictions on growth temperature. They also need to be resistant to diffusion of moisture and other ambient impurities, and in some applications they need to possess optical properties such as UV transparency. The top surface needs to be hard in order to provide scratch protection to the underlying devices. Often a stack of different dielectrics is deposited by CVD techniques to produce films satisfying each requirement. Phosphorus-doped glass could be used to prevent water diffusion, then a nitride or oxynitride film could be deposited on top to prevent impurity diffusion.

9.2. CVD OF SILICON NITRIDE

Thermal nitridation of silicon produces a self-limiting film thickness of less than 100 nm at temperatures as high as 1 000°C.[5] However, deposition of silicon nitride by thermal CVD and by plasma CVD techniques for bulk as well as thin film applications has been well established since the 1970s.[6] The nitride film is noted for its high hardness, high dielectric constant (compared to SiO_2) and low permeability to mobile metallic ions. Silicon nitride films find wide application in IC manufacture, with property requirements at each level being quite diverse, as discussed in the previous section. For instance, the nitride film is used at the isolation level as a sacrificial film to prevent oxidation of silicon; at the transistor level as the gate dielectric in some optical devices, and as interpoly dielectric; at the passivation level as a moisture and sodium diffusion barrier; in photolithography during the patterning of multilayer resists; and in the fabrication of X-ray lithography masks.

Bulk Si_3N_4 is a hexagonal crystal with a density of 3.18 g/cm^3. However, the thin film in its applications in IC manufacture is amorphous with a density often as low as 2.6 g/cm^3. Compositional purity and stoichiometry of the film strongly affect its properties, and the method of growth significantly influences both purity and stoichiometry.

9.2.1. Growth Chemistry and Thermodynamics

Silicon sources for the nitride vary from SiH_4 (the most common) through SiH_2Cl_2 and $SiCl_4$ to SiF_4. Silicon iodide has also been attempted. Nitrogen sources are NH_3 and N_2, used either individually or together. The commonly used reaction is

$$3SiH_4 + 4NH_3 \rightarrow Si_3N_4 + 12H_2$$

The reaction produces acceptable nitride films at temperatures higher than 800°C. More economical and lower temperature films have been obtained through the following reaction using dichlorosilane:

$$3SiH_2Cl_2 + 4NH_3 \rightarrow Si_3N_4 + 6HCl + 6H_2$$

where rates of about 100 Å/min can be obtained around 700°C in a tube reactor.[7] At higher temperatures, NH_4Cl is also found in the exhaust. Growth is usually limited by the surface reaction. Films grown by thermal CVD have high stresses and spontaneous delamination is seen over 2 000 Å.

Morasanu and Segal compare the free energy changes as a function of temperature for reducing SiH_2Cl_2 by NH_3 and N_2 (see Figure 4.4).[8] Up to about 900 K, the NH_3 reaction has a smaller negative free energy of formation, even though both reactions are feasible. However, in the more useful temperature regime above 900 K, the NH_3 reaction is more favorable. Equilibrium constants for all the partial reactions involved in the overall reaction have been well established, and CVD phase diagrams for this reaction are also available.[9]

9.2.2. Thermal and Plasma CVD Processes

Atmospheric pressure reactors using silane and LPCVD tube reactors with both SiH_4 and SiH_2Cl_2 have been utilized in producing Si_3N_4 films by thermal CVD. In tube reactors, wafer spacing needs to be larger to overcome mass transfer limitations, which lower the growth rate at the center of the wafer even at reduced pressures.[10] Figure 9.3 shows how the wafer spacing in an LPCVD tube reactor increases the growth rate at the center of the wafer.

In order to reduce the temperature of deposition, especially for the passivation applications, plasma enhanced CVD was introduced in the

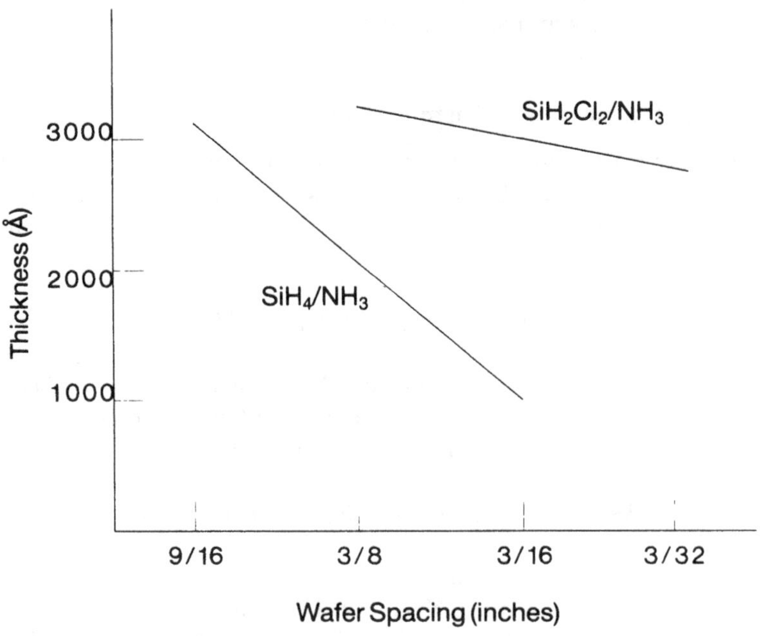

FIGURE 9.3. The effect of wafer spacing in an LPCVD tube reactor in determining the film growth uniformity at the center of the wafer. Reprinted, by permission, from Richard Rosler, *Solid State Technology*, April 1977, p. 67.

TABLE 9.2 Properties of silicon nitride

	Property	APCVD 900°C	PECVD 300°C	Bulk
1.	Structure	amorphous	amorphous	hexagonal
2.	Density (g/cc)	2.3–3.1	2.5–2.8	3.2
3.	Refractive index	2–2.1	2–2.1	2.0
4.	Dielectric constant	6–7	6–9	8.4
5.	Dielectric strength (V/cm)	1×10^7	6×10^6	1.2×10^7
6.	Resistivity (Ω cm)	10^{15} to 10^{17}	10^{15}	
7.	Young's modulus (Msi)			45
8.	Thermal expansion (10^{-6} K^{-1})	4	4–7	3.28
9.	Specific heat (cal g^{-1} °C^{-1})			0.171 (20°C)
10.	Etch rate in 49% HF (Å/min)	80 A/min	1 500–3 000	
11.	IR Si–N absorption peak (cm^{-1})	~870	~830	
12.	IR Si–H absorption peak (cm^{-})	none	2 180	none
13.	Na^+ penetration (Å)	< 100	< 100	
14.	Water permeability	none	low or none	

mid-1970s and has since been the more popular method for producing silicon nitride. Batch and single-wafer reactors for PECVD are available; throughput considerations favor the batch reactors. Plasma deposited nitride is more correctly represented as Si_xN_yH, where the hydrogen content is on the order of 20 at%, and the silicon to nitrogen ratio is close to 1. Table 9.2 compares the properties of thermally and plasma grown silicon nitride films.

Notice in general that the slower-growing, more stoichiometric thermal CVD film possesses properties closer to the bulk values. This is in part due to the normally higher temperatures needed for thermal CVD. A particularly important property to note is the etch rate of the film in HF solutions. The etch rate varies with the hydrogen content of the film and is generally considered to be a measure of the quality of the film.

9.2.3. Growth–property Relationships in Plasma CVD SiN_x

Interactions between the plasma input variables, such as power, pressure, frequency, gas flow rates, pumping speed, substrate temperature, reactor geometry, and electrode spacing, strongly affect the structure and composition and hence the properties of the nitride film. Higher power results in increased electron density, consequently increasing the concentration of the active

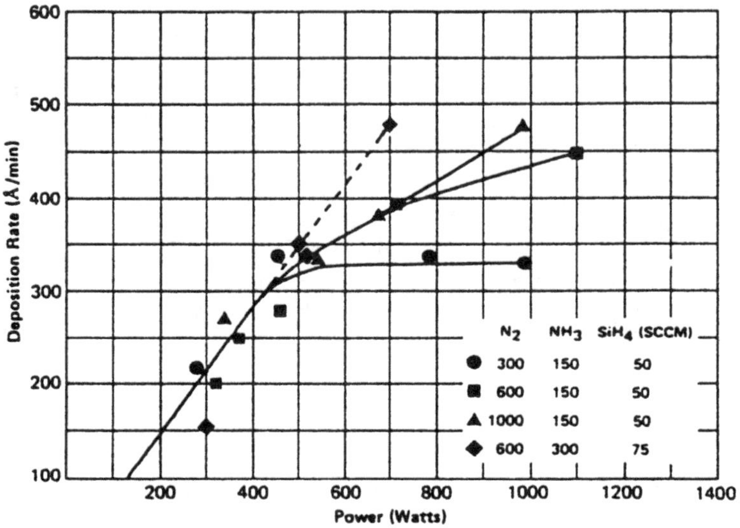

FIGURE 9.4. Nitride growth rate as a function of input plasma power. Reprinted, by permission, from E. P. G. T. van de Ven, *Solid State Technology*, April 1981, p. 168.

species. An increase in power generally increases the rate of deposition until transport begins to control the rate. At this point, the rate is influenced by the pumping speed and the gas feed rates. For nitride films the transition from power dependence to transport dependence occurs at about 0.07 W/cm^2.[12] Up to this point, change in deposition rate with input power depends on the geometry of the reactor. Rate increases as high as 0.4 nm min^{-1} W^{-1} can be observed in parallel-plate reactors. Figure 9.4 shows how the rate of nitride film growth depends on the input plasma power. Note that power density rather than power is the proper input variable, even though the reactor geometry is not available in most reports in the literature.

In general, substrate temperature results in a lowering of the deposition rate (Figure 9.5). Thermal CVD of Si_3N_4 follows a conventional Arrhenius type dependence with a transition from surface reaction control to mass transfer control at around 700°C. But PECVD shows a negative temperature dependence, often attributed to adsorption control on the surface.[13] As the temperature of the substrate is increased, the desorption rate increases, reducing the concentration of the active species on the surface. The mechanism is not well understood for nitride deposition, but it has been suggested the decreased incorporation of hydrogen in the higher temperature film might play a role in reducing the rate.[13]

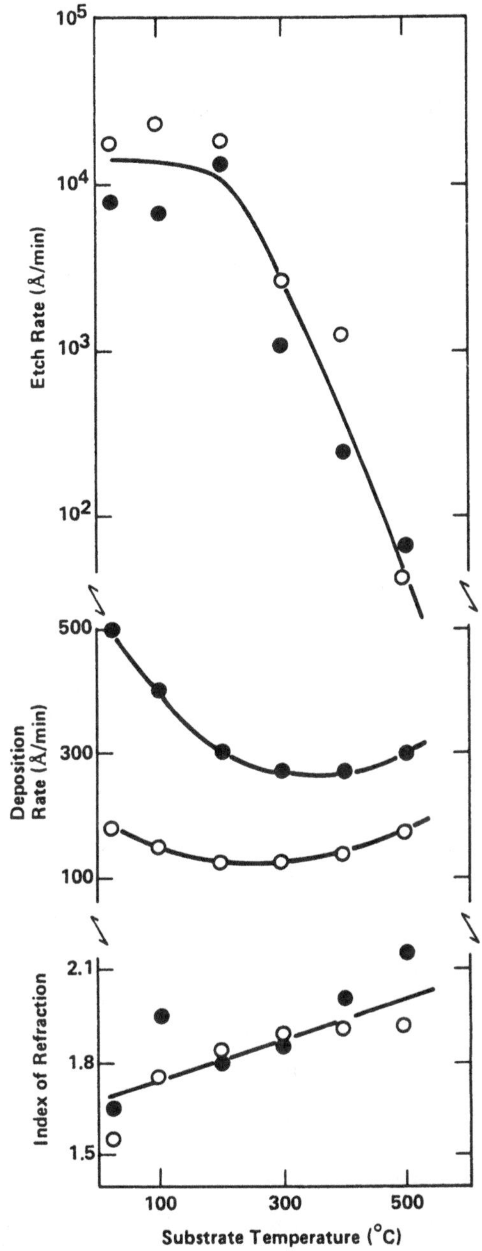

FIGURE 9.5. Effect of substrate temperature in reducing growth rate of plasma nitride. This often indicates adsorption phenomena to be rate-controlling. Other film properties are also shown. C. Blaauw, *J. Electrochem. Soc.* **131**(5), 1114 (1984). Reprinted by permission of the publisher, The Electrochemical Society, Inc.

The frequency of the applied electric field influences the mechanical state of the film. At low frequencies the stress is compressive; at higher frequencies the stress is tensile. Above 0.5 MHz the massive ions have too much inertia to move with the rapidly changing field but electrons do respond. However, as the frequency is reduced below 0.5 MHz, the ions begin to track the field. This results in low energy ion bombardment and ion implantation in the growing film, making it compressive. Dual frequency reactors take advantage of this phenomenon in ensuring film integrity by simultaneously applying a high frequency field to sustain the plasma and a low frequency field to control the film stress. Figure 9.6 shows the effect of the low frequency component in changing the film stress.[14]

The hydrogen content of the film is influenced by the deposition temperature, the power density, and the frequency. Hydrogen could exist as both Si–H and N–H bonds, even though most of the hydrogen is introduced from ammonia. The concentration of the Si–H bond affects the ultraviolet transparency of the film and has to be kept to a minimum in such applications. Increasing the frequency decreases the Si–H content and increases the N–H content near the surface of the film. Infrared spectroscopy is the commonly used technique in determining the concentration of the Si–H and N–H bonds.

The silicon to nitrogen ratio in PECVD films is variable, whereas thermal CVD films tend to be more stoichiometric. Increasing the silicon content increases the refractive index of the film and decreases the resistivity. In general, an increased silicon content degrades the dielectric properties. It lowers the breakdown strength and increases the power dissipation. The silicon to nitrogen ratio is strongly affected by the partial pressures of the silicon- and nitrogen-bearing species in the feed gas.

Thermal CVD films show a thin layer of SiO_2 at the nitride–silicon interface, probably introduced by the native oxide on silicon.[15] PECVD films show incorporation of oxygen in the bulk, introduced from the system background. They also show an increased oxide content near the surface, probably introduced by surface oxidation. The presence of oxygen tends to lower the refractive index of the nitride.

9.2.4. Device Performance

Stress in the nitride film is critical in all device applications. Even as a sacrificial film in the LOCOS isolation scheme, high nitride stress leads to damage of the semiconductor surface, resulting in interface-trapped charge. At the passivation level, high compressive stresses lead to stress voiding of underlying aluminum interconnects; high tensile stresses lead to hillocking

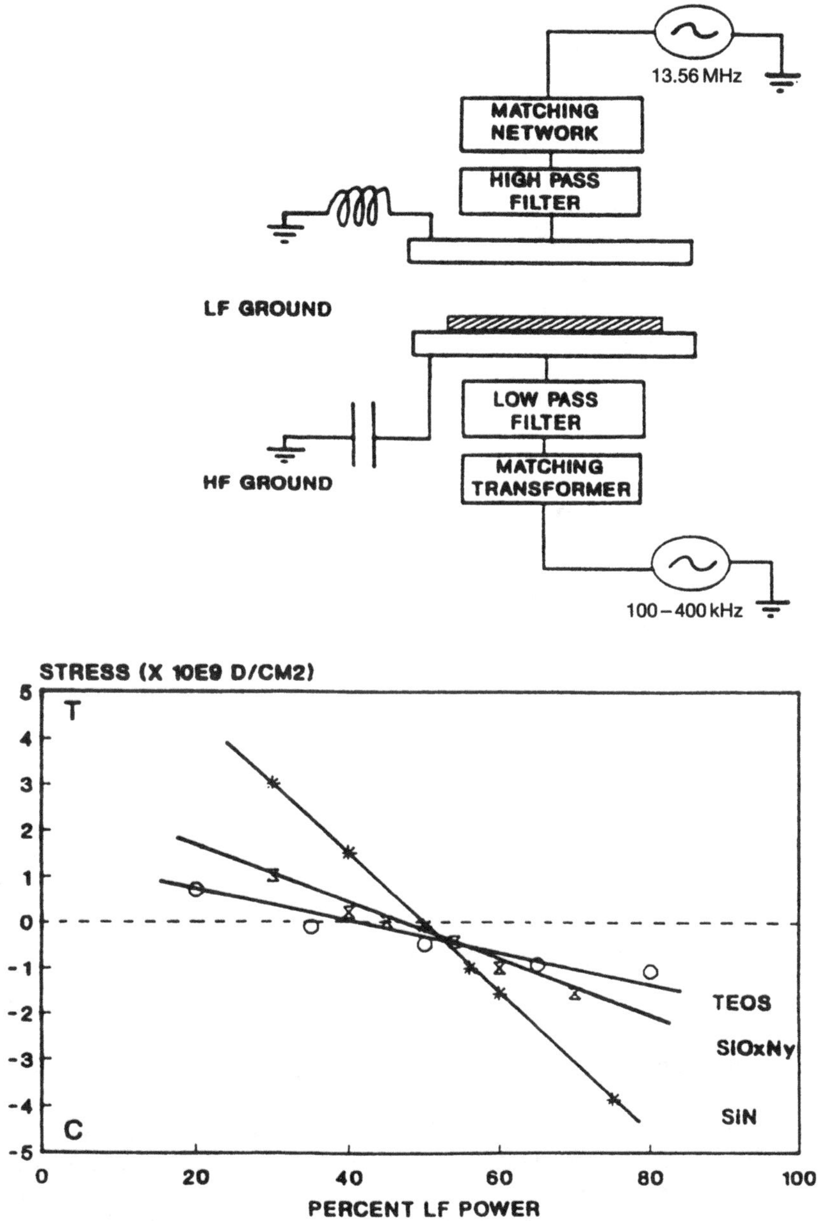

FIGURE 9.6. A dual frequency reactor setup to control the mechanical properties of the nitride film. The lower figure shows the effect of the low frequency power input in controlling ion bombardment and hence the film stress. Reprint from the 1990 Proceedings of the VLSI Multilevel Interconnection Conference.

of the metal. Interconnect yield is a strong function of the mechanical state of the passivation film.

Step coverage of the film and compositional uniformity on vertical surfaces have also led to concerns about PECVD nitride. For instance, Figure 7.8 shows a profile of a typical PECVD nitride over a narrow topography.[16]. Even if the step coverage were adequate, HF etch rate studies show differing nitride etch rates on the vertical surface, the bottom surface, and a horizontal surface, indicating differences in film quality. Films deposited with nitrogen without ammonia also exhibit pinholes.

However, the excellent barrier and scratch protection properties of the film improve the reliability of VLSI devices. Compared to devices passivated with phosphorus-doped glass, PECVD nitride passivation increases the mean time to failure of a VLSI device five times in temperature–bias–humidity testing.

9.3. CVD OF SILICON DIOXIDE

Silicon dioxide is the most commonly used dielectric in integrated circuits. Growth techniques for each of its applications and the properties of the dielectric itself vary significantly to suit the specific needs of each application. For this reason, each will be treated individually. We will explore the applications, study the growth chemistry, and examine the effects of growth conditions on film properties for low temperature oxides (LTOs), plasma oxides, doped glasses, and high temperature oxides (HTOs). Silicon- and oxygen-bearing precursors for these films also have seen considerable development in the last decade. We will study the modifications needed to conventional thermal and plasma CVD reactors to suit the reaction of choice for each application.

9.3.1. Structure and Properties of Silicate Glasses

In its bulk crystalline form, SiO_2 takes many forms, including quartz. As a thin film it is an amorphous glass. The structure of the glass is similar to a long-chain polymer, with cross-linking at intervals.[17] The glassy state has no extended periodicity or symmetry. The Si^{4+} cations are surrounded by oxygen ions in the form of triangles or tetrahedra. The oxygen ions are either bridging oxygen ions, which link two tetrahedra, or nonbridging oxygen ions, which belong to only one tetrahedron. Other cations, such as boron, phosphorus, arsenic, and germanium, also form such a network with oxygen and are called glass formers. These can substitute the silicon ions in the SiO_2 network isomorphically. Other large cations, such as sodium, potassium, and tin, can exist in the large holes between the oxygen polyhedra, where they

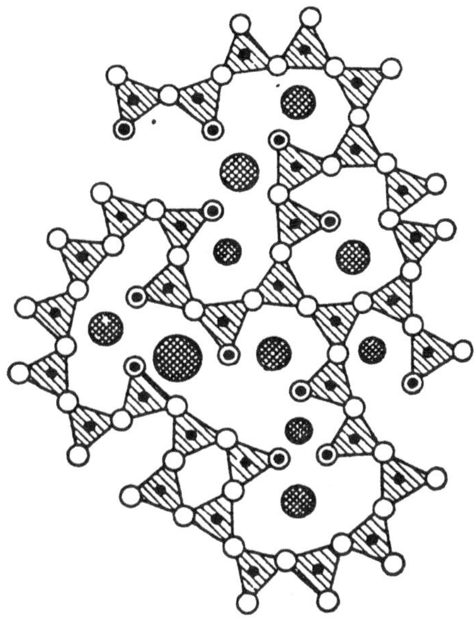

FIGURE 9.7. A glassy network consisting of silicon atoms (dark circles), bridging oxygen atoms (open circles), nonbridging oxygen atoms (concentric circles), and network formers (large circles).

compensate for the excess negative charge of the nonbridging oxygen ions. These ions modify the tetrahedral network and can result in significant changes to the properties of the glass. Figure 9.7 shows a schematic representation of the glass structure, indicating the two kinds of oxygen ions and the network modifiers.

The glassy state is a nonequilibrium condition that resembles a supercooled liquid in structure. Slow cooling of some glasses from melting can produce the equilibrium crystalline state. Rapid cooling *freezes in* the liquid structure, resulting in a vitreous solid. Figure 9.8 shows the transformations occurring when molten glass is cooled either rapidly or slowly. Slow cooling shows a typical melting point and a stable state. Rapid cooling shows a delayed break in the curve, the transition from the supercooled liquid to the vitreous solid at the glass transition temperature. Notice how there is no clearly defined melting temperature. The glass transition temperature is particularly important for doped glasses in IC applications, where the glass is flowed after deposition to achieve planarity.

9.3.2. Reaction Chemistry and Precursors

Silane has been the most common silicon-bearing precursor for various silicon-containing films in integrated circuits. Silane is a pyrophoric gas

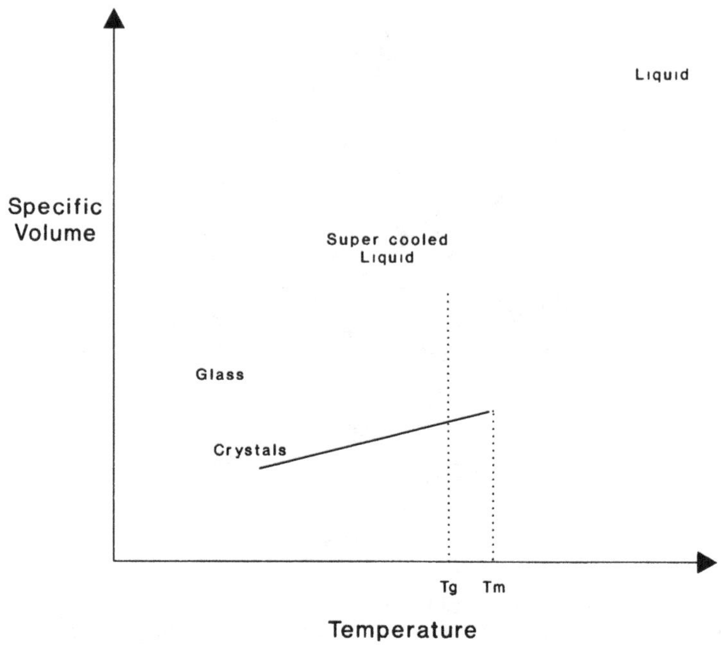

FIGURE 9.8. Transformations during cooling from a liquid. Slow cooling produces a crystallline state whereas rapid cooling produces a glossy solid. In the case of the glass, note the absence of a clear melting point.

requiring extreme care in handling. It is primarily for this reason that organic silicon sources such as tetraethylorthosilicate (TEOS) are replacing silane in many applications.[18] But it is also because some silicon sources lead to better film conformality. Conformality is also the reason for the use of halogenated silicon hydrides such as SiH_2Cl_2. We will examine many of these chemistries and their limitations in this section.

Oxidation of Silane

Formation of glassy SiO_2 films through the oxidation of diluted silane in the presence of oxygen has been accomplished by atmospheric pressure, reduced pressure, and plasma enhanced CVD processes. The reaction can be represented as

$$SiH_4(N_2) + O_2 \rightarrow SiO_2 + 2H_2$$

and

$$SiH_4(N_2) + 2O_2 \rightarrow SiO_2 + 2H_2O$$

The latter reaction is favored at higher silane concentrations and at higher temperatures. The nitrogen in parentheses indicates dilution of the silane.

Excess silane causes gas phase nucleation, forming a cloud of silicon dust. Under controlled gas compositions (an O_2/SiH_4 ratio of 3:1 at 325°C and 60:1 at 515°C), the reaction proceeds by surface adsorption and surface catalysis to produce amorphous SiO_2 at the surface.[19] Gas composition and temperature are closely related. At high oxygen contents, oxygen molecules preferentially occupy surface reaction sites to the exclusion of other reactants, thus slowing the reaction. At high silane concentrations, homogeneous reactions lead to particle problems. Both processes are thermally activated, leading to interplay between concentration and temperature. Diluent gases such as nitrogen and argon reduce the silane partial pressure, so a safe level can be maintained irrespective of the reactor pressure.

Above 800°C, N_2O has been used as the oxygen source, since stoichiometric or silicon-rich films can be produced with this reaction.[20] The overall reaction is as follows:

$$SiH_4 + 2N_2O \rightarrow SiO_2 + 2H_2O + 2N_2$$

At these temperatures, gas phase decomposition of N_2O initiates the reaction. A series of radical chain reactions then leads to the formation of stoichiometric oxide when the N_2O/SiH_4 ratio is about 100. At lower ratios, the reaction of SiH_2 with N_2O produces silicon-rich films. And this reaction is commonly used in PECVD of silicon dioxide, carried out at temperatures between 200°C and 400°C with O_2 as an additive to control stoichiometry.

Oxidation of Halogenated Silanes

Oxidation of chlorine-bearing silanes such as dichlorosilane and hexachlorodisilane with N_2O and O_2 mixtures produces SiO_2. Dichlorosilane-based reactions at over 900°C produce excellent quality HTO films[21] according to the following equation:

$$SiH_2Cl_2 + 2N_2O \rightarrow SiO_2 + 2HCl + 2N_2$$

Oxygen addition lowers the incorporation of chlorine in the film. This reaction has also been used in PECVD oxide deposition at reduced reaction temperatures. The use of hexachlorodisilane is of interest since the source gas is not pyrophoric, unlike other silanes. However, residual chlorine in the film can have adverse effects on film properties.

Organic Silane Substitutes

We have repeatedly mentioned a safety hazard associated with silane, it autoignites on contact with air. Safety has been primarily responsible for an industry-wide shift away from SiH_4 chemistries to safer alternatives. TEOS provides conformal films and is a safer alternative. The reaction occurs as follows:

$$Si(OC_2H_5)_4 \rightarrow SiO_2 + 4C_2H_4 + 2H_2O$$

Reaction temperatures are generally in excess of 700°C. At lower temperatures, H_2 forms in the place of H_2O. Earlier applications of the TEOS reaction were in the formation of borophosphosilicate glass (BPSG) films, given the high temperature required for pyrolysis. With proper dopant additives, this reaction is still one of the most popular atmospheric pressure BPSG reactions.

However, for the production of undoped SiO_2, the reaction has two disadvantages: (a) the film stress is relatively uncontrolled and (b) the reaction temperature is still very high for aluminum metallization. Addition of oxygen controls the film stress but does not alter the deposition temperature. This is accomplished by adding up to 10% ozone into the oxygen stream.[22] Plasma reactions based on TEOS have also become very popular for intermetal dielectric depositions at temperatures below 400°C.

TEOS is a liquid at room temperature and hence needs special delivery systems to provide the gaseous mix in the reactor. Several other organic precursors have been developed with the intent of improving on the successes of TEOS. These include diacetoxyditertiarybutoxysilane (DADBS), 2,4,6,8-tetramethyltetrasiloxane (TMCTS), and 2,4,6,8-tetraethylcyclotetrasiloxane (TECTS).[23] Reduced temperature, reduced pressure thermal depositions have been reported from all these sources.

9.3.3. Low Temperature Oxides (LTOs)

Low temperature oxides find use in spacers between the gate and diffusion in lightly doped drain (LDD) structures (see Figure 9.1), in passivation schemes, and as sacrificial oxide layers. Thickness control across the wafer diameter and an ability to survive high temperature operations without cracking are essential properties of this layer. LTO deposition has been performed in atmospheric pressure reactors, where the wafer is held flat; low pressure reactors, tube reactors where the wafers are held upright and close together; and in plasma reactors, considered in Section 9.3.5. Both hydride and organic sources have been used, even though the $SiH_4 + O_2$ reaction is the most popular.

As discussed earlier, the reaction is constrained within a window formed by the growth temperature and the partial pressures of hydride and oxygen. The process is limited by the adsorption of oxygen and hence observed activation energies for the reaction are dependent on process pressure; activation energy increases with decreasing pressure.[24]

Commercially available reactors have been optimized for this reaction with specially designed quartz fixtures. The process is harsh on the vacuum pumps used in the LPCVD reaction, since the silane/oxygen mixtures produce silicon dust in the exhaust tubes.

9.3.4. Doped Glasses

Doping of the SiO_2 network by network modifiers and other glass-forming oxides serves the following purposes: (a) dopant additives tend to lower the melting point of the glass, (b) they alter the viscosity of the glass at high temperatures, and (c) below certain concentrations they introduce the properties of the glass former without phase segregation from the SiO_2 matrix. For instance, boron oxides provide high temperature stability to the glass, whereas phosphorus oxides make the glass hygroscopic. Dopants also alter the glass transition temperature. For instance, ordinary window glass has sodium-based network modifiers included in the SiO_2 structure. To retain high temperature properties, the oven-to-table commercial glasses contain glass formers such as B_2O_3.

The common additives to SiO_2 for semiconductor applications are boron and phosphorus oxides for poly–metal dielectrics, and arsenic for doping applications. Boron and phosphorus additions tend to lower the melting temperature, reduce the intrinsic stress in the glass, allow for better glass flow due to reduced viscosity, and phosphorus additions getter sodium ions. Phosphorus doped-glass (PSG) finds application as a passivation film to protect the device against sodium. Boron-doped glass (BSG) is used as a boron dopant source.

When used together as borophosphosilicate glass (BPSG), they are the most commonly used dielectric between polysilicon and interconnect metal. Their superior flow characteristics at temperatures above 800°C allow them to be planarized after deposition through a viscoelastic flow process. They also protect the substrate silicon from mobile contaminant ions. BPSG films in this application need to possess good conformality as deposited and should completely cover all features without surface asperities or sharp edges after they are densified. Hence the flow characteristics of the glass are crucial to the application. Figure 9.9 shows the as-deposited and flow characteristics of BPSG films deposited with silane + O_2 and TEOS + O_3 reactions.[25]

In general, the higher the combined dopant concentration in the glass, the better the flow characteristics of the BPSG film. Up to 10 combined weight percent of boron and phosphorus has been used in BPSG, with the individual concentrations of the boron or phosphorus ranging up to 7 wt%. At

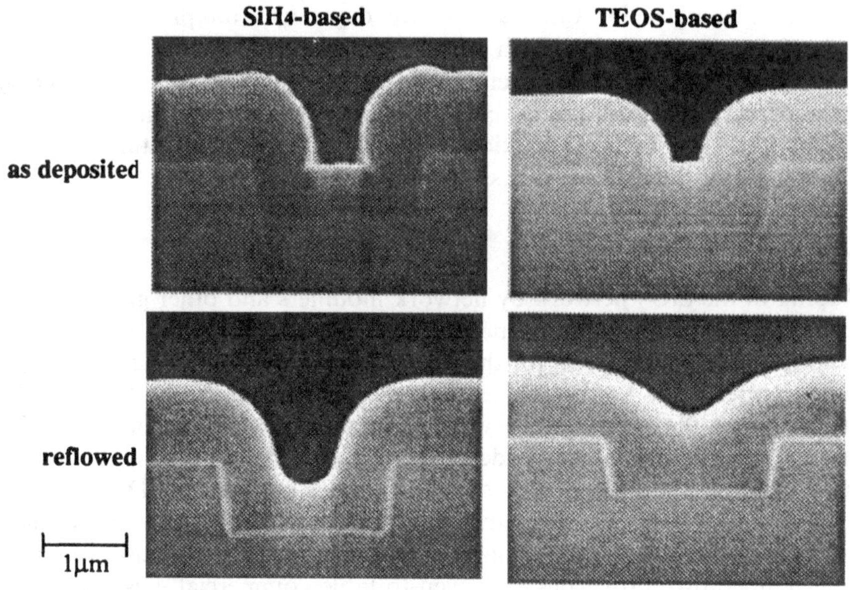

FIGURE 9.9. Comparison of the step coverage and flow characteristics of SiO_2 films produced by silane and TEOS processes. TEOS films consistantly produce better BPSG step coverage. Reprint from the 1992 Proceedings of the VLSI Multilevel Interconnection Conference, p. 19.

concentrations higher than this level, phase segregation and local devitrification at the surface occur. For instance, in the presence of moisture, BPSG films often precipitate BPO_4, B_2O_3, and P_2O_5 crystallites. Such precipitation is deleterious to film performance since (a) the particles, although soluble in water, leave pits in the glass that later processing steps decorate and (b) they leave local areas of low dopant concentration that affect the flow performance of the glass.

Table 9.3 shows the commonly used organic and inorganic CVD sources for the films. In general, thermal CVD is the popular method of growth for BPSG films. The organic sources provide better conformality in as-deposited films compared to inorganic films. This is attributed to the enhanced surface mobility of the TEOS molecule, resulting from its low sticking coefficient.

Atmospheric pressure deposition of BPSG, especially from hydride sources such as SiH_4 suffers from gas phase nucleation problems. These can be suppressed by proper dilution of the feed gases and by ensuring well-defined exhaust flow. Often boron and phosphorus incorporation rates have a strong interaction factor, i.e., the presence of one affects the rate of incorporation of the other. This is often a function of the temperature and pressure of

TABLE 9.3 Precursors for BPSG Films

Element	Inorganic sources	Organic sources
Silicon	SiH_4, SiH_2Cl_2	TEOS, DADBS
Oxygen	O_2	O_2, O_3
Phosphorus	PH_3	Trimethyl phosphite (TMP-ite), Trimethyl phosphate (TMP-ate)
Boron	B_2H_6, BCl_3	Trimethyl borate (TMB), triethyl borate (TEB)

deposition and the type of source gas used. The presence of the dopant gases also affects the growth rate of the BPSG film itself.

All BPSG films need to be densified after deposition at temperatures above 800°C, preferentially in the presence of moisture (called pyroflow) or oxygen. This ensures spatial dopant uniformity and homogeneity of the glass. The presence of oxygen during the flow compensates for any oxygen deficiency in the film. However, some device applications might preclude the use of oxygen as there is oxidation of the substrate during the densification process.

9.3.5. Plasma Oxides

The primary motivation for the plasma deposition of SiO_2 is the reduction in deposition temperature. However, some unique modifications to the plasma process, such as deposition/sputter etch/deposition sequences, and high density sources, such as the ECR deposition process, produce films with excellent gap-filling characteristics. SiH_4–N_2O–O_2 and TEOS–O_2 gas mixtures are the most common commercially used plasma oxides. Their applications include low temperature oxides as spacers adjacent to the gate structure and as intermetal dielectrics. They are also used as passivation films for compound semiconductor devices.

Parallel-plate, tube, and ECR configurations are the commonly used plasma reactor types. In general the deposition rate is linearly dependent on power, increases with increasing substrate temperature (Figure 9.10) for the hydride reactions, and exhibits mixed behavior with substrate temperature for organic sources.[26] The composition of the film is dependent on the gas flow ratios and the substrate temperature. The silicon to oxygen ratio in plasma deposited films can vary between 0.5 and 0.9. Poor growth conditions can result in nitrogen incorporation up to 7% and carbon incorporation up to 10%. Both are minimized by the addition of oxygen to the reaction. Hydrogen concentrations in the film range from 2% to 9%.[27]

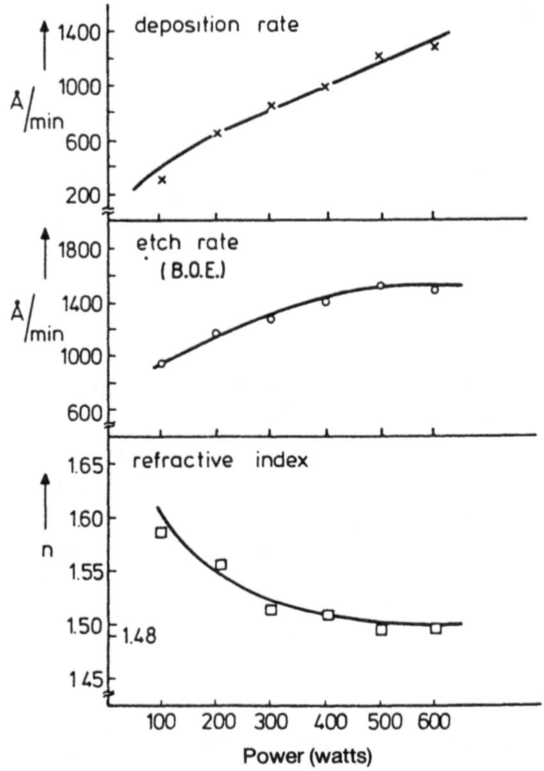

FIGURE 9.10. Deposition rate of SiO_2 with substrate temperature. Related film properties are also shown. Reprinted, by permission, from E. P. G. T. van de Ven, *Solid State Technology*, April 1981, p. 169.

An easy predictor of the silicon to oxygen ratio is the refractive index of the film. Compared to thermally oxidized SiO_2, which has a refractive index of 1.46, silicon-rich films have a refractive index above 1.5. Refractive index, like the silicon to oxygen ratio is modulated by the N_2O/SiH_4 ratio. The dielectric constant of the film is also high when the silicon content is high and can vary between 4 and 7 for plasma deposited films. Both refractive index and dielectric constant are modulated by nitrogen incorporation.

Step coverage of the film is a crucial property for interlevel dielectric applications. Without modifications to the deposition process, the plasma deposited film is not conformal and leaves voids in narrow structures and cracks at sharp corners. Figure 9.11 shows a deposition method where alternate plasma depositions and sputter etching combine to produce films with good step coverage.[28] Sputter etching preferentially etches off the

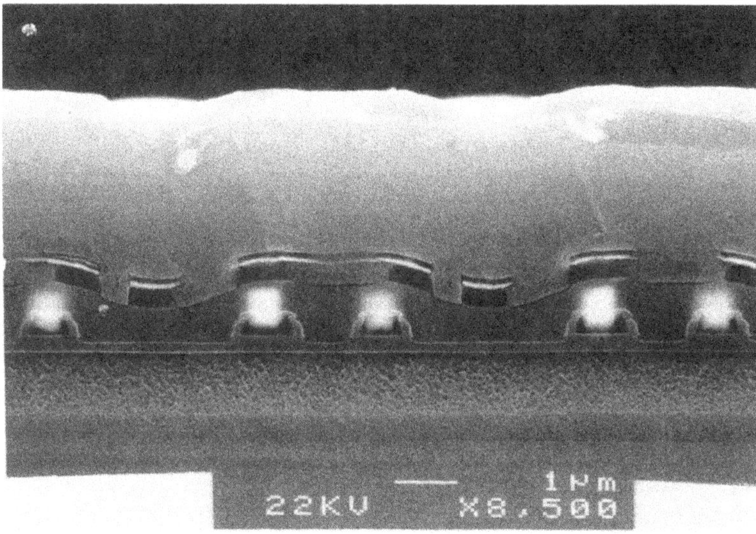

FIGURE 9.11. Alternate deposition/etch/deposition sequences produce good step coverage. Photo courtesy F. Moghadam, Intel Corp.

corners and rounds the profile. Sputtering yield is maximized at a 45° incident angle. Alternate deposition and etching sequences thus produce an acceptable step coverage.

Although very good at ensuring film step coverage, such a process often suffers from two major deficiencies: the overall process throughput is small and the sputter etch can often result in plasma-related damage to the underlying devices. Electron cyclotron resonance plasmas (see Chapter 7) solve this problem through a high density plasma source, which accomplishes deposition and etching simultaneously. Figure 7.13 shows very narrow, high aspect ratio gaps filled using the ECR plasma process.

9.4. CVD OF SILICON OXYNITRIDES

Silicon oxynitrides find application in the passivation of MOS devices, where they combine useful properties of both silicon oxides and nitrides. EPROMs use ultraviolet irradiation to erase their contents. In EPROMs and other devices that require UV transparency, nitride-based passivation is inadequate. Silicon oxynitrides are hard and UV transparent. They have a wide process window that allows tuning of film characteristics such as stress and dielectric properties.

Since the temperature limitation on the passivation film is stringent, the film is often grown by plasma enhanced CVD. The commonly used reactants are SiH_4, NH_3, and N_2O or O_2. Dual frequency reactors similar to those described in Section 9.2 have been used to control film stresses and UV transparency, which are strongly affected by the hydrogen content.

References

1. W.R. Runyan, and K. E. Bean, *Semiconductor Integrated Circuit Processing Technology*, Addison-Wesley, Reading, Mass., 1990.
2. S. J. Hillenius, in *VLSI Technology* (S. M. Sze, ed.), 2nd Ed., p. 466, McGraw-Hill, New York, 1988.
3. T. Hosaka, in *Proc. 10th Int. Conf. on CVD* (Cullen, ed.), Vol. 87–8, p. 518, The Electrochemical Society, Pennington, N.J., 1987.
4. Y. S. Hisamune, et al., in *Proc. IEEE IEDM,* 1989, p. 583.
5. R. V. Giridhar, Ph.D. thesis, Rensselaer Polytechnic Institute, Troy, N.Y., 1984.
6. W. Baerg, S. Sachdev, P. R. Engel, and P. A. Gargini, in *Silicon Nitride Thin Insulating Films* (Kapoor and Stein, eds.), p. 476, The Electrochemical Society, Pennington, N.J., 1983.
7. K. Jensen, and W. Kern, in *Thin Film Processes II* (Vossen and Kern, eds.), p. 330, Academic Press, New York, 1991.
8. C. E. Morosanu, and E. Segal, *Thin Solid Films*, **88**, 339 (1982).
9. K. E. Spear, and M. S. Wang, *Solid State Technol.* **23**(7), 63 (1980).
10. R. S. Rosler, *Solid State Technol.* **20**(4), 63 (1977).
11. J. R. Hollahan, M. T. Wauk, and R. S. Rosler, in *Proc. 6th Int. Conf. on CVD* (Donaghey, Rai-Chaudhury, and Tauber, eds.), p. 224, The Electrochemical Society, Pennington, N.J., 1977.
12. C. Blaauw, *J. Electrochem. Soc.* **131**(5), 1114 (1984).
13. E. van de Ven, *Solid State Technol.* **24**(4), 169 (1981).
14. E. van de Ven, I. Connick, and A. S. Harrus, in *Proc. IEEE VMIC*, 1990, p. 194.
15. F. H. P. M. Habraken, et al., *J. Appl. Phys.* **53**(1), 404 (1982).
16. Novellus Inc., private communications.
17. L. Holland, *The Properties of Glass Surfaces*, p. 5, John Wiley, New York, 1964.
18. F. S. Becker, D. Pawlik, H. Anzinger, and A. Spitzer, *J. Vac. Sci. Technol.* **B5**(6), 1555 (1987).
19. N. Goldsmith, and W. Kern, *RCA Review* **28**, 153 (1967).
20. A. J. Learn, and R.B. Jackson, *J. Electrochem. Soc.* **132**, 2975 (1985).
21. D. W. Freeman, *J. Vac. Sci. Technol.* **A5**(4), 1554 (1987).
22. K. Fujino, in *Proc. 2nd Ann. Dielectrics and CVD Metallization Symp.,* J. C. Schumacher, San Diego, Calif., 1989.
23. G. Smolinsky, in *Proc. 10th Int. Con. on CVD* (Cullen, ed.), p. 490, The Electrochemical Society, Pennington, N.J., 1987.
24. C. S. Pai, R. V. Knoell, C. L. Paulnack, and P. H. Langer, *J. Electrochem. Soc.* **137**, 971 (1990).
25. H. Kotani, et al., in *Proc. IEEE VMIC*, 1992, p. 15.
26. A. C. Adams, F. B. Alexander, C. D. Capio, and T. E. Smith, *J. Electrochem. Soc.* **128**(7), 1545 (1981).
27. I. T. Emesh, G. D'Asti, J. S. Mercier, and P. Leung, *J. Electrochem. Soc.* **136**, 3404 (1989).
28. H. Kotani, et al., *J. Electrochem. Soc.* **130**, 645 (1983).

Chapter 10

CVD of Semiconductors

Semiconducting films such as Si and GaAs used in ICs depend on their crystallographic perfection to produce the device performance demanded of them. In theory, the growth of single-crystal thin films to match the substrate lattice, or epitaxy, is possible through most common methods of thin film deposition, such as evaporation, PVD, and CVD. However, CVD, liquid phase epitaxy and molecular beam epitaxy are the most commonly used methods of epitaxial thin film growth. In this chapter, we will examine the role of the chemistry, process conditions, and the reactor, that promote epitaxial growth of the semiconducting thin film.

The chapter covers the CVD of two broad sets of semiconductors: elemental semiconductors, where we will study silicon, both in its polycrystalline form and as an epitaxial layer; and compound semiconductors such as group III–V materials and group II–VI materials. Even though there is a certain amount of overlap between the two sets of technologies, we will treat them independently.

We will begin the section on elemental semiconductors with polycrystalline silicon (polysilicon), its applications, structure, and how choice of deposition process affects an application. We will then follow the transition from polycrystallinity to epitaxial growth and the structural consequences of epitaxy on semiconductor processing (for instance, pattern shift). This leads us on to the applications and processes for epitaxial silicon.

The second portion of the chapter will concentrate on the technology of metal–organic chemical vapor deposition of GaAs, GaAlAs, and InP. Their application in widely different roles as lasers, solar cells, phototransistors, photocathodes, and field effect transistors, take us away from the mainstream of microelectronics, so we will study their applications only where it is relevant to specific properties developed during deposition. Instead we will focus on the deposition chemistries, precursors, and processes that occur during deposition.

10.1. POLYCRYSTALLINE SILICON

Over its evolution, microelectronic processing has moved away from the first transistor made of germanium to being essentially silicon-based. Almost all forms of crystallinity in silicon find specific uses: single-crystal and polycrystalline silicon in mainstream logic, memory, and linear ICs; amorphous silicon in solar cells; and microcrystalline silicon in certain active matrix displays. In most of its applications in MOS and bipolar devices, polycrystalline silicon (or polysilicon, as we shall call it) is used more as a conductor than a semiconductor by degenerately doping it with phosphorus or boron. Together with the gate oxide, the polysilicon film forms the heart of a MOS transistor. It is also widely used in bipolar technology as a contact to the emitter and base regions. We will review these two applications in some detail before discussing the CVD of polysilicon from silane-based chemistries.

10.1.1. Applications of Polysilicon

Polysilicon-based discrete devices such as sensors, diodes and transistors find wide applications. However, the bulk of polysilicon use in ICs is in the gate electrode in MOS devices, and as poly-emitter/base contacts in bipolar devices. Novel applications such as trench capacitors for DRAMs, and polysilicon-based trench isolation schemes are also becoming popular.

MOS Devices

Self-aligned source and drain technology for MOS transistors minimizes the overlap of the gate electrode required over the source and drain diffusion regions.[1] Figure 10.1 illustrates the self-aligned gate structure. Reduction in the silicon real estate required per device is achieved by patterning the gate oxide using the polysilicon as the mask. For this technology to work, the

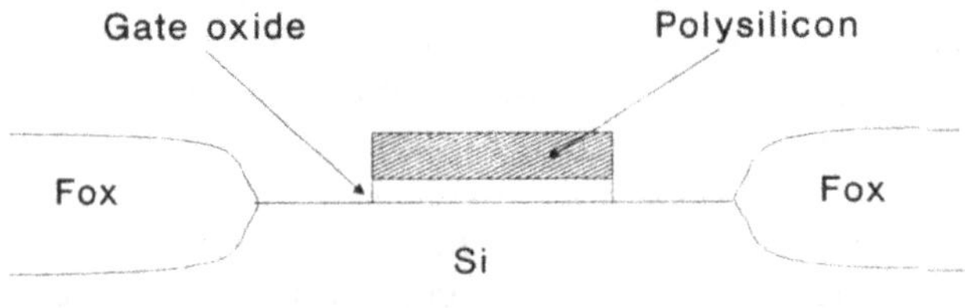

FIGURE 10.1. A self aligned gate structure utilizing polysilicon as a mask for etching gate oxide. This technique saves considerable amounts of silicon real estate.

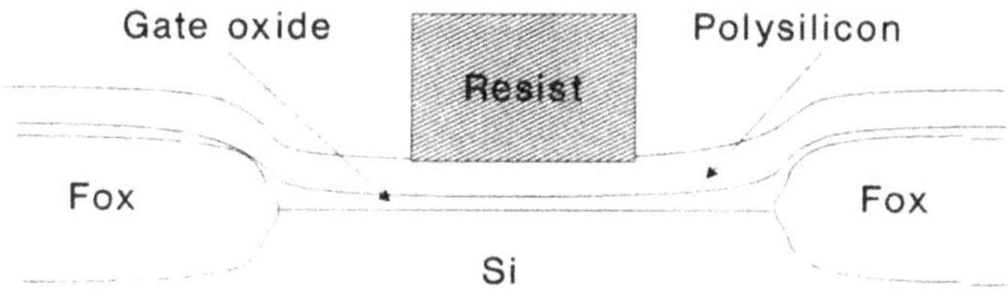

FIGURE 10.2. Process integration of the self-aligned gate structure: (1) deposit polysilicon on the gate oxide, (2) pattern the photoresist, (3) etch the polysilicon, (4) use the polysilicon as a mask to etch the gate oxide, (5) remove the photoresist and clean.

essential requirements for the polysilicon films are its conformality over the gate region and its ability to withstand the oxide etch chemistry without undercutting. The film also needs to possess high temperature stability to withstand subsequent oxidation temperatures. Figure 10.2 shows the sequence of operations in the fabrication of the self-aligned polysilicon gate.

Once the gate is formed, the voltage at which the gate turns on or off is determined by the threshold voltage of the device. The threshold voltage can be defined as the voltage applied between the gate and source of a MOS device below which the drain to source current drops to zero (or the device switches).[2] The threshold voltage is a function of the difference in the work functions of the conductor and the substrate. In the case of heavily doped polysilicon (which acts as the conductor over the substrate silicon), the work function term can be written as:[3]

$$\Phi_s = \chi + E_c - E_f \qquad (10.1)$$

where χ is the energy difference between vacuum and the conduction band edge, and the subscripts c and f on E denote the energy at the edge of the conduction band and the Fermi level, respectively. The Fermi level is modulated by the presence of the dopants and hence the work function of the polysilicon is a strong function of the type and concentration of the dopant species. To achieve maximum conductivity of the polysilicon, it is usually degenerately doped, making the Fermi level close to the valence band edge for N-type dopants such as phosphorus and close to the conduction band edge for P-type dopants such as boron. It is thus critical that as the MOS gate, the polysilicon film is uniformly doped and activated. It also needs to provide a consistant interface to the underlying oxide.

Other applications for polysilicon films in MOS transistors include (a) local interconnects, where they are heavily doped, (b) floating charge retention gates in some memory devices, where their lack of surface asperities and ability to maintain a clean interface to a tunneling oxide are critical, (c) load resistors in NMOS logic structures and static RAM devices, where they are doped according to the resistance needed, and (d) programmable

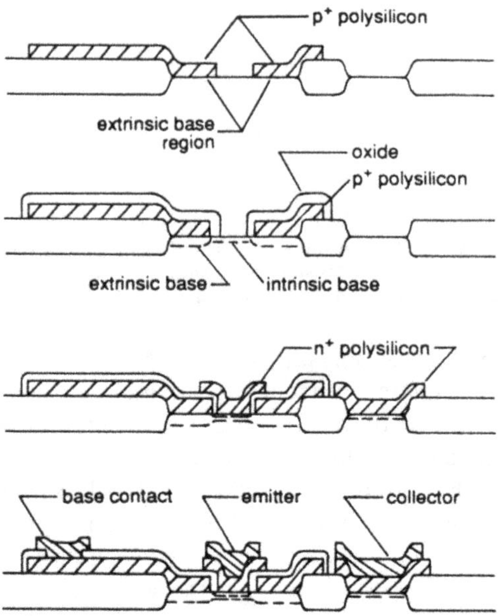

FIGURE 10.3. Schematic fabrication sequence for the polyemitter structure commonly used in bipolar and biCMOS structures. Reprinted from Ref. 5 with permission from Kluwer Publishing Co.

connectors in certian ROMs, where they are fused together by subsequent laser treatments.[4]

Bipolar Devices

In high performance bipolar devices, device speed can be significantly enhanced by the reduction of parasitic elements. Polysilicon emitter transistors have become a very popular alternative to structures where metal makes direct contact to intrinsic semiconductor. This structure, illustrated in Figure 10.3, provides the advantages of low base resistance, low parasitic capacitance, and provides mask alignment margin.[5] To take full advantage of the reduction in capacitance, and the increased metal–poly area, step coverage and dopant control of the film are critical. The dopant, usually boron, is also used to dope the single-crystal silicon below it.

Polysilicon is used extensively in bipolar device isolation. Radiation hard devices are fabricated in single-crystal silicon tubs, buried in thick polysilicon and separated from it by a thin layer of oxide.[6] In trench isolation, now becoming popular in many CMOS devices, deep trenches in silicon are filled with polysilicon after a thin isolating layer is grown on the sidewalls.[7] In CMOS and BiCMOS structures, this isolation is beneficial in reducing latch-up.

In summary, in most polysilicon applications in MOS and bipolar devices, the polysilicon needs to exhibit conformality during deposition, uniformity in doping, high temperature stability during oxidation, surface smoothness to prevent field accumulation, and a consistant interface to SiO_2. The deposition chemistry and conditions during deposition play a key role in ensuring consistent film properties.

10.1.2. Polysilicon Deposition Chemistry

Pyrolysis of silane to produce polycrystalline silicon has been in use since the beginning of the integrated circuit industry. The following overall reaction

$$SiH_4 \rightarrow Si + 2H_2$$

is still the common method of polysilicon deposition. The temperature regime used is 600–900°C. Growth rate in this regime is exponentially dependent on the temperature. Atmospheric and reduced pressure reactors have been used; however, volume manufacturing techniques have adopted low pressure CVD as the deposition thickness and better step coverage.[8] Photolysis, and plasma deposition techniques have also been attempted and are not currently commercially popular.

In the gas phase, at high gas phase temperature and silane partial pressure, the following reaction is thermodynamically favorable:

$$SiH_4 \rightarrow SiH_2 + H_2$$

Spectroscopic measurements of the gas phase do show the presence of SiH_2. When conditions are favorable (high pressures and temperatures), homogeneous gas phase reaction can lead to the formation of hydrogenated silicon powder.[9]

Under subatmospheric pressures and temperatures in the regime 600–800°C, the following sequence of reactions is thought to lead to the deposition of polysilicon:[10]

$$SiH_4 + 2^* \rightarrow SiH_2^* + H_2^*$$

$$SiH_2^* + {}^* \rightarrow Si^* + H_2^*$$

where the *s represent surface adsorption sites. The sites that occur at kinks and ledges on the surface are more favorable for the adsorption of the gas phase silane species.

The growth reaction is strongly controlled by adsorption of the reactants at higher temperatures and desorption of adsorped hydrogen at lower temperatures. Competition for the surface adsorption sites drastically lowers the growth rate. For a hot wall, tube reactor at differential conversion, the

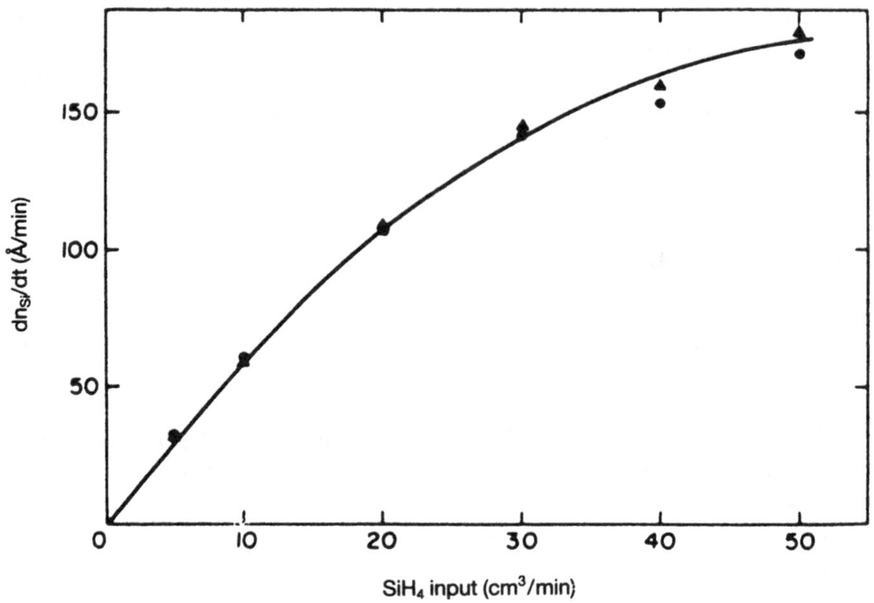

FIGURE 10.4. Model prediction and experimental verification for polysilicon growth. Reprinted, by permission, from M. L. Hitchman, J. Kane, and A. E. Widmer. *Thin Solid Films* **59**, 231–247 (1979).

expression for growth rate becomes that due to the surface concentration of the adsorbate and can be written as

$$\text{Rate} = \frac{n_0}{An_0 + B} \tag{10.2}$$

where n_0 is the input concentration of silane and A and B are constants for a reactor.[11] Figure 10.4 shows the experimental and model fits for this model of polysilicon growth. Partial pressure of silane increases the rate to the point where the surface coverage of silane is no longer limited by input concentration.

The deposition of doped films involves the addition of dopant-bearing gases, such as PH_3 or B_2H_6, to the gas stream. Both phosphorus and boron additions modify the growth process: phosphorus lowers the growth rate while boron increases the growth rate. The change in growth rate with addition of the doping gases is related to the growth mechanism. In the case of phosphine, trivalent phosphine preferentially attaches itself to the surface to the exclusion of the tetravalent silicon.[12] The resultant drop in silane surface concentration results in the drop in growth rate. In the case of boron, a new parallel silicon deposition pathway opens up due to the presence of adsorbed boron atoms.[13]. Boron atoms on the surface catalyze the decomposition of the SiH_2 molecule resulting in enhanced deposition rate. Phosphorus and arsenic atoms make the surface more negative while boron makes it more positive. And these electrostatic properties have also been used to explain the difference in condensation rates of the slightly negative SiH_x molecule.

The rate increase in the presence of boron has been used in depositing polysilicon at reduced temperatures where otherwise the deposition rate would not be viable. Similarly, disilane has been used to increase the rate of phosphorus-doped deposition.

More interestingly, the presence of dopants degrades the step coverage of otherwise very conformal polysilicon films.[14] The different surface reaction rates of different dopant atoms, and the different surface mobilities of the different species result in the degredation of the conformality.

10.1.3. Reactors for Polysilicon Growth

Most of the polysilicon deposition in IC manufacturing utilizes horizontal, low pressure, hot wall, tube furnaces. The tubes are similar to oxidation and diffusion furnaces. They are usually three-zone furnaces; gases enter at the source end and are pumped from the exhaust end of the tube. Resistive heating coils are used for maintaining precise temperature control in the three zones

Wafer spacing between vertically placed wafers limits the mass transfer from the forced convection gas stream (due to the pumping) flowing normal to the wafers. Wafer spacing is critical in determining the load size of the tubes.[15]

Other tube configurations are becoming popular, such as vertical furnaces where the furnace footprint is much smaller than large horizontal furnaces. Special loading stations are needed for vertical furnaces. Figure 10.5 shows a low pressure vertical furnace reactor made by the Silicon Valley Group.

Compared to these batch reactors, recently a single-wafer polysilicon reactor has been introduced. The Applied Materials Centura Polysilicon deposition system shown in Figure 10.6 makes use of a high temperature wafer transport capability along with multiple chambers to match the throughput of tube furnaces. Both doped and undoped films can be grown by this reactor, which claims excellent thickness and dopant uniformities at a growth rate of approximately 2 000 Å/min for undoped polysilicon films. The reactor operates between 10 and 200 torr, at process temperatures varying between 550°C and 750°C.

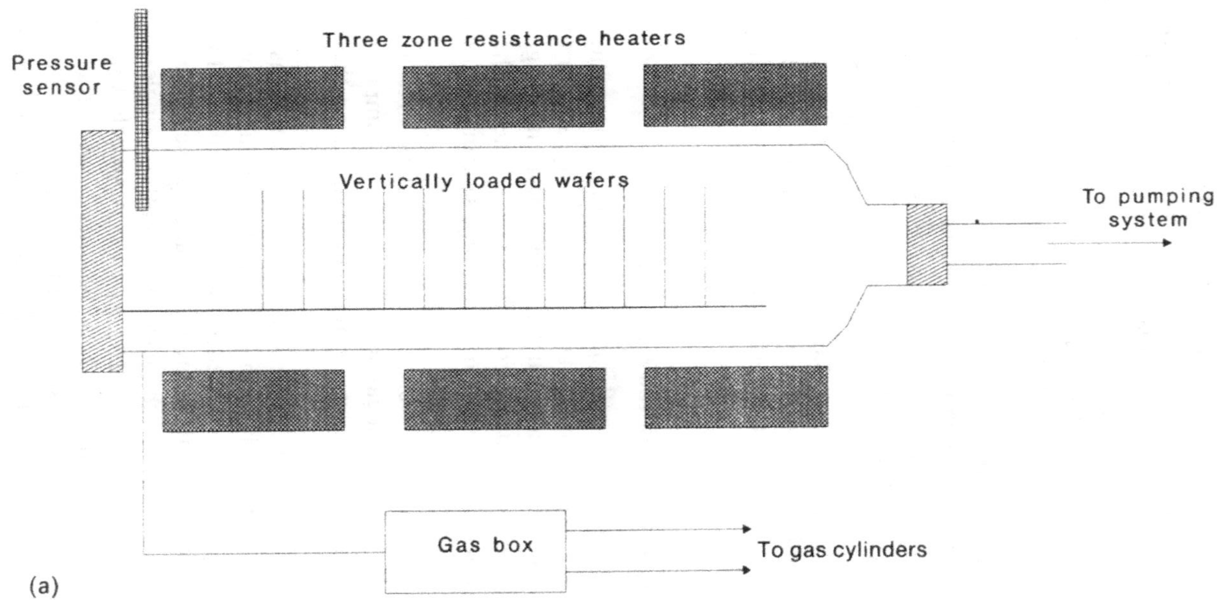

(a)

FIGURE 10.5. (a) Schematic of conventional LPCVD furnace for the deposition of polysilicon. (b) LPCVD vertical furnace made by the Silicon Valley Group. Notice the significant reduction in floor space and the loading mechanisms. (b) Reprinted with permission from SVG, Inc.

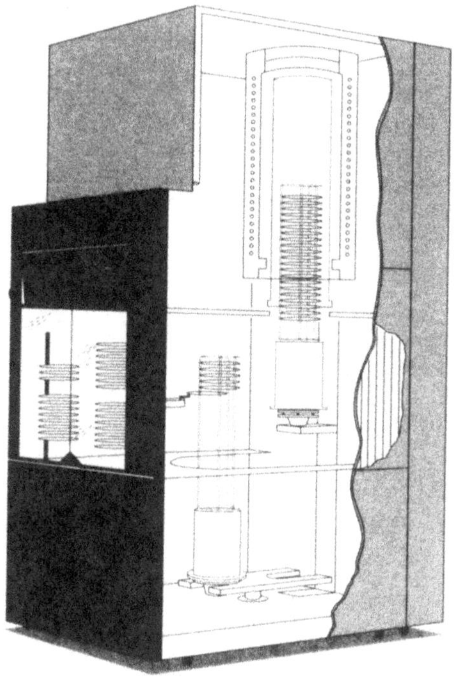

(b)

FIGURE 10.5. *Continued*

10.1.4. Structure and Properties of Polysilicon

The transition from amorphous to polycrystalline silicon films is strongly affected by temperature. Nucleation of crystallites and their subsequent growth is treated in Chapter 2 and is classically applicable to the case of polysilicon. Pressure can affect the transition temperature.[16] At atmospheric pressure and below 550°C, amorphous films are seen. At pressures on the order of 0.1 torr, the temperature needs to be lower than 500°C for the deposition of amorphous films. At very low pressures, there is evidence of crystallinity at temperatures below 400°C. The substrate surface also plays a role in the development of crystallinity. For instance, films grown on epitaxial silicon tend to possess additional low energy kinks or steps for nucleation to occur.

As-deposited undoped polysilicon films have columnar grains, which are the result of growth from a saturation density of critical nuclei at a given temperature and pressure. Growth from these critical nuclei follows the classical pattern of minimizationm of surface free energy, resulting in films

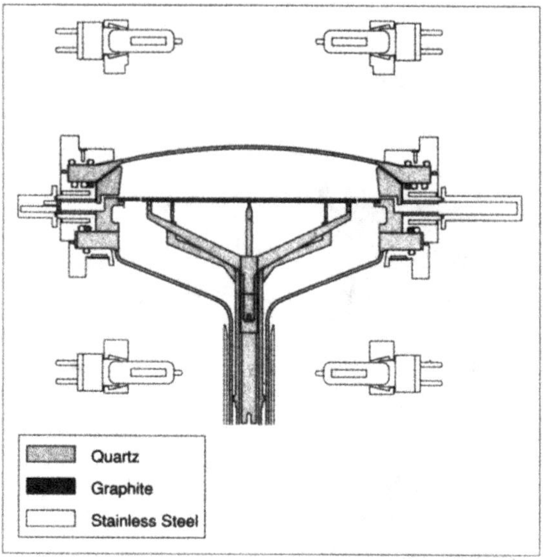

FIGURE 10.6. The Centura low pressure, single-wafer reactor for polysilicon deposition developed by Applied Materials, Incorporated. Reprinted with permission from Applied Materials, Inc.

strongly textured along the [1 1 0] directions.[17] Further heat treatment does not affect grain growth significantly till above 1 100°C, since the activation energy for the nucleation of new crystals for recrystallization is high. Above this temperature, grain growth occurs by nucleation and growth of fault-free new crystals at the expense of the faulted deposition grains.

Doping changes the grain morphology by altering the free energy associated with the grain surfaces. Reduction of surface free energy in the presence of phosphorus results in larger, equiaxed grains in phosphorus-doped films.[18] Boron does not result in similar grain growth. Annealing of the doped films further promotes the growth of equiaxed grains, with sizes approximately equal to the film thickness. Addition of boron in phosphorus-doped films can slow down the grain growth. Figure 10.7 shows a series of TEM micrographs that show grain growth during deposition and during further heat treatment of doped and undoped films.

Other properties of the polysilicon films are strongly affected by structure. For instance, the refractive index of the film can vary continuously between the single-crystal value of 3.55 and the amorphous film value of around 3.95. Optical absorption of undoped polysilicon decreases with grain size. Grain size strongly moderates the electrical conductivity, decreasing with increasing

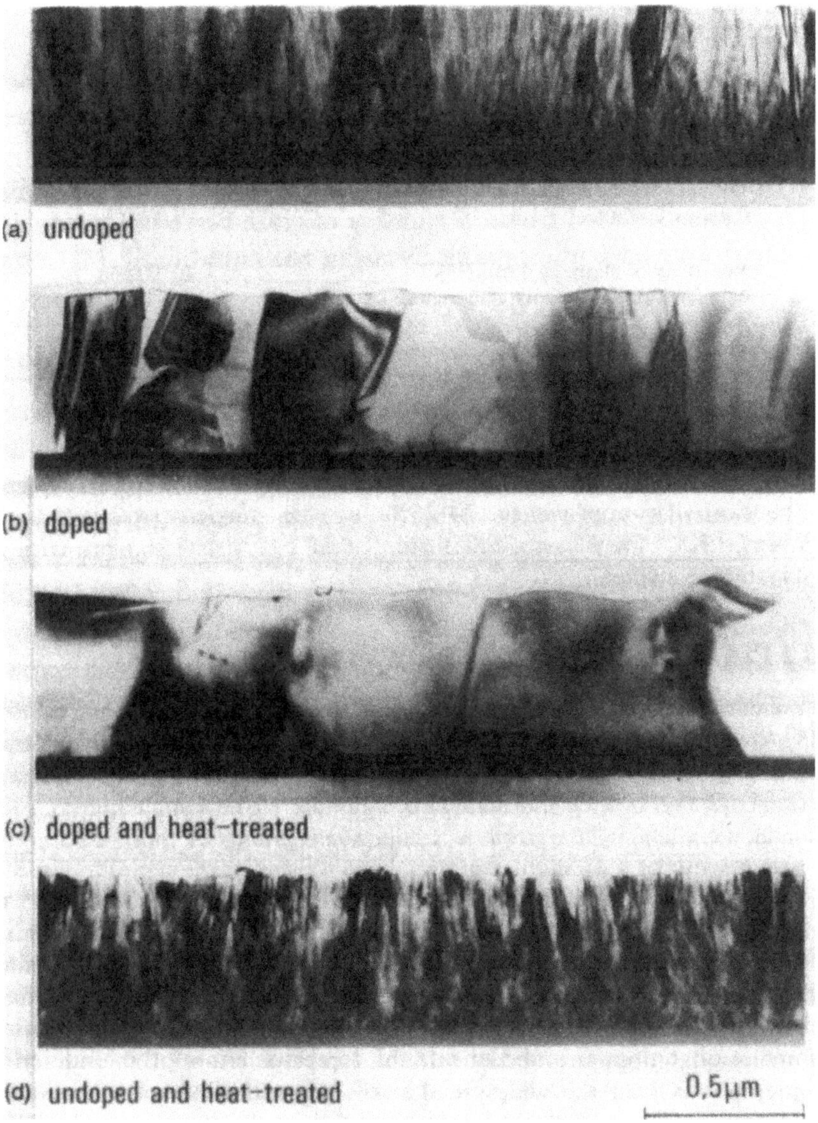

FIGURE 10.7. Grain growth during annealing of undoped and phosphorus doped polysilicon films. R. Falckenberg, E. Doering, and H. Oppolzer, *Extended Abstracts 570, Electrochemical Society Fall Meeting,* Los Angeles, October 1979, pp. 1429–1432. Reprinted by permission of the publisher, The Electrochemical Society, Inc.

TABLE 10.1 Properties of Polysilicon

	Property	Value
1.	Activation energy for growth from silane	1.6 eV
2.	Index of refraction	~3.55 (up to 1.8 μm)
3.	Thermal conductivity	0.3–0.35 W cm^{-1} K^{-1}
4	Diffusion coefficients and activation energy for dopants at 900°C	As: 2×10^{-14} cm^2/s, 3.2 eV P: 1×10^{-12} cm^2/s, 3.4 eV B: 3×10^{-13} cm^2/s, 3.3 eV
5.	Resistivity of film grown over 750°C	5×10^5 Ω cm
6.	Minimum doped resistivity after anneal	B, As: ~2 000 $\mu\Omega$ cm P: ~400 $\mu\Omega$ cm
7.	Electron mobility in heavy phosphorus-doped poly	30 cm^2 V^{-1} s^{-1}

grain size and increasing dopant concentration. For columnar grains, the conductivity approaches 20–25% of the single-crystal value of 1.5 W cm^{-1} K^{-1} for fine-grained films. Table 10.1 lists the properties of polycrystalline silicon.

10.2 EPITAXIAL SILICON GROWTH

We mentioned the concept of epitaxy in Chapter 2, with reference to thin films growth. The deposition of a film on a single-crystal substrate such that the growing film has the same crystallographic orientation as the substrate is termed epitaxy. If the substrate and the film are of the same material, such as silicon on silicon, the growth is called *homoepitaxy* or more commonly *epi.* The growth of a different material, such as silicon on saphire, is called *heteroepitaxy.* We also discussed some of the conditions of deposition that favor epitaxial growth during CVD: higher substrate temperature, higher surface mobility of the adatoms, cleanliness of the substrate, and crystallographic quality of the substrate surface. In the deposition of epitaxial silicon for IC applications, we also need to be concerned about the incorporation of intentional dopants, and the faithful reproduction of the underlying topography. Given the widespread use of epitaxial silicon in the microelectronics industry and the complexity of the requirements placed on the deposition, this is one of the widely studied ares of chemical vapor deposition.

10.2.1. Applications of Epitaxial Silicon

The use of epitaxial layers of silicon in integrated circuits arises essentially from the ability to grow lightly doped silicon layers over heavily doped bulk

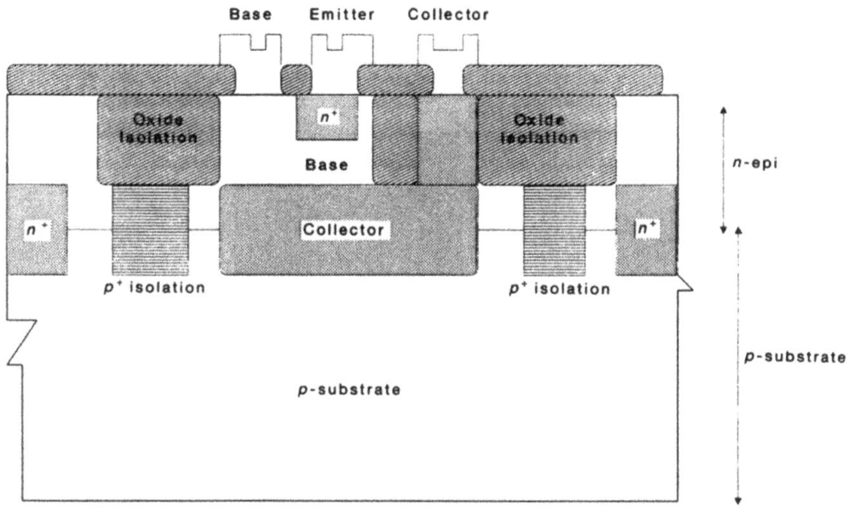

FIGURE 10.8. A high speed *n–p–n* transistor showing N^+ epitaxial layer in a *p*-substrate. *p*- and *n*-wells are formed in the epi layer for fabricating the devices. Reprinted, by permission, from G. R. Srinivasan, *Solid State Technology*, November 1981, p. 101.

substrates.[19] This translates into abrupt transitions in doping between layers, with excellent control in thickness, both of which are essential for high performance circuits. Early applications for epi layers pertained to the reduction in the collector series resistance in bipolar devices.[20] In a grown junction, the high resistivity material extended all the way from the collector–base junction to the collector contact leading to poor high current performance. By building a high resistance epi mesa on a low resistivity bulk material, the effective collector resistance was reduced significantly.

Epi films are now used both in bipolar and MOS applications. Figure 10.8 shows epitaxial silicon as commonly used in a high speed *n–p–n* bipolar transistor.[21] An abrupt *n*-doped epi layer is grown on the bulk *p*-substrate. p^+ and n^+ buried layers below the epi layers are grown prior to the epi growth and we will see in a subsequent section the problems introduced by heavily doped substrates. The structure shown in Figure 10.8 uses oxide isolation combined with diffusion in the epi layer for device isolation. This structure provides high device density, low device parasitic capacitance, and low wiring capacitance.

Figure 10.9 shows a typical CMOS transistor built on an epi layer.[22] The advantages of using the lightly doped epi layer over the heavily doped substrate arise from a CMOS device phenomenon called latchup, caused by adjacent *n*, *p*, and *n* regions.[23] The presence of the high resistivity epi layer

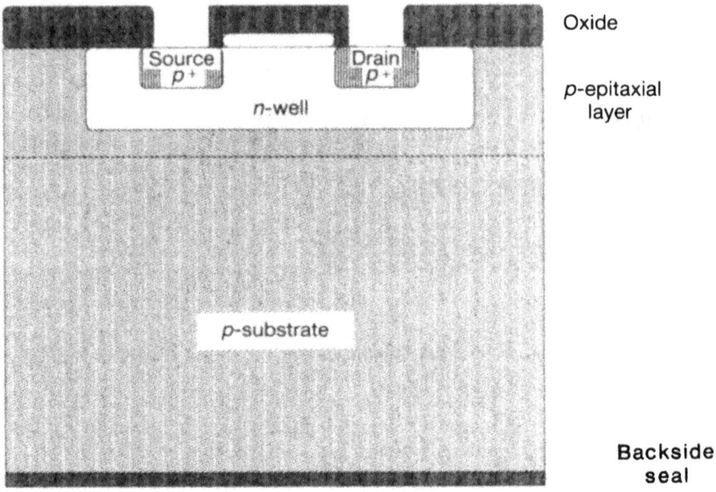

FIGURE 10.9. A typical CMOS device built on a grown epi layer. The biggest advantage of such a structure is the reduction in latch-up.

alleviates the latch-up related breakdown. In dynamic RAMs, the epi layer also helps reduce soft error problems, caused by alpha particles generating electron–hole pairs in the semiconductor. And epi layers help to reduce parasitic capacitances.[24] Other novel applications include raised source–drain structures, where selective epitaxial growth is used to increase the junction depth in the contact areas. Simultaneous polysilicon and single-crystal depositions have also been attempted in some unique structures. Addition of epitaxial silicon to CMOS processing adds to the cost of the device, which has to be carefully balanced against its improved performance.

Even though CVD is by far the most popular means of growing epitaxial silicon films in integrated circuits, silicon molecular beam epitaxy, and liquid phase epitaxy have also been tried. However, wide commercial acceptance of these techniques in volume IC manufacture is limited.

10.2.2. Growth Chemistry for Epitaxial Silicon CVD

The most common chemistry used for thermal CVD of single-crystal silicon involves chlorosilanes ($SiCl_xH_{4-x}$ where $x = 0$ to 4) and hydrogen. The net reaction can be written as

$$SiH_4 \rightarrow Si + 2H_2$$

$$SiH_2Cl_2 \rightarrow Si + 2HCl$$

$$SiCl_4 + 2H_2 \rightarrow Si + 4HCl$$

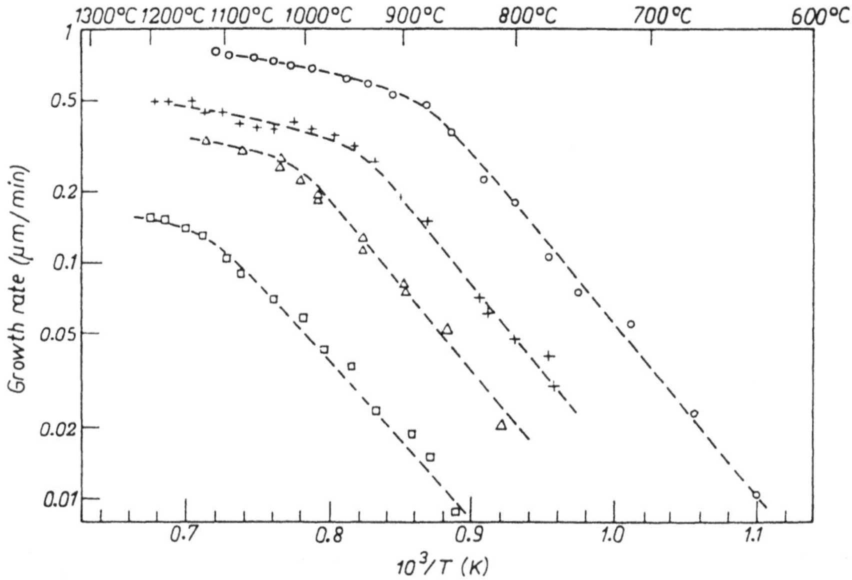

FIGURE 10.10. Arrhenius plots for the growth of epitaxial silicon from various chlorosilanes. The transition from reactor control to diffusion control occurs at progressively higher temperature as the chlorine content increases. Curves correspond to increasing Cl content going from top to bottom. Reprinted from Ref. 27 with permission from Academic Press.

depending on the precursor species used. Other halogens such as silicon iodide have been attempted, but ease of operation of the source materials has favored the chlorides.

Figure 10.10 shows the Arrhenius plots for the reaction between hydrogen and $SiCl_xH_{4-x}$ where $x = 0$, 1, 3, and 4. In all cases the activation energy is approximately 40 kcal/mol; the transition to diffusion-controlled growth occurs at higher temperatures as x increases.[25] The minimum temperature at which epitaxy occurs also increases with x; below this transition temperature, polycrystalline silicon growth is seen. The minimum temperature for epitaxial growth is an important concept because temperature of deposition is constrained by integration of the process. To circumvent the minimum epitaxy temperature other techniques are being sought, such as PECVD.

The effect of reduced pressure on lowering of the deposition rate is quite dramatic below 100 torr, as shown in Figure 10.11.[26] Above 100 torr, the rate is only weakly dependent on increasing pressure. This phenomenon is the basis of reduced pressure epi growth, which offers significantly reduced autodoping without compromising rate, as we will see in a subsequent section. Growth rate increases with reactant flow till the transition to the diffusion-

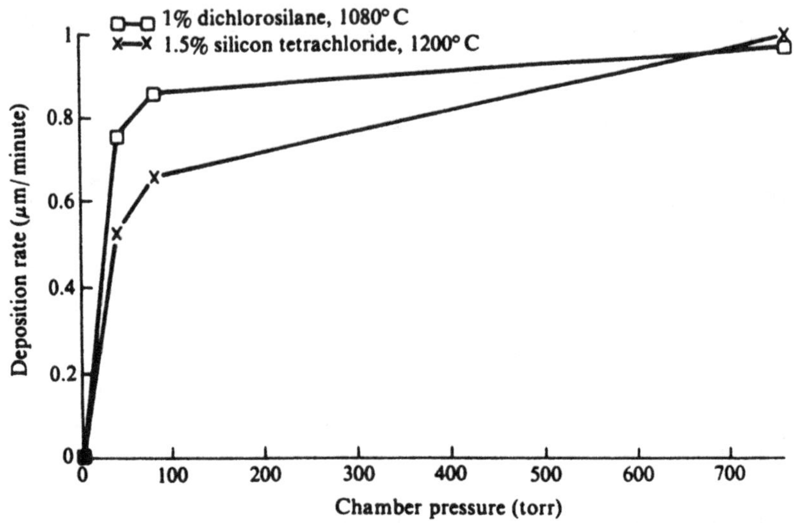

FIGURE 10.11. Effect of silicon growth rate on the pressure. Above 100 torr, the rate is not very strongly influenced by pressure. Walter Runyan and Kenneth Bean, *Semiconductor Integrated Circuit Process Technology* (pp. 130, 341), © 1990 by Addison-Wesley Publishing Company, Inc. Reprinted by permission of the publisher.

controlled growth occurs as in Figure 10.10. The rate varies close to the square root of the flow, as expected.

With chlorosilanes, the reaction product is HCl and under proper conditions (such as high $SiCl_4$ flow), the reverse reaction can dominate, resulting in the etching of substrate silicon by HCl:

$$Si + 2HCl \rightarrow SiCl_2 + H_2$$

This reaction is quite useful as the initial cleaning step for subsequent epitaxial growth and is often the first step in an epi process. The overall silicon deposition reaction is thought to proceed in the following steps.[27]

Homogenous Decomposition of the Input Gas:

$$SiH_4 \rightarrow SiH_2 + H_2$$

$$SiH_2Cl_2 \rightarrow SiCl_2 + H_2$$

With increasing chlorine content in the gas, this reaction becomes more difficult in the gas phase, requiring a heterogeneous surface catalyst at lower gas temperatures. For $SiCl_4$, the decomposition of the precursor on the surface can be the rate-controlling step.

Surface Adsorption

$$SiH_2 + * \rightarrow SiH_2*$$

$$SiCl_2 + * \rightarrow SiCl_2*$$

Both species preferentially adsorb on ledges, kinks, and other sites of reduced surface energy. This might result in adsorption to multiple adjacent surface sites (SiH_2*_2, $SiCl_2*_2$). This is particularly important when the substrate surface has been cleaved at a small angle away from the crystallographic plane. We will discuss this further in Section 10.2.4.

Surface Reaction

$$SiH_2*_2 \rightarrow Si* + H_2 + *$$

$$SiCl_2*_2 \rightarrow Si* + 2HCl + *$$

where Si* = * or crystalline silicon.

Surface reaction has been shown to be the rate controller in the deposition process. For hydrochlorosilanes above 700°C and for $SiCl_4$ above 1 000°C, the overall enthalpy of the reaction is about 40 kcal/mol. The rate controller has been postulated to be the breaking of the Si–Cl bonds of the adsorbed species. In the case of SiH_4, the rate controller is the desorption of molecular hydrogen from the surface.

10.2.3. Incorporation of Dopants

Since the aim of the epi layer is to form abrupt layers with different doping concentration than the substrate, both intentional and unintentional doping of *n*- and *p*-type species needs to be characterized. *p*-type dopants in epi layers are almost exclusively boron. Phosphorus, arsenic, and antimony are used as n-type dopants. Phosphorus is the most common of the three and has been studied extensively.

For CVD epitaxial growth, the dopants are introduced as their hydrides (B_2H_6, PH_3, AsH_3, and SbO_3), diluted in either H_2 or Ar. BCl_3 is also used as a boron source. The behavior of the dopants in affecting growth rate was discussed in Section 10.1. In general, phosphorus doping tends to lower the growth rate by a mechanism where there is a competition for surface sites. The reduction in rate becomes insignificant as the temperature of deposition is increased.

Concentration of dopants in the growing film is controlled by the partial pressure of the dopant in the feed gas. The concentration of phosphorus in the film as a function of its concentration in the gas phase is shown in Figure 10.12. The slope of the line in Figure 10.12 is called the segregation coefficient,

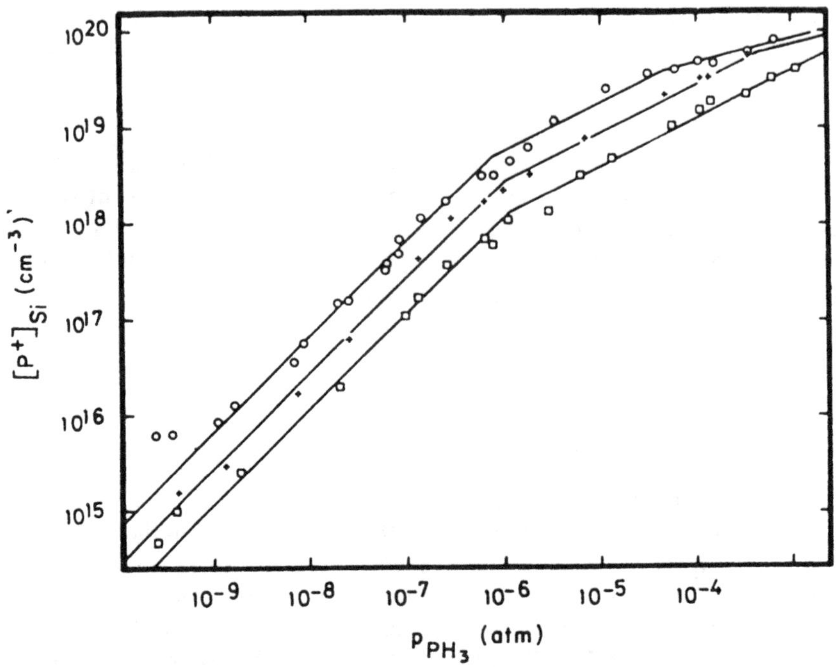

FIGURE 10.12. Segregration coefficient of phosphorus at different temperatures. The figure shows the concentration of phosphorus in the film as a function of gas phase concentration at 1 400 K (top), 1 200 K (middle), and 1 000 K (bottom). Reprinted from Ref. 27 with permission from Academic Press.

the ratio of atomic dopant in the solid to its partial pressure in the gas phase.[28] For phosphorus, the segregation coefficient increases with increasing deposition temperature.

Seldom do we obtain an ideally abrupt dopant concentration change from one substrate to another. Often there is a transition dictated by solid-state diffusion from the substrate and by a phenomenon called autodoping. Figure 10.13 shows a typical transition and the contributions of the different phenomena to the transition region.[29] Solid-state diffusion from the substrate is affected by the time the substrate is maintained at the deposition temperature, and the diffusion coefficient of the dopant from the substrate, which in turn is dependent on the temperature. Hence lower growth temperature and faster growth time (either thinner films or higher growth rates) slow down solid-state diffusion from the substrate into the film.

Autodoping occurs through transfer of the dopant from the heavily doped substrate to the gas phase prior to incorporation back into the growing film. Before deposition, perhaps during HCl cleaning or flow stabilization, dopants

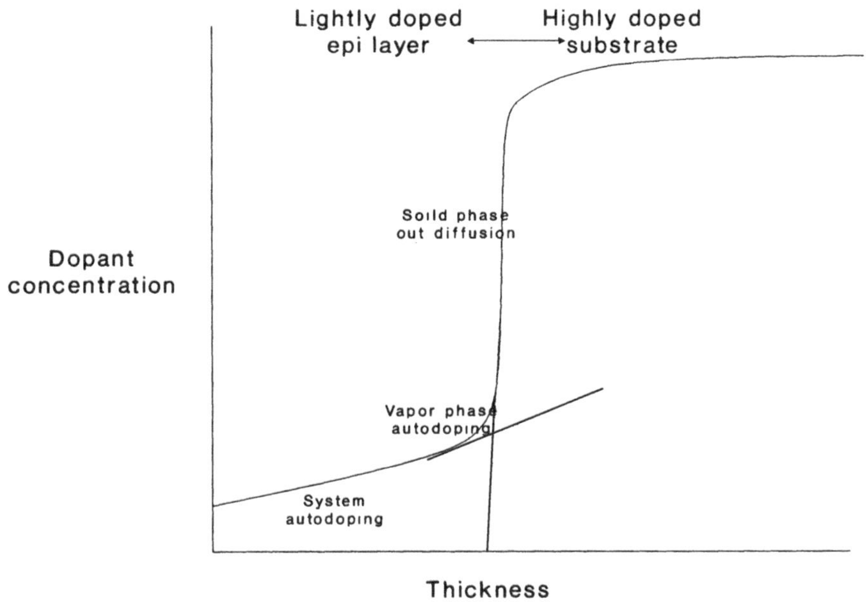

FIGURE 10.13. Deviation from abrupt dopant profile in an epi layer. Autodoping causes doping of the film.

evaporate from the substrate and enter the gas stream. If the gas molecules have a long residence time in the reactor, the incorporation back into the film is dictated by the segregation coefficient. This effect is particularly prevalent in areas adjacent to heavily doped regions on a single wafer; in a horizontal reactor into the downstream wafers; when the edges and the back of a heavily doped wafer are not sealed to prevent outdiffusion; and when the reactor is not optimally designed, resulting in local vortices that are difficult to pump. The following pointers help reduce the problem of autodoping.[30]

a. Reduction in deposition pressure from atmospheric CVD to as low as 40 torr reduces deposition rates by about 20% while reducing autodoping by almost two orders of magnitude (Figure 10.14).
b. Autodoping in both the vertical and horizontal directions needs to be considered, as shown in Figure 10.15. Sealing the back and sides of the reactor can reduce vertical autodoping but areas adjacent to buried regions will still suffer from it. Ion implantation, as opposed to diffused buried layer, can help tailor the surface concentration of the substrate.
c. Flow and temperature stabilization time prior to actual deposition needs to be minimized—even a thin layer of deposition drastically

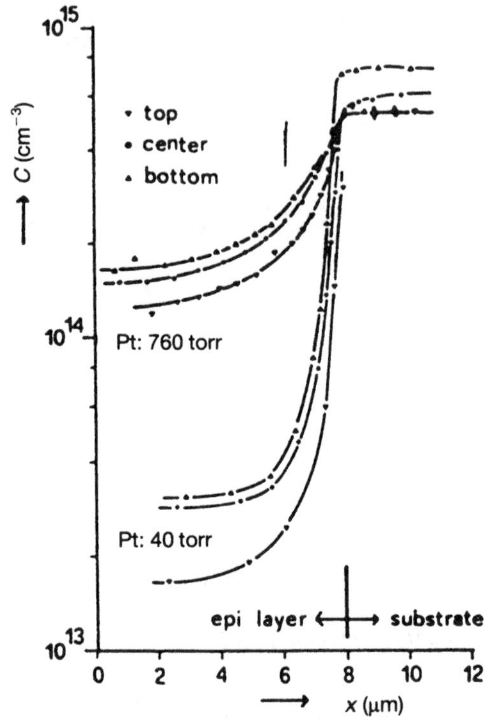

FIGURE 10.14. Reduction in autodoping as a function of pressure. Note that growth rate itself is not seriously affected by this change in pressure. Reprinted, by permission, from Eike Krullman and Walter Engl, *IEEE Trans. Electron Devices* **ED29**(4), 491 (1982). © 1982 IEEE.

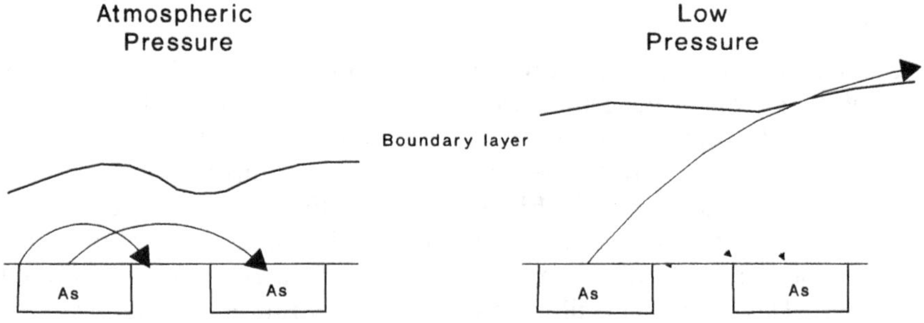

FIGURE 10.15. Schematic illustration of vertical and horizontal autodoping and the influence of pressure.

reduces autodoping. Less time at elevated temperature also helps in better junction control.

d. Both tube and radial flow reactors suffer autodoping. Increased flow and decreased residence time are necessary.
e. With respect to autodoping, antimony is better than arsenic, and arsenic is better than phosphorus. Different dopants behave differently with changing temperature and flow.

Proper characterization of the deposition process and careful process integration, taking into account the substrate characteristics, can minimize autodoping and ensure nearly abrupt dopant transitions.

10.2.4. Nucleation and Crystallographic Quality

Substrate silicon wafers are often produced from bulk crystal growth techniques where wafers are sliced from the ingot and polished to the wafer specifications. The surface orientation of the polished wafer varies from (1 1 1) in bipolar applications to predominantly (1 0 0) in MOS applications. However, epitaxial growth on a true (1 1 1) silicon surface is extremely slow, whereas (1 0 0) offers acceptable rates.[31] The origin of this difficulty with the (1 1 1) surface is the low density of bonds available between the surface double layers. This problem is solved by polishing the surface to be 3–5° away from the (1 1 1) plane towards the (1 1 0) plane. This in essence produces a series of atomic steps that act as ledges for the nucleation of the film.

In Chapter 2, we mentioned the role of the surface in reducing the critical nucleus size. The ledges and kinks modify disk-shaped nuclei so they have a larger nucleus to substrate interface and a lower overall surface free energy. Growth along these atomic steps (even (1 0 0) surfaces are often produced with a slight slope) in theory extends the steps with perfect atomic alignment from the substrate. The resulting islands of silicon all have similar orientations, and when the islands coalesce to form a continuous film, reorientation of growth directions is not required.

A clean substrate is essential to maintain its crystallographic quality. Oxide inclusions on the surface can lead to film surface roughness. Similarly, even small oxygen partial pressures in the feed gas (often in the form of moisture) can lead to poor epi quality.

Crystallographic defects in silicon can be electronically active. For instance, the strain field around a dislocation can produce local distortion of the energy bands by as much 0.1 eV.[33] Stacking faults in the silicon can also be electronically active, while simultaneously attracting point defects which decorate the fault. Dislocations and stacking faults are the most common process-induced defects in epi films. The quality of the substrate surface, thermal stresses, and the crystallographic quality of the substrate can cause these defects. Nonuniform heating of the wafer (either poor contact between wafer and susceptor, or unoptimized induction heating in the reactor) can produce thermal stresses. When these stresses exceed the yield strength of silicon, dislocations are produced. Preexisting dislocations often propagate into the growing film, they might get annihilated by conventional processes or extend all the way to the film surface, depending on the thickness of the film.

Stacking faults are created as a result of growth on surface steps produced by thermal stresses.[34] Oxide inclusions on the surface also produce the same effect. Impurities in the gas feed, such as carbon and moisture, also lead to stacking faults by the same mechanism. Oxidation-induced stacking faults can getter metallic precipitates at the interface. Figure 10.16 shows electron micrographs that illustrate the crystallographic defects in epitaxial silicon.[35] Minimization of these defects is crucial for achieving device performance specifications.

10.2.5. Reactors for Epitaxial Growth

Epitaxial silicon reactors in principle follow the general guidelines in Chapter 5. However, variations on the theme have been restricted only by the imagination of the researchers, leading to multiple choices in reactor chamber design. The following special precautions apply to the design and construction of epi reactors:

1. The dopants used are almost always toxic; the gases are pyrophoric. Hence safety and effluent treatment are important criteria in the choice of a reactor.
2. Depletion of reactants passing from the leading wafer edge to the trailing wafers can be severe and have to be compensated; similarly downstream wafer surfaces see increased autodoping.
3. Thickness control and uniformity of the coating are important not only on the wafer but also on the hot surfaces of the reactors. Peeling and popping of the deposit on the chamber walls can lead to particulate problems.
4. Vertical wafer positioning lowers particles on the wafer; but depletion from the leading edge to the trailing edge needs to be accounted for. The converse is true for horizontal wafer positioning.
5. Induction heating causes the susceptor to heat first and transfers the energy to the wafer through a combination of conduction, convection, and radiation. Heating the wafer from one side creates a temperature gradient within the wafer, warping it into a dish shape and producing dislocations and slip.

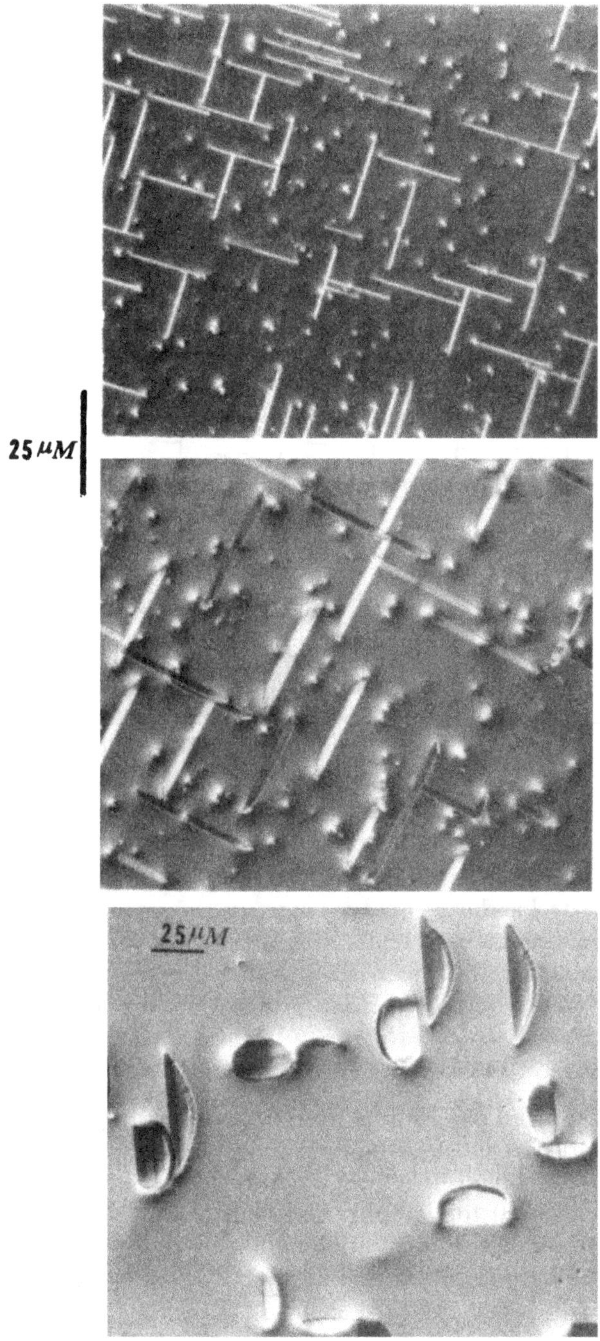

FIGURE 10.16. Oxidation induced stacking faults in an epi layer. The lower figures show longer etching time to decorate the faults. Such defects can be electronically active, affecting the performance of the device. Reprinted from Ref. 35 with permission from John Wiley & Sons.

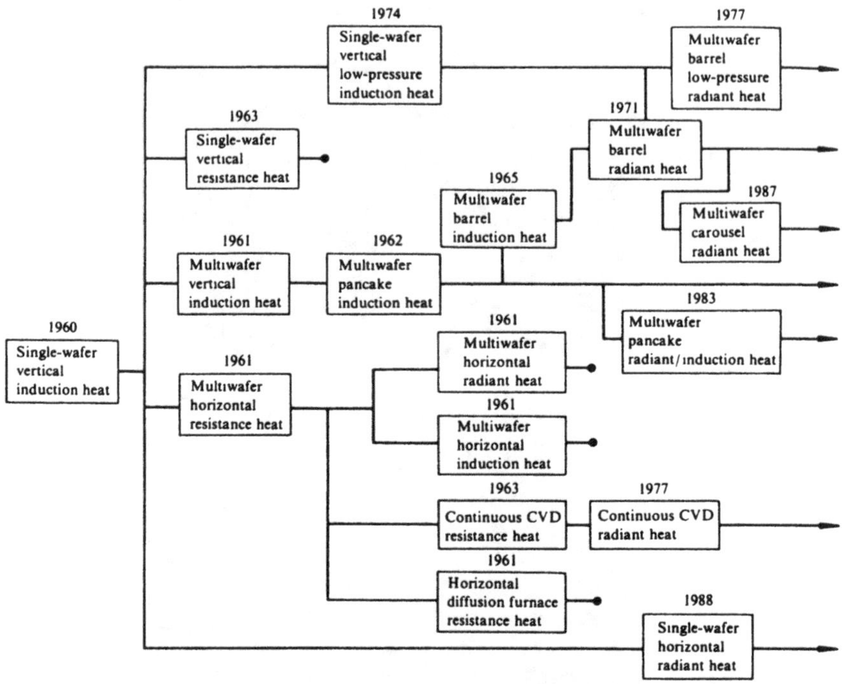

FIGURE 10.17. Chronology of the evolution of epitaxial reactors. Reprinted from Ref. 31 with permission from *Semiconductor International*.

After significant evolution (Figure 10.17), the popular choices for epitaxial CVD are the barrel and carousel type reactors shown in Figure 5.1.[36] These are radiantly heated, batch reactors with wafers held vertically. They can be operated at atmospheric or reduced pressures. Processes on these systems offer slip-free epi layers with excellent within-wafer uniformity control.

10.2.6. Integration of Epitaxy into IC Processing

Integration of any unit process into overall IC processing requires that some common concerns be satisfied: minimization of processing temperature to conserve the thermal budget; control of particulates; faithful reproduction of underlying or overlaid topography; and control of cost. We will consider three important aspects of integrating epi layers in a process sequence; (a) pattern distortion, (b) gettering, and (c) selective epitaxial growth.

Pattern Distortion

Pattern distortion during epitaxial growth pertains to the change in dimension or location of an underlying feature when an epi film is grown on the feature:

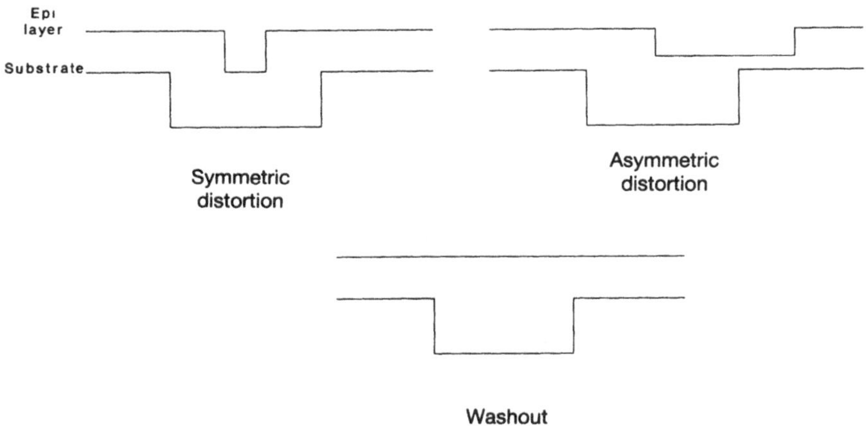

FIGURE 10.18. Pattern distortion during epitaxial growth. Distortion occurs due to the different silicon growth rates on different crystal planes.

feature dimensions shrink or grow, circular features become ovals, squares turn into rectangles, and some feature edges get no growth. Pattern distortion arises because the growth rates of epitaxial layers on different crystallographic planes can be different, and an underlying feature on a single-crystal layer will be exposing different crystal planes on its faces.

Pattern distortion, illustrated in Figure 10.18 can be classified as symmetric distortion, asymmetric distortion without size change, asymmetric distortion with size change, or pattern washout.[37] In all cases, pattern distortion is a function of epi layer thickness: the smaller the thickness, the lower the distortion. Pattern shift is also lowered for higher deposition temperature. However, the most important parameter in containing pattern distortion is the substrate crystallographic orientation. It has been shown that a 3° shift away from the true $\langle 1\,1\,1 \rangle$ direction and precise $\langle 1\,0\,0 \rangle$ direction for the surface normal of the substrate give the least pattern distortion.

The importance of pattern shift arises from the fact that, during lithography, elevated or recessed features on the substrate are used as alignment marks for the imaging tool to align one mask layer to the next. Washout or lateral shift of the pattern during epi growth can cause misregistration of the next layer mask resulting in shorts and opens.

Gettering

Since the device is built on the epitaxial semiconductor layer, impurities in the epi layer can result in performance loss in the device. In particular, species that affect the carrier lifetime, such as gold, copper, nickel, or iron atoms incorporated into the film from either the feed gas or from the reactor itself, can result in degradation of the device characteristics. Similarly, electrically active crystal defects such as stacking faults and dislocations can also alter device performance. Gettering is used to clear the electrically active epilayer of such contaminants and defects and gather them in the single-crystal bulk, so they do not affect the device.

Two distinct techniques have been used in clearing the epi layer of contaminators and defects; extrinsic gettering attracts contaminants by imposing an external strain field near the epi–substrate interface;[38] intrinsic gettering creates a denuded zone by utilizing the thermal cycling of device processing to nucleate and attract the defects towards the rear of the wafer.[39] Intrinsic gettering is particularly useful for removing oxygen precipitation, whereas extrinsic gettering helps in the removal of metallic contaminants.

Extrinisic gettering techniques include the use of a thick polysilicon layer on the wafer backside, germanium doping of the interface between the substrate and the epi layer, and multilayer strain fields at the interface by using dopants to create misfit dislocations. Intrinsic gettering uses different heat cycles to nucleate oxygen precipitates, stacking faults, and discloations; to diffuse the nuclei away from the epi layer; and to create a zone of fault-free silicon close to where the device is being fabricated. Figure 10.19 shows a typical heat cycle for intrinsic gettering.

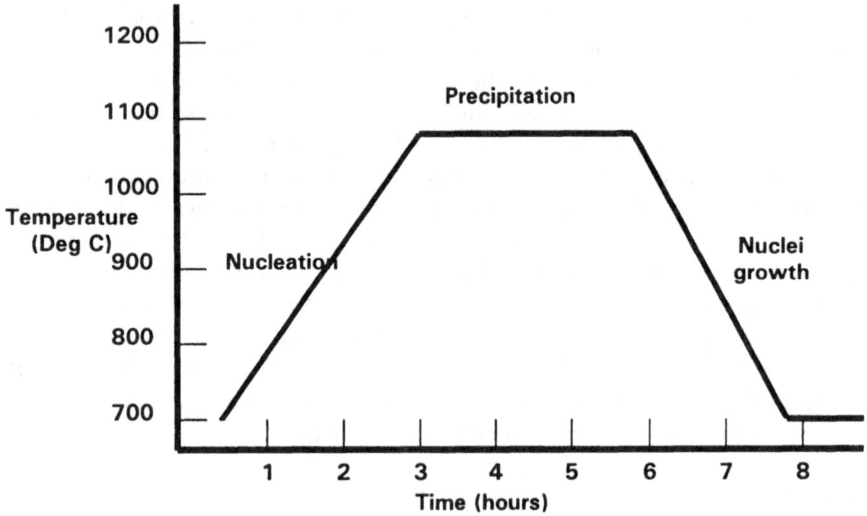

FIGURE 10.19. A typical intrinsic gettering cycle. The defects are nucleated and diffused away to form a denude zone. This heating cycle would be part of a regular process sequence, such as oxidation or dopant diffusion.

Selective Epitaxial Growth

We saw earlier that nucleation of epitaxial layers preferentially occurs at kinks and steps on the single-crystal substrate layer. We also saw the dependence of growth rate on different crystal faces. If we mask portions of the substrate with an amorphous layer such as SiO_2, nucleation will preferentially occur only on exposed crystalline faces, producing selective epitaxial growth.

Interest in selective epitaxial grwoth for CMOS integrated circuits comes from a desire to raise the source and drain regions from the original interfaces such that problems of shallow junctions can be alleviated, as shown in Figure 10.20.[40] Even though selectivity in epitaxial growth has been known for some time, along with other selective CVD processes, selective epitaxial growth is yet to find widespread commercial application.

Growth chemistry plays a significant role in determining selectivity. Even though $SiCl_4$ shows the best selectivity, the high deposition temperature needed is incompatible with IC processing, and the trend is to use SiH_2Cl_2. SiH_4 is the least selective. Addition of HCl to the feed gases improves selectivity. Lowering the temperature also has a profound effect on improving selectivity.

10.3. COMPOUND SEMICONDUCTORS

Compound semiconductors find windespread use in optoelectronics and high speed devices, in such applications as solar cells, lasers, phototransistors, photocathodes, and field effect transistors (FETs). Chemical similarity among Group IIIA elements and also among Groups IIB, IVA, and VA leads to the existence of a wide variety of compound semiconductors. Of these, the compounds GaAs, $Al_xGa_{1-x}As$ and $In_xGa_{1-x}As_yAs_yP_{1-y}$ are commercially very important. Varied methods of growth such as liquid phase epitaxy, molecular beam epitaxy, vapor phase epitaxy (VPE), and metallorganic CVD (MOCVD) have been used in the growth of compound semiconductors. Metallorganic CVD, also called organometallic CVD (OMCVD) or organometallic vapor phase epitaxy (OMVPE), is commonly used for the deposition of III–V and II–VI compound semiconductors since precursors for most of the elements of interest are readily synthesized, and since the free energy of formation of the compounds from these precursors is highly negative.[41] Also the chemical similarity between the precursors of different elements of the same groups assists in easy compositional changes in the substitutional alloys. Figure 10.21 shows the relevant portion of the periodic table useful in compound semiconductor applications.

GaAs and AlGaAs have been deposited by MOCVD for lasers, light emitting devices, light-detecting devices, photocathodes, high speed FETs,

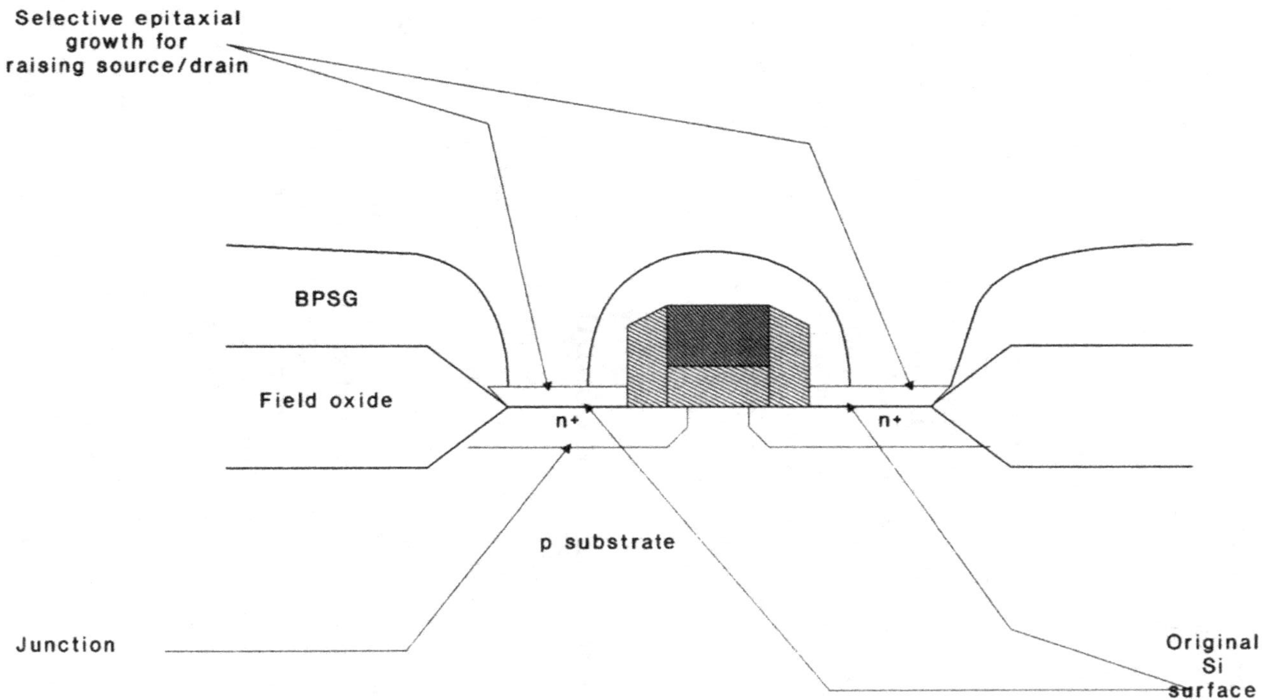

FIGURE 10.20. Application of selective epitaxial growth to raise the source and drain regions above and away from the shallow junction.

	Group	IIB	IIIA	IVA	VA	VIA
	2		B	C	N	O
	3		Al	Si	P	S
Period	4	Zn	Ga	Ge	As	Se
	5	Cd	In	Sn	Sb	Te
	6	Hg	Tl	Pb	Bi	Po

FIGURE 10.21. Section of the periodic table relevant to compound semiconductor MOCVD.

and heterojunction bipolar transistors. InP and InGaAs have been used for longer wavelength optical devices, and HgCdTe for infrared detectors. Due to the diversity of applications and requirements, we will treat only the CVD process. Device requirements will be mentioned only where processing is a key variable in determining a device property.

10.4. GALLIUM ARSENIDE AND $Al_{1-x}Ga_xAs$

GaAs is a direct band gap semiconductor, with an energy gap between the valence and conduction band extrema of 1.42 eV, compared to the indirect 1.08 eV for silicon. The addition of Al allows the band gap to be changed continuously without altering the interatomic spacing. Substitutional impurities from groups II, IV, and VI of the periodic table act as donors and acceptors in the GaAs structure. GaAs has a mass density of 5.31 g/cc and crystallizes into the zinc blende structure. The structure consists of two interpenetrating face centered cubic lattices, one containing Ga and the other As. The two sublattices are separated by 2.44 Å along the body diagonal of the unit cube. The structure is similar to that of silicon (which is diamond cubic), but with the two different atoms in the lattice. The presence of different types of atoms results in different chemical activity on the opposite (1 1 1) crystal planes; one is all Ga and the other is all As. Table 10.2 summarizes the physical, electrical and mechanical properties of GaAs.[42]

$Al_{1-x}Ga_xAs$ is a substitutional compound where Al substitutes for the

TABLE 10.2. Properties of GaAs

Property	Value	Units
Density	5.3174	g cm^3
Lattice parameter	5.654	Å at 300 K
Structure	zinc blende	
Young's modulus (1 0 0)	75.5×10^{10}	dyne/cm^2
Poisson's ratio	0.55	
Band gap	1.42	eV
Acceptors	Zn, C, Mg, Cd, Ge	
Donors	Si, S, Se, Te	
Refractive index	3.45	at 1 eV

Ga in the lattice sites. The importance of $Al_{1-x}Ga_xAs$ comes from the fact that the lattice parameters of AlAs and GaAs are very close to each other at 5.661 Å and 5.654 Å respectively. This allows for heteroepitaxy of various Al substitutional compositions on GaAs substrates. The structural characteristics of $Al_{1-x}Ga_xAs$ are as expected very similar to those of GaAs.

10.4.1. Vapor Phase Epitaxy (VPE) of GaAs

VPE from inorganic precursors, especially halide transport, was a popular technique for the growth of GaAs, due to its simplicity and because the impurity content of early MOCVD films was not acceptable.[43] However, growth of GaAlAs using MOVCD has always been the accepted technique due to the difficulty in incorporating inorganic Al sources. Precursors for Ga and As for VPE are either the chlorides or the hydrides. The following reaction has been successfully used in the deposition of GaAs:

$$GaCl(g) + \tfrac{1}{4}As_4(g) \rightarrow GaAs(s) + HCl(g)$$

It is carried out in a hot wall, quartz reactor in a hydrogen ambient.[44] In the hydride process, GaCl is produced by passing HCl over liquid Ga, and As_4 is produced by the pyrolysis of AsH_3 (arsine). In the chloride process, $AsCl_3$ is passed over liquid Ga. A GaAs crust forms on the hot Ga. The transport of GaCl and As_4 to the substrate is achieved by the reaction of the GaAs crust with HCl according to the reaction

$$GaAs(g) + HCl(g) \rightarrow GaCl(g) + \tfrac{1}{4}As_4(g) + \tfrac{1}{2}H_2(g)$$

The temperature range for all of the reaction is 700–850°C, above which increased vacancy concentration in the grown crystal limits its usefulness.

The requirement for a hot wall reactor, which increases the particulate concentration, and the inability to transport aluminum has led to a move away from this technology in favor of MBE and MOCVD. The rest of this chapter will concentrate only on MOCVD techniques.

10.4.2. MOCVD of GaAs and GaAlAs

The term MOCVD and the earliest demonstration of the process can be found in a series of articles by Manasevit starting in 1969, where he showed that triethylgallium (TEGa) and AsH_3 deposited single-crystal GaAs in an open tube reactor.[45] Despite its long history, the exact series of reactions that lead to the deposition from the organic precursors is not well understood. Of all the epitaxial techniques for compound semiconductors, including MBE, liquid phase epitaxy and vapor phase epitaxy, MOCVD has demonstrated the widest variety of III-V materials with excellent uniformity, monolayer thickness and transition control.

We will examine in detail the properties of the precursors, the chemical reaction in the gas phase and on a solid surface, the reactors, the growth parameters and the properties of the grown film. This should set up the discussion for the growth of other films such as InP. We will not treat the other films in detail but will suggest device-specific reviews for each of them.

Organometallic Precursors

The availability of high vapor pressure, pure precursors for a wide variety of elements that are of interest in compound semiconductor deposition and the relatively large negative free energy of deposition has led to the popularity of MOCVD. Most of the precursors are colorless liquids with vapor pressures between 1 and 100 torr around room temperature. Some are white solids which sublime readily. In general, they are pyrophoric and react adversely with moisture. Hence they need to be enclosed in special gas cylinders, which also act as bubblers for liquids and sublimators for the solids. The containers need to have welded joints, Diaphragm or bellows valves prevent leakage and moisture contamination. Figure 10.22 is a schematic of the equipment needed for the effective confinement and use of MOCVD precursors.[46]

For liquid sources, the vapor can be considered to be in thermodynamic equilibrium with the liquid (for solids this approximation is not true). The vapor pressure of the organometallic compound can be expressed in terms of the following relation:

$$\log P = B - \frac{A}{T} \tag{10.3}$$

where P is the pressure, T the temperature, and A and B are constants. The

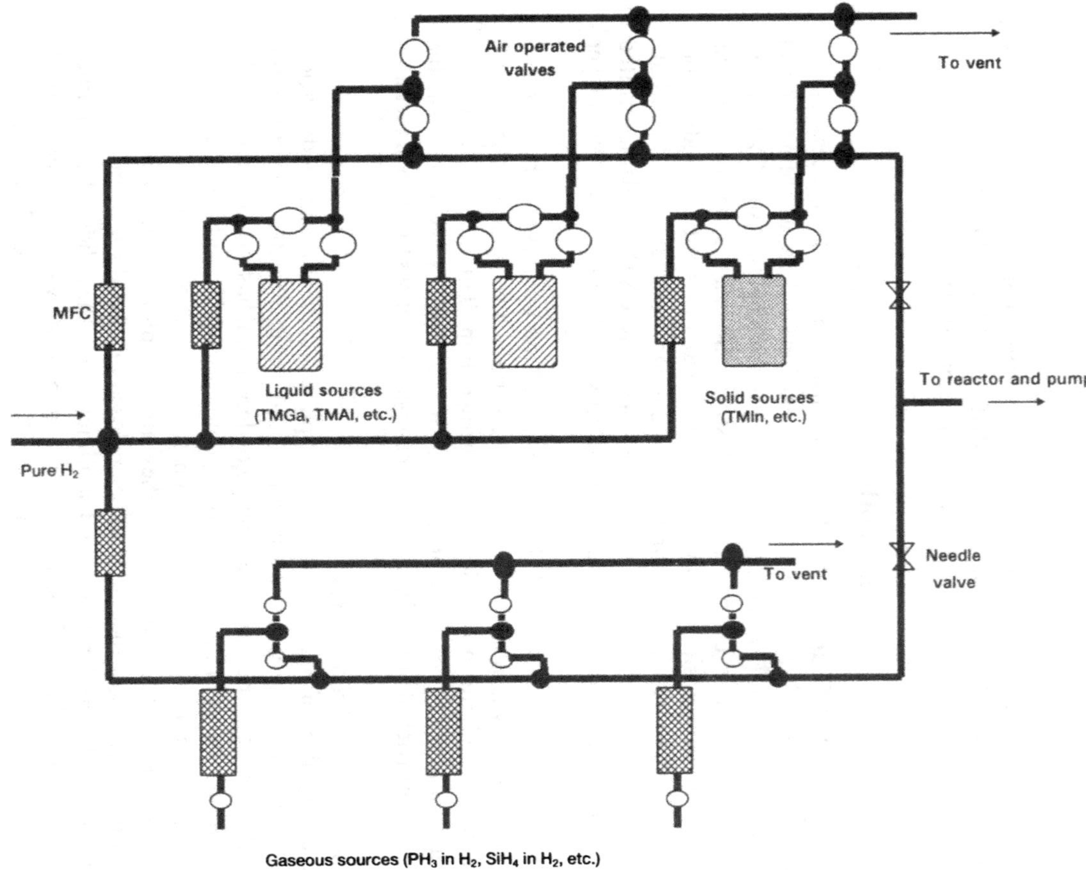

FIGURE 10.22. Apparatus needed for the effective confinement and use of organometallic precursors. In all cases safety is of prime concern.

TABLE 10.3 MOCVD Precursors

Element	Precursor	State*	Boiling point (°C)	Vapor pressure in torr (at temperature in °C)	Name
Al	Triisobutyl Al	L	130	1 (50)	TIBAL
	Triethyl Al	L	194	4 (80)	
As	Triethyl As	L	140	15.5 (37)	TEAs
Ga	Trimethyl Ga	L	55.7	65.4 (0)	TMGa
	Trimethyl Ga–Trimethyl As adduct	S	121		
In	Trimethyl In	S	133.8	7.2 (30)	TMIn
P	Trimethylphosphine	L	37.8		TMP
Sn	Tetramethyl tin	L	78	10 (−21)	TMSn

* L = liquid; S = solid.

vapor is transported from the containers to the reaction chamber in a carrier gas, often hydrogen or nitrogen, which is bubbled through the container.

Reaction between gases during transport needs to be prevented. If reaction between gases during transport produces volatile products, no harm is done, since the film deposition reaction still proceeds as desired. Nonvolatile reaction products can be prevented by the use of adducts which have lower vapor pressure than either of the two gases.[47] In general, the sources are used below ambient temperature in order to prevent condensation on the walls of tubing. However, if the source has a very low vapor pressure, tubes and all connections can be heated to avoid condensation on the walls. The reactors have controlled wall temperatures and the dissociation of the source occurs when it encounters the hot susceptor with the wafers. Table 10.3 lists some of the common MOCVD precursors and their properties.

Alkyls of group III elements and hydrides of group V elements are used as precursors in MOCVD. For GaAs, the reaction can be described as

$$(CH_3)_3Ga + AsH_3 \rightarrow GaAs + 3CH_4$$

Ignoring intermediate steps, this reaction can be generalized for reactions involving III–V compounds as

$$R_3M + EH_3 \rightarrow ME + 3RH \tag{10.4}$$

where M is a group III element, such as Ga or Al, and R is an organic group, such as C_nH_{2n+1}. E is P, As, or Sb.[49] The presence of more than one group III element results in a substitutional alloy of the form $Ga_xAl_{1-x}As$.

Impurities in the precursors can result in significant property changes in the semiconductor. For instance ppm levels of Zn or Si in TMG can lead

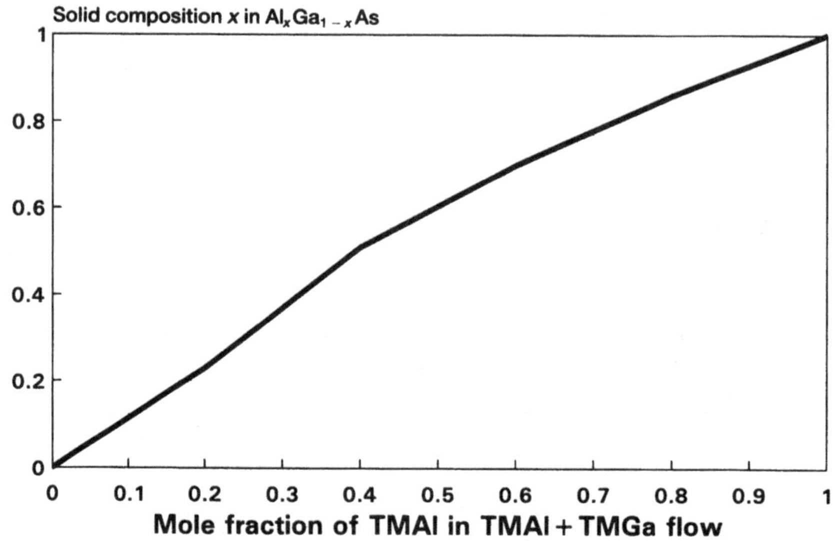

FIGURE 10.23. Concentration of Al in the solid as a function of the gas phase concentration of TMAl. Data from multiple temperatures and reactors. Reprinted from Ref. 50 with permission from Noyes Publications.

to highly compensated GaAs. Electronic and optical characteristics of the devices are the most sensitive tests for detecting impurities. Synthesis of the precursor involves classical purification techniques, such as distillation and sublimation, along with adduct purification schemes.

Film Growth from Organic Precursors

Growth of GaAs using the chemistry in equation (10.4) results in a rate that is directly proportional to the partial pressure of the group III alkyl and is independent of temperature and substrate orientation. It is also relatively insensitive to the concentration of AsH_3 above a threshold. The growth of $Al_xGa_{1-x}As$ has very similar parameters to the growth of GaAs. Simple substitution of TMAl for TMGa leads to inclusion of Al in the place of Ga. Figure 10.23 shows the variation of composition of the solid with the fractional concentration of TMAl in the feed gas.[50] At a given alloy composition, the growth rate is proportional to the total alkyl concentration in the reactor, and largely independent of substrate temperature.

Intentional doping of GaAs and $Al_{1-x}Ga_xAs$ with donor and acceptor atoms is accomplished by the addition of dopant gases along with the gas feed. Zn and Mg are the common *p*-type dopants and Si, Ge, and Sn are the *n*-type dopants. High vapor pressure dopants such as Zn are incorporated under mass transport control, whereas low vapor pressure dopants such as

Si are incorporated under surface reaction control. Alkyl precursors for IIA and IIB elements and hydrides for group IVA elements are commonly used.

The details of the actual partial reactions in MOCVD are still not well understood. It has been suggested that TMG decomposition starts in the gas phase with the loss of sequential methyl radicals. Arsine decomposes heterogeneously with the loss of hydrogen on the surface. If TEG is used, more complicated gas phase reactions involving the elimination of H and C_2H_5 radicals have been noticed. TMG adsorbs on the surface with the loss of a methyl radical. At higher temperatures (around 500°C), complete decomposition of the TMG occurs on the surface, leading to Ga droplets. No residual carbon is found on the surface.

The useful temperature range of GaAs growth is 600–850°C. Growth in this temperature range is mass transport controlled by the diffusion of TMGa through the boundary layer, in the presence of sufficient AsH_4. Even though the growth rate in this regime is unaffected by the partial pressure of AsH_4, many electrical properties of the film are modulated by the AsH_4/TMGa ratio. The free carrier background doping changes from *p*-type to *n*-type as this ratio is increased. As the growth pressure is dropped to below 0.5 torr, the concentration of background carbon increases and this has been suggested as the reason for carbon as the residual acceptor dopant in the film.

In the growth of both GaAs and $Al_xGa_{1-x}As$, substrate cleaning forms a crucial part of epitaxial growth. Even with the use of pregrowth, wet chemical cleaning, in situ cleaning prior to growth is essential. Substrates are heated in the growth ambient with flowing hydrogen to reduce the native oxide. In situ pregrowth anneal has also been used to remove native oxides.

One of the most important device properties affected by growth conditions is photoluminescence (PL). Poor PL intensity in $Al_xGa_{1-x}As$ is often due to incorporation of oxygen from moisture residual in the reactor. Use of molecular sieves in the AsH_3 line, or the growth of gettering layers prior to deposition have been useful in reducing the oxygen partial pressure in the reactor. Simultaneous presence of oxygen and aluminum produces deep traps in the semiconductor, also leading to poor PL intensity.

Film Characteristics

GaAs has been grown homoepitaxially, and heteroepitaxially in applications requiring sharp doping profiles, unobtainable by conventional diffusion or implantation techniques. Pseudomorphic growth where the substrate and film are lattice mismatched is also possible. The resulting film has a strained lattice which matches the substrate orientation. Even though the growth rate itself is not a strong function of the substrate crystallographic orientation, the morphological defect density of the epi layer strongly depends on the crystal orientation. (1 0 0) is the preferred growth plane and leads to the smallest defect density in the film. (1 1 1) films result in higher densities of stacking faults.

Heteroepitaxial layers of $Al_{1-x}Ga_xAs$ offer the advantage of *band-gap engineering*. Incorporation of Al in the GaAs lattice produces a very small change in the lattice parameter. However, the band gap is increased rapidly when Al is substituted by Ga. This allows us to grow films with the quality of homoepitaxy without extended misfit dislocations at the same time as altering an important characteristic of the semiconductor.

Pseudomorphic growth is usually limited in thickness, dictated by the maximum strain energy the distorted lattice can store. Above a certain thickness, defects such as dislocations begin to form to relieve the stored strain in the lattice.[51]

Reactor Modifications

CVD reactors described in Chapter 5 need to be modified to accommodate the requirements of the MOCVD precursors. The overall reactor consists of the precursor source (bubbler or sublimator), a getter or filter to remove trace moisture and oxygen, gas manifolds for mixing and metering gases, the reaction chamber and the exhaust. Diluent and carrier gases, such as H_2 and N_2, flow through these components, along with the reactants. For sub-atmospheric reactors, vacuum systems need to be included. The exhaust needs to be able to handle toxic and pyrophoric effluents from the reactor.

Common reaction chamber configurations are vertical and horizontal. In the vertical reactor, gases are introduced at the top of the reactor and flow normally to a susceptor, which holds the wafers. Radial uniformity is maintained by rotating the susceptor or by swirling the inlet gas flow. The horizontal reactors are similar in concept to the tilted susceptor reactor discussed in Exercise 5.1. The gas flow is parallel to the wafer, and gas depletion is compensated by increasing flow velocity. Other reactor configurations are the pancake and the barrel.

Similar to the conventional tilted susceptor reactor, the region immediately close to the susceptor develops a boundary layer through which the reactant species need to diffuse to get to the wafer surface. Above this layer is a well-mixed zone of gases, often turbulent due to thermal, free convection effects. If the reactant gases have widely different diffusivities through the boundary layer, concentration gradients in the axial direction will produce a compositional gradient in the solid, even with a tilt in the susceptor. Hence we need to match the reactants to have similar molecular weights to avoid composition changes. One method we have already seen to eliminate diffusion control is the use of lower pressures. Many commercial reactors operate at sub-atmospheric pressures, as low as 0.5 torr.

Residence times of the reactants in the chamber (including the section of manifolds downstream of isolation valves) have the effect of degrading abrupt junction profiles. High diluent flow rates decrease species residence time and the avoidance of local vortices and dead spots is crucial in obtaining sharp concentration profiles.

Wafers are heated by contact with the susceptor, which in turn is either inductively or radiatively heated. All reactors tend to develop a coating of As wall deposition.[52] Maintenance of electrical feedthroughs is difficult, hence the reactors are seldom resistively heated. Inductive heating requires the susceptor to be electrically conducting, so it is often made of graphite. The chamber walls are made of quartz, which is chemically inert to the organometallic reactants.

10.5. MOCVD OF OTHER SEMICONDUCTORS

So great is the array of binary, ternary and quartenary compounds, MOCVD-grown from elements in groups IIIA, IVA, and VA, it is impossible for us to deal with them individually. The reader is directed to excellent reviews such as the one by Ludowise[49] for detailed bibliographies on each of the compounds.

In general, the growth kinetics are similar to the case of $Al_{1-x}Ga_xAs$, and the precursors for each element are described in Table 10.3. In their useful ranges of deposition, the growth processes are mass transport controlled by the concentration of the group III alkyl.

In the case of indium-containing compounds, such as InP and $Ga_{1-x}In_xAs$, the reaction chemistry is complicated by parasitic reactions between the indium alkyls and the group V hydrides at room temperature in the gas phase.[53] Adducts have become a popular solution, since separation of the two species as a means of preventing the reaction also leads to film nonuniformity. The reactor design and contaminants in the source gases play key roles in determining the efficiency of incorporation of indium into the film from the gas phase, as opposed to the parasitic reaction. Indium-bearing ternaries and quarternaries, such as $Ga_{1-x}In_xAs$ and $Ga_{1-x}In_xP_yAs_{1-y}$, are important as fiber-optic materials and short wavelength LEDs.

Compounds containing antimony are unique because their growth conditions are not simple extensions of the other MOCVD compounds. TESb and TMSb are the preferred sources over the hydride but in excessive concentrations they lead to the formation of whiskers and droplets inside the reactor.

In general, the purity of the MOCVD source limits the purity of the epitaxial film produced. Along with a lack of understanding of the thermodynamics of some of these reactions, detailed chemical pathways through which film growth proceeds are not clearly known. These factors are limiting the controllability and the precision with which films are synthesized for optical and electronic applications.

References

1. F. Faggin, and T. Klein, *Solid State Electron.* **13**(8), 1125 (1970).
2. R. S. Muller, and T. I. Kamins, *Device Electronics for Integrated Circuits*, p. 443, John Wiley, New York, 1986.
3. S. M. Sze, *Physics of Semiconductor Devices*, 2nd Ed., John Wiley, New York, 1981.
4. M. E. Lunnon, and D.W. Greve, *J. Appl. Phys.* **54**, 3278 (1983).
5. T. Kamins, *Polycrystalline Silicon for Integrated Circuit Applications*, p. 218, Kluwer Academic, Boston, 1988.
6. U. S. Davidsohn, and F. Lee, *Proc. IEEE* **57**, 1532 (1969).
7. R. D. Rung, H. Momose, and Y. Nagakubo, *IEEE IEDM Techn. Dig.* 236 (1982).
8. R. S. Rosler, *Solid State Technol.* **21**(4), 63–70 (1977).
9. C. H. J. van den Brekel, and L. J. M. Bollen, *J. Cryst. Growth* **54**, 310 (1981).
10. M. L. Hitchman, J. Kane, and A. E. Widmer, *Thin Solid Films* **59**, 231 (1979).
11. T. Kamins, *Polycrystalline Silicon for Integrated Circuit Applications*, p. 27, Kluwer Academic, Boston, 1988.
12. B. S. Meyerson, and M. L. Yu, *J. Elecrochem. Soc.* **131**(5), 2366 (1984).
13. L. H. Hall, and K. M. Koliwad, *J. Electrochem. Soc.* **120**(10), 1438 (1973).
14. F. Moghadam, Intel Corporation, private communications.
15. R. S. Rosler, *Solid State Technol.* **21**(4), 63–70 (1977).
16. M. Matsui, Y. Shiraki, and E. Maruyama, *J. Appl. Phys.* **53**(2), 995 (1982).
17. T. I. Kamins, *J. Electrochem. Soc.* **127**(3), 686 (1980).
18. S. Nakayama, I. Kawashima, and J. Murota, *J. Electrochem. Soc.*, **133**(8), 1721 (1986).
19. P. S. Burggraff, *Semiconductor International*, Oct. 1983, pp. 45–51.
20. W. R. Runyan, and K. E. Bean, *Semiconductor Integrated Circuit Processing Technology*, p. 295, Addison-Wesley, Reading, Mass., 1990.
21. G. R. Srinivasan, *Solid State Technol.* **24**(11), 101 (1981).
22. P. S. Burggraff, *Semiconductor International*, Oct. 1983, pp. 45–51.
23. N. Weste, and K. Eshraghian, *Principles of CMOS VLSI Design*, p. 58, Addison-Wesley, Reading, Mass., 1985.
24. D. S. Yaney, J. T. Nelson, and L. L. Vanskike, *IEEE Trans. Electron Devices* **ED26**(1), 10 (1979).
25. J. Bloem, and L. J. Giling, in *VLSI Electronics Microstructure Science*, Vol. 12, (Einspruch and Huff, eds.), p. 91, Academic Press, Orlando, Fla., 1985.
26. W. A. P. Claassen, and J. Bloem, *Phillips J. Res.* **36**, 124 (1981).
27. J. Bloem, and L. J. Giling, in *VLSI Electronics Microstructure Science*, Vol. 12, (Einspruch and Huff, eds.), p. 117, Academic Press, Orlando, Fla., 1985.
28. J. Bloem, L. J. Giling, and M. W. M. Graef, *J. Electrochem. Soc.* **121**, 1354 (1974).
29. H. B. Pogge, in *Handbook of Semiconductors* (Keller, ed.), Vol. 3, p. 335, North Holland, Amsterdam, 1980.
30. M. L. Hammond, *Solid State Technol.* **21**(11), 68 (1978).

31. K. E. Bean, W. R. Runyan, and R. G. Massey, *Semiconductor International,* May 1985, p. 136.
32. W. R. Runyan, and K. E. Bean, *Semiconductor Integrated Circuit Processing Technology,* p. 305, Addison-Wesley, Reading, Mass., 1990.
33. W. Guth, *Phys. Status Solidi b* **51**, 143 (1972).
34. K. V. Ravi, C. J. Varker, and C. E. Volk, *J. Electrochem Soc.* **120**, 533 (1973).
35. K. V. Ravi, *Imperfections and Impurities in Semiconductor Silicon*, John Wiley, New York, 1981.
36. K. E. Bean, W. R. Runyan, and R. G. Massey, *Semiconductor International,* May 1985, p. 136.
37. S. P. Weeks, *Solid State Technol.* **24**(11), 111 (1981).
38. A. S. Salih, H. J. Kim, R. F. Davis, and G. A. Rozgonyi, *Appl. Phys. Lett.* **46**(4), 419 (1985).
39. C. Y. Tan, *Appl. Phys. Lett.* **30**, 175 (1977).
40. J. Manoliu, *Semiconductor International*, April 1988, pp. 90–92.
41. R. D. Dupuis, *Proc. Electrochem. Soc.* **83**, 175 (1983).
42. M. J. Howes and D. V. Morgan, *Gallium Arsenide*, John Wiley, New York, 1985.
43. J. R. Knight, *Solid State Electronics,* **8**, 178 (1965).
44. J. C. Hong, and H. H. Lee, *J. Electrochem. Soc.* **132**, 427 (1985).
45. H. M. Manasevit, *Appl. Phys. Lett*, **12**, 156 (1968); also see M. J. Ludowise, *J. Appl. Phys.* **58**(8), R31 (1985) for a complete set of papers by Manasevit.
46. K. F. Jensen, *J. Crystal Growth* **98**, 148 (1989).
47. K. W. Benz, H. Renz, J. Wiedlein, and M. H. Pilkuhn, *J. Electron. Mater.* **10**, 185 (1981).
48. T. F. Kuech, and K. F. Jensen, in *Thin Film Processes II* (Vossen and Kern, eds.), p. 378, Academic Press, New York, 1991; M. J. Ludowise, *J. Appl. Phys.* **58**(8), R33 (1985).
49. M. J. Ludowise, *J. Appl. Phys.* **58**(8), R31 (1985).
50. J. L. Zilko, in *Handbook of Thin Film Deposition Processes and Techniques* (Schuegraf, ed.), p. 234, Noyes Publications, Park Ridge, N.J. (1988).
51. R. H. Moss, and P. C. Spurdens, *J. Cryst. Growth* **68**, 96 (1984)
52. See bibliography of Indium compounds in M. J. Ludowise, *J. Appl. Phys.* **58**(8), R31 (1985).
53. C. B. Cooper, III, R. R. Saxena, and M. J. Ludowise, *J. Electron. Mater.* **11**, 1001 (1982).

Chapter 11

Emerging CVD Techniques

The bulk of the discussion so far has concentrated on the two main modes of CVD film growth for microelectronics: thermal and plasma CVD. However, to suit novel applicatons, new CVD techniques are being developed where the energy for the forward progress of the CVD reaction is supplied by sources other than heat and electrical power. In these techniques, the reactant molecules are raised to excited states by direct absorption of energies from sources including photons, electrons and ions.

Emerging techniques such as photochemical CVD, laser CVD and focused ion beam CVD, have been used for the deposition of conducting, semiconducting and insulating films. They tend to avoid the broad disadvantages of thermal and plasma CVD, such as temperature limitations in thermal CVD and radiation and ion bombardment related effects in plasma CVD. These techniques tend to be used in niche applications rather than general volume manufacturing; applications include mask repair, post fabrication circuit modification, and connections in certain types of gate arrays. As the scope of their acceptance in the mainstream suggests, each technique has its own special advantages and disadvantages.

Due to their wide range and rapidly changing nature, we will choose for our study some of the more mature techniques. We will illustrate the technique, its advantages and its potential applications through the deposition of a representative film. There are vast bibliographies for more detailed information on individual film depositions using these techniques.

11.1. PHOTOCHEMICAL CVD

Photochemical CVD is different from laser CVD, discussed in the next section. It utilizes a broadbeam ultraviolet source such as a mercury vapor lamp to excite gaseous species prior to deposition.[1] Its advantages are a reduction in deposition temperature and the absence of any charged species. Its disadvantage is a relatively low deposition rate.

There are two major photo-CVD mechanisms; direct photolysis of gaseous reactants in the presence of UV irradiation (laser CVD is an extension of this technique), and more effective energy absorption using mercury vapor as a sensitizer to absorb energy from the incident irradiation. Using a sensitizer produces almost an order of magitude increase in deposition rate compared with direct photolysis.

Following the work of Peters et al.[1] let us study the deposition of silicon nitride using mercury vapor sensitized photo-CVD. Photochemical CVD, by definition, proceeds only if at least one of the reactants has absorbed energetic photons. The low limit in the emission spectrum of the commonly used quartz-mercury vapor lamp is 2 000 Å. In the reaction to form Si_3N_4 from ammonia and silane, neither of the reactants absorb UV radiation significantly above 2 200 Å. To overcome the relatively narrow wavelength range in the spectrum that will be absorbed by the reactants, mercury vapor is added to the reactants to a partial pressure of approximately 1 millitorr. Nearly all the irradiated photons are absorbed by the mercury vapor which, in turn, excites the gas phase reactants SiH_4 and NH_3 through gas phase collisions. The sequence of reactions is as follows:

$$h\nu + Hg \rightarrow Hg^*$$

$$Hg^* + SiH_4 \rightarrow SiH_4^* + Hg$$

$$Hg^* + NH_3 \rightarrow NH_3^* + Hg$$

$$3SiH_4^* + 4NH_3^* \rightarrow Si_3N_4 + 12H_2$$

All the reactions occur at room temperature or slightly elevated temperatures, much lower compared with thermal CVD. The gaseous excited species are all neutrals without any charged particle generation.

The same setup has been used for the deposition of SiO_2 from N_2O and SiH_4 using Hg sensitization. Using photochemically enhanced nitride, oxynitride, and oxide formation techniques, MOS isolation and passivation schemes have been proposed.[2] Growth rates of oxide on the order of 15 nm/min on 150 mm wafer areas (where the irradiation window area is on the order of 650 cm^2) have been achieved. Figure 11.1 shows the schematic of a reactor used for oxide deposition designed by Hughes Research Laboratory.

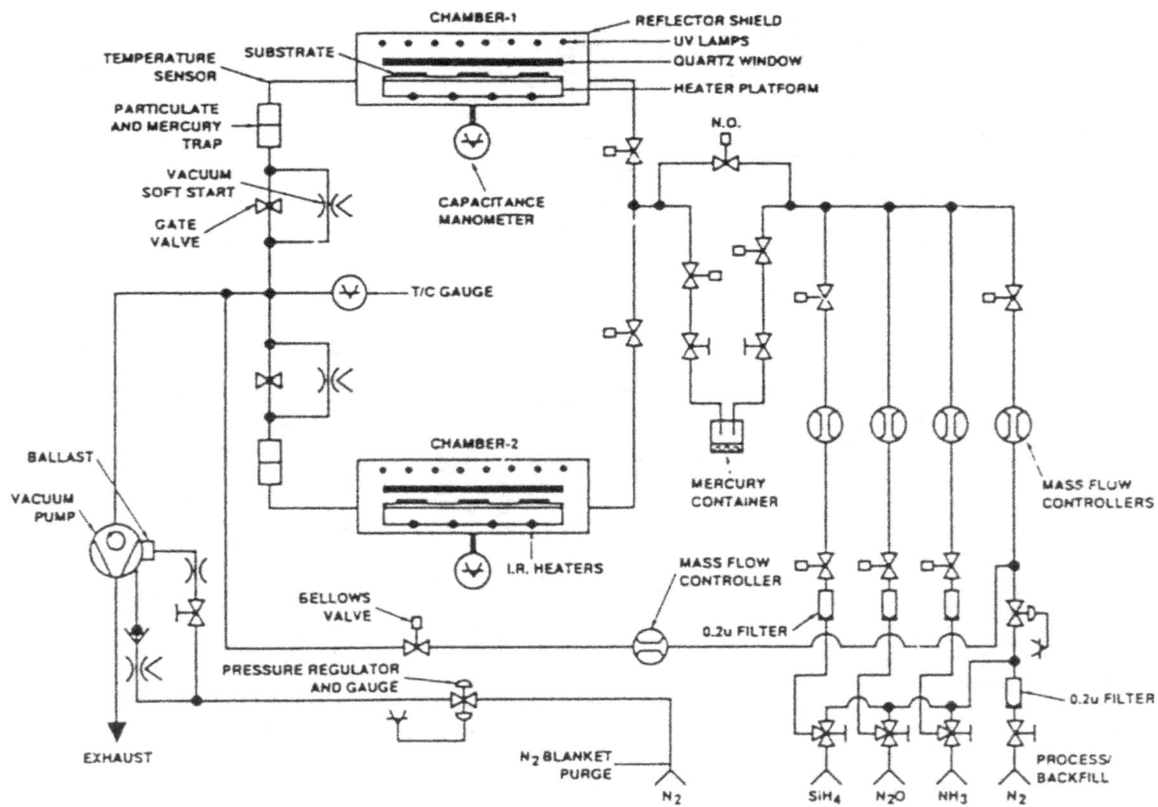

FIGURE 11.1. Setup by Hughes Research for depositing dielectric films using mercury vapor enhanced photo CVD. Reprinted from Ref. 1 with permission from *Solid State Technology*.

Similar processes have been used for the deposition of doped and undoped silicon films from silane–again the rate is very low–and from disilane, where the rate is close to being acceptable.[3]

11.2. LASER CVD

Laser CVD (LCVD) differs from photolysis because it uses a narrow, coherent, monochromatic light source instead of a wide-band mercury vapor lamp. Laser enhanced evaporation of a reactant species and substrate heating also fall within the definition of LCVD. Since laser photons have a narrow distribution of energy (0.1–7 eV, with >7 eV available in commercial UV lasers), selective excitation of a species to a specified excited state is possible.[4] Figure 11.2 shows the distribution of energies between conventional thermal CVD and a laser beam.

The reaction volume in the case of conventional reactors is the complex shape of the reaction chamber, whereas laser CVD possesses a defined and controllable reaction volume. The reaction volume is defined by the cylinder of the laser beam interacting with the gas phase. The advantages of controlling reaction volume are utilized in applications that involve direct writing and

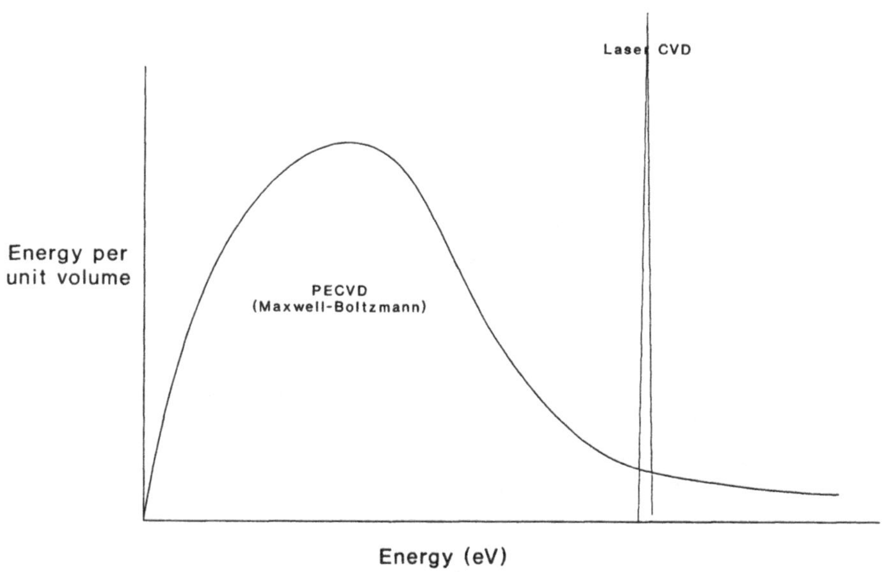

FIGURE 11.2. Laser CVD shows a narrow spectrum of input photon energies compared to plasma CVD. The laser helps minimize reaction volume. Adapted from Ref. 4 with permission from *Solid State Technology*.

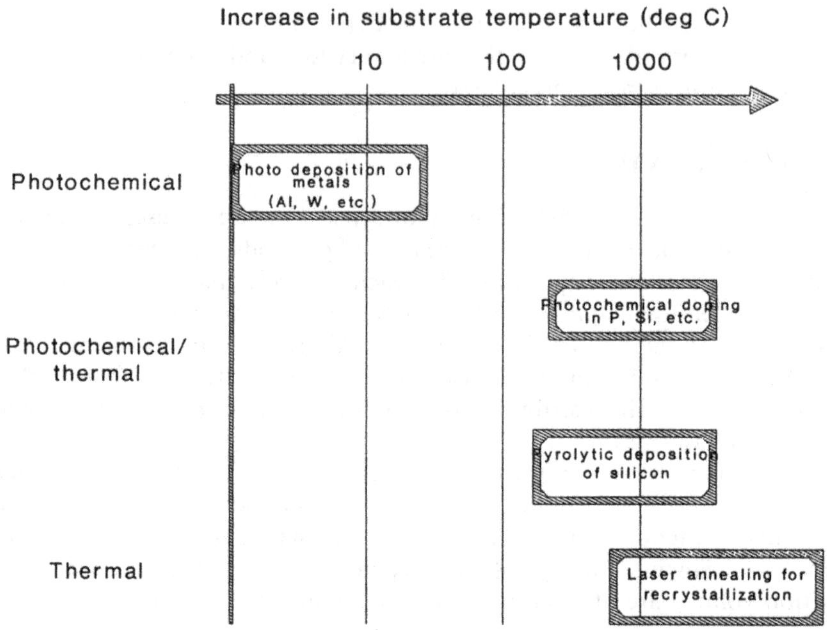

FIGURE 11.3. Operative mechanisms in laser CVD as the substrate temperature is increased. Direct heating by the laser beam leads to new applications at higher temperatures where gas phase photon absorption leads to reaction at lower temperatures.

circuit modifications. Other advantages include reduced particulate generation, since there is little deposition on chamber walls.

Both pyrolytic and photolytic mechanisms can be utilized for deposition from lasers. Pyrolytic mechanisms involve controlled substrate heating whereas photolytic mechanisms use a photochemical process.[5] Figure 11.3 illustrates the operative mechanisms as a function of the temperature increase of the substrate.

The two major applications of laser CVD are in flood-exposed nucleation and direct writing. We discussed the deposition of aluminum films using prenucleation in Chapter 4. Using wide area laser beams various conductive films have been deposited from relatively conventional precursors. Table 11.1 compares the deposition characteristics of various metallic films deposited by broadbeam LCVD.

Laser direct writing is popular both in mask repair and in post fabrication device modification.[7] A pulsed Ar^+ laser can be used to directly deposit polysilicon links between Si and Al gap structures. Figure 11.4 illustrates the

TABLE 11.1 **Properties of LCVD Metallic Films**

Film	Precursor	Laser power density (W/cm^2)*	Resistivity Film	Resistivity Bulk	Deposition rate (Å/min)
Mo	MoF_6, $Mo(CO)_6$	3	36	5.2	2 500
W	WF_6, $W(CO)_6$	3	40	5.6	1 700
Cr	$Cr(CO)_6$	3	56	12.9	2 000
Al	TMA, AlI_3	1.6	9	2.6	500

* Power density at 248 nm.

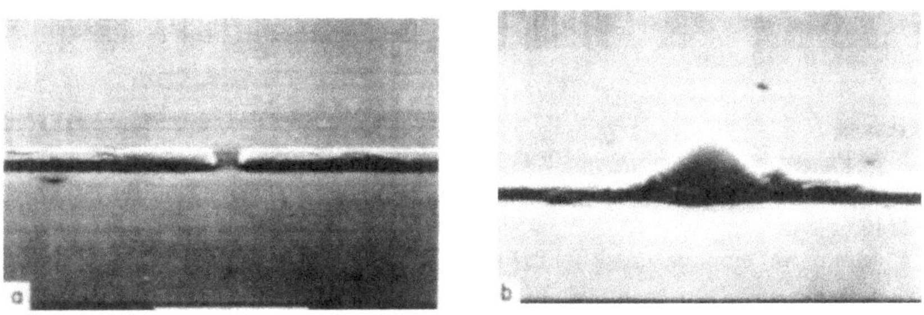

FIGURE 11.4. A typical application of direct laser CVD spot deposition. A connection is made between the two poly line sections by a doped polysilicon link. Reprinted from Ref. 7 with permission from Academic Press.

application. The precursors used in this example are $B(CH_3)_3$ and SiH_4. Except for spot depositions, continuous fine-line deposition is generally throughput-limited and is not seriously considered for volume manufacture.

11.3. FOCUSED ION BEAM AND ELECTRON BEAM CVD

Extending the concept of controlled reaction volume, we can expect a controlled volume plasma deposition involving charged species to be effective in eliminating large volume and wall effects. An electron beam CVD setup, for instance, produces a sharply defined sheet of plasma, about 20–25 mm wide and 1–2 mm above the substrate.[8] The electron beam is produced from an electron gun and this configuration has been used in the production of SiO_2 films from N_2O and 5% SiH_4 in N_2.[9] The deposition rate is not a strong function of the electron beam current density and the substrate temperature for deposition is comparable with conventional plasma CVD. The dependence on electron density is to be expected, since radical excitation is not limited by electron collison. However, confinement of the plasma to such a small volume reduces unwanted wall interactions and energetic ion bombardment seen in plasma CVD.

Akin to laser direct writing, a focused beam of ions can also be used for deposition.[10] Since it is possible to focus ions to a finer spot than is possible with photons, such a technique can be used for fine-line direct deposition. Moreover, focused ion beams are already in use in mask repair, where they etch away unwanted bridges in the mask chrome.[11] The focused ion beam can be used locally to excite precursors, resulting in fine-line CVD. So far, depositions of aluminum from TMA, tungsten from WF_6, carbon from certain hydrocarbons and gold from dimethyl gold hexafluoroacetylacetonate have been reported.[12] The focused ion beam is usually Ga^+ ions with a Gaussian profile beam. The average energy is about 15 eV. The main disadvantages of the method are the low throughput and the presence of high concentrations of carbon in the film.

References

1. J. W. Peters, F. L. Gebhart, and T. C. Hall, *Solid State Technol.* **23**(9), 121 (1980).
2. J. Y. Chen, R. C. Henderson, J. T. Hall, and J. W. Peters, *J. Electrochem. Soc.* **131**(9), 2146 (1984).
3. T. Inoue, M. Konagai, and K. Takahashi, *Appl. Phys. Lett.* **43**(8), 774 (1983).
4. R. Solanki, C. A. Moore, and G. J. Collins, *Solid State Technol.* **28**(6), 220 (1985).
5. D. J. Ehrlich, and J. Y. Tsao, *J. Vac. Sci. Technol.* **B1**, 969 (1983).
6. R. Solanki, P. K. Boyer, and G. J. Collins, *Appl. Phys. Lett.* **41**, 1048 (1983).
7. D. J. Ehrlich, and J. Y. Tsao, in *VLSI Electronics* (Einspruch, ed.) Vol. 7, Ch. 3, Academic Press, New York, 1983.
8. L. R. Thompson, J. J. Rocca, K. Emery, P. K. Boyer, and G.J. Collins, *Phys. Lett.* **43**, 777 (1983).
9. K. Emery, et al., Proc. Materials Research Society Symposium, Boston, Mass., 1983.
10. G. M. Shedd, H. Lezec, A. D. Dubner, and J. Melngailis, *Appl. Phys. Lett.* **49**(23) 1584 (1986).
11. J. Melngailis, C. R. Musil, E. H. Stevens, M. Utlaut, E. M. Kellogg, R. T. Post, M.W. Geis, and R. W. Mountain, *J. Vac. Sci. Technol.* **B4**, 176 (1986).
12. K. Gamo, N. Takakura, N. Samoto, R. Shimizu, and S. Namba, *Jpn. J. Appl. Phys.* **23**, L293 (1984).

Appendix

Vacuum Techniques for CVD

The operating pressures for both low pressure and plasma enhanced CVD vary between 100 millitorr and several hundred torr. Some emerging techniques such as molecular beam CVD operate under high vacuum conditions. However, in order to minimize the system background gases (in particular oxygen, moisture and fluorine-bearing species), CVD reactors are often pumped to many orders of magnitude lower than the operating pressure, often as low as 10^{-7} torr. Even though this is quite a high pressure compared with modern ultrahigh vacuum systems, design of the vacuum system is an integral part of reactor design.

A.1. FUNDAMENTALS OF VACUUM SYSTEM DESIGN

Consider a vacuum pump removing gas from an enclosed chamber through a short tube connecting the pump to the chamber. If A is the area of cross section at the mouth of the pump, and v the velocity of the gas through the cross section, pumping speed S is defined as

$$S = Av \tag{A.1}$$

S can also be defined as the volumetric removal rate of the gas from the chamber and be written as

$$S = dV/dt \tag{A.2}$$

Commonly used units for pumping speed are liters per second or cubic feet per minute (CFM). Note that S is defined independent of pressure, even though it can drop off significantly near the ultimate pressure limit for an actual pump.

The throughput Q of a pumping system is defined as the product of the pumping speed and the inlet pressure

$$Q = PS \tag{A.3}$$

Since S is defined independent of pressure, the product PS carries units such as torr liters/s or sccm/s. From the definition of S in terms of gas velocity and P in terms of molecules per unit volume, Q can be written as

$$Q = kTdN/dt \tag{A.4}$$

where dN/dt is the rate of removal of molecules from the volume V. The throughput of a serial system at steady state has to be constant across the various elements of the vacuum system. In Figure A.1, it is constant at the neck of the chamber being pumped, through the pump and into the exhaust.

Q might consist of several components of gas load that need to be pumped. These include: (a) the gas originally in the chamber, (b) gas desorbing from the chamber walls, (c) leaks or other external intentional gas introductions such as through a mass flow controller, (d) gas reemitted or backstreaming from the pump. The pumping speed of the pump that fits in a vacuum system is determined by the throughput that needs to be evacuated and the time in

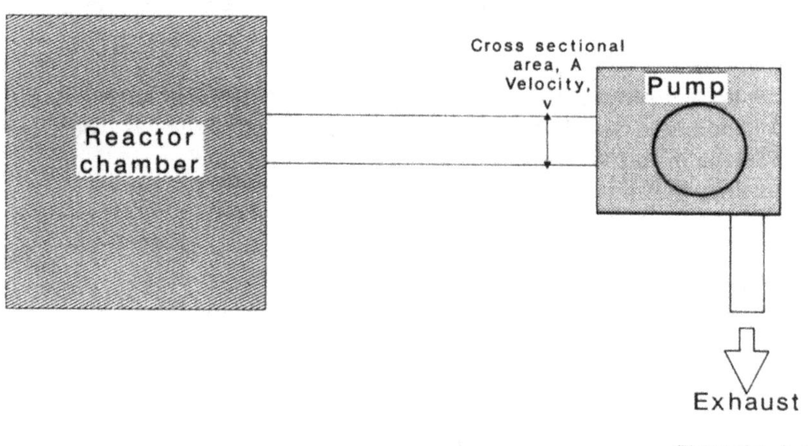

FIGURE A.1. Schematic showing a pump exhausting a reactor chamber through a connecting pipe. The outlet of the pump is to atmosphere.

which the ultimate pressure P_u is to be reached. From equation (A.3), the pumping speed needed can be written as

$$S = \frac{\sum Q}{P_u} \tag{A.5}$$

Integrating this equation, with proper substitutions, we can arrive at an expression for instantaneous pressure in the system.

$$P = P_0 \exp(-St/V) + P_u \tag{A.6}$$

The system will reduce its pressure by about 67% in a time S/V. The larger the pumping speed, the faster the rate at which the ultimate pressure is reached. Hence the size of the pump needs to be slightly larger than to handle precisely the gas load, so that pumping can be done in a reasonable time.

P_u is not a function of pumping speed. It is an intrinsic characteristic of the pumping system itself. We will discuss this further when we examine actual hardware.

The impedance offered by any element of a vacuum system is often characterized in terms of its conductance C, which is defined as

$$C = Q/(P_i - P_o) \tag{A.7}$$

where P_i and P_o are the pressures upstream and downstream of the element. It is analogous to electrical conductance (Q is analogous to current and the pressure difference to potential difference) and has the same units as pumping speed. Conductance is a strong function of the regime of flow under which the system is operating. For instance, C for any cross section is highest in the viscous flow regime and lowest in the molecular flow regime. The conductance of a circular tube in viscous flow can be derived as

$$C_v = \frac{\pi D^4}{128 \eta L} P_{ave} \tag{A.8}$$

Notice the tube diameter D very strongly influences the conductance of the tube. L is the length of the tube and η is the viscosity of the fluid. Also, at low average pressures, the flow converts from being viscous to transition flow or molecular flow and equation (A.8) is no longer valid. From equation (A.8) the throughput during viscous flow through a circular pipe is given by

$$Q = C(P_1 - P_2) = \frac{\pi D^4}{128 \eta L} P_{ave}(P_1 - P_2) \tag{A.9}$$

When the mean free path of the gas is on the order of the tube diameter, a new component of the flow along the tube walls becomes important. Roughness of the chamber walls can determine if the molecules impinging on the wall are reflected back to the gas stream or suffer multiple collisions on the rough walls, resulting in a net flow along the wall. This region of flow, as we have seen in Chapter 5, is called the transition flow regime. A term, χ, called the coefficient of slip, corresponding to reflections of the molecules along the rough tube walls is defined as

$$\chi = \frac{2}{3}\lambda\frac{2-f}{f} \tag{A.10}$$

where $f = 0$ for specular reflection from the walls and $f = 1$ for diffuse reflections from rough walls. λ is the mean free path of the gas. The throughput Q_{tr} in transition flow can then be written as

$$Q_{tr} = \left(\frac{\pi D^4}{128\eta L} + \frac{\pi D^3\chi}{64\eta L}\right)P_{ave}(P_1 - P_2) \tag{A.11}$$

Notice the throughput (and hence the conductance C_u) contains two terms. The first corresponds to viscous flow, the same as equation (A.8). The second includes the coefficient of slip and corresponds to flow along the surface.

When the pressure drops further and the mean free path is much larger than the tube diameter, the flow becomes molecular. In this case tube length is of particular importance. A tube of finite length has to be characterized as a tube in series with an orifice, since there is a definite probability that a molecule in the short tube can jump back into the chamber through the mouth of the tube. The conductance C_m of the tube and the orifice can be written as follows

$$C_{m,t} = \frac{D^3}{6L}\sqrt{\frac{2\pi kT}{m}} \tag{A.12}$$

and

$$C_{m,o} = \frac{\pi D^2}{16}\sqrt{\frac{8kT}{\pi m}} \tag{A.13}$$

The overall conductance is then written as the sum of the two conductances in series as

$$\frac{1}{C_{\mathrm{m}}} = \frac{1}{C_{\mathrm{m,t}}} + \frac{1}{C_{\mathrm{m,o}}} \tag{A.14}$$

in an analogous fashion to electrical conduction.

For the system shown in Figure A.1, the net pumping speed at the inlet to the chamber can be written in general terms as

$$\frac{1}{S} = \frac{1}{S_{\mathrm{p}}} + \frac{1}{C} \tag{A.15}$$

where S is the pumping speed of the pump and C is the conductance of the other elements in the system. It can thus be seen that if the rate of pressure drop in the chamber is being limited by the conductance of the various elements, increasing the pumping speed through the use of a larger pump does not lead to a shorter pump downtime. We need to identify the element whose conductance is limiting the pumping speed in order to decrease pumping time.

Some pumps used in vacuum system cannot exhaust to the atmosphere. Hence they need to be used in conjunction with other pumps which can exhaust to the atmosphere. This is termed pump staging. Since the throughput of such a system has to remain constant, the following equality can be written:

$$S_1 P_1 = S_2 P_2 = S_3 P_3 \tag{A.16}$$

or

$$\frac{P_2}{P_1} = \frac{S_1}{S_2}, \text{ etc.} \tag{A.17}$$

Suppose the first pump is a turbomolecular pump, which can pump at very low pressures but cannot exhaust to the atmosphere. And suppose the second pump is a roughing pump such as a rotary vane pump, which can exhaust to atmosphere but has a high ultimate pressure. If the first pump has a high outlet-to-inlet pressure ratio, the pumping speed of the second pump can be relatively small. When staging a pumping system, we need to make sure that all the pumps in series can handle the overall system throughput at their respective inlet and exhaust pressures.

A.2. TYPICAL HARDWARE CONFIGURATIONS

Any vacuum system contains the following typical components: the enclosure that is being evacuated, the pump, gauges for pressure and other system properties, feedthroughs to communicate with the inside of the system mechanically and electrically, and valves to isolate various components. Since the explosion in the use of high vacuum systems for many applications, many different kinds of pumps, gauges and feedthroughs have become commercially available. In this section, we will discuss some typically used pumps and gauges which find wide applications in CVD reactors.

A.2.1. Rotary Vane Pumps

Rotary vane pumps are the most common of the mechanical pumps which use the compression and expansion of the gas to achieve pumping. Figure A.2 shows the operating principle of the pump, which consists of a cylinder rotating eccentrically within a cylindrical block. Spring-loaded vanes projecting from the rotor provide the vacuum sealing from the inlet to the exhaust. A film of oil lowers the friction between the vanes and the stator,

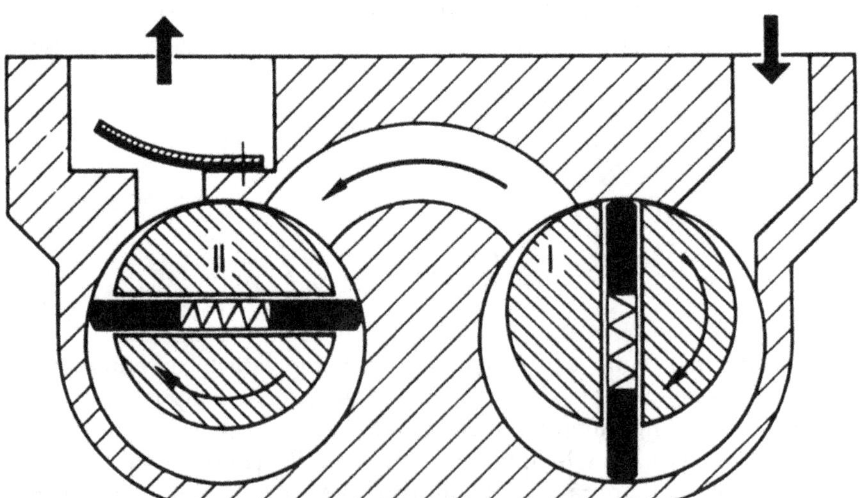

FIGURE A.2. Schematic of two-stage rotary vane pump from Leybold-Heraeus. Reprinted with permission.

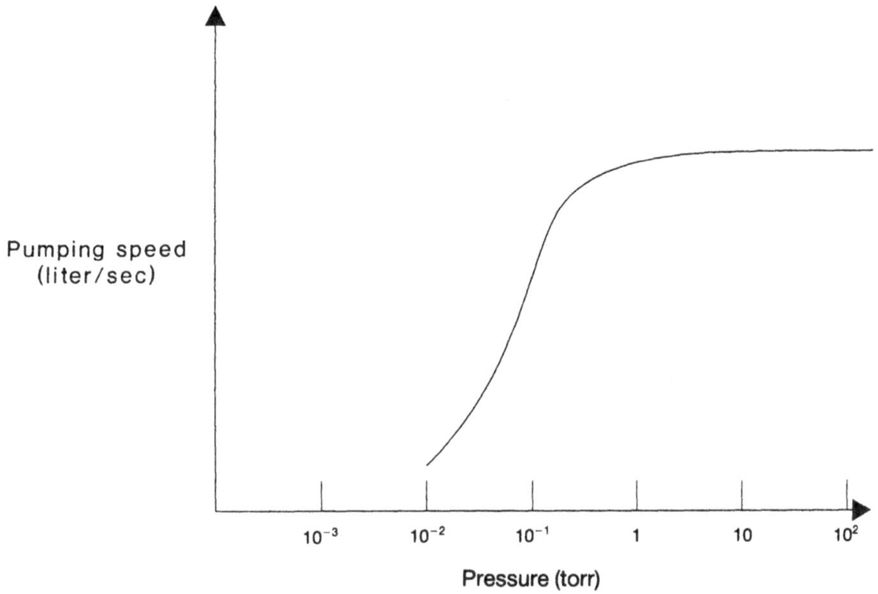

FIGURE A.3. Typical pumping speed characteristics of a single-stage rotary pump. The pumping speed drops when the pressure approaches the ultimate pressure of the pump.

and assists in the establishment of the seal. As the pump turns, a volume of gas trapped between the rotor and the stator is compressed to above atmospheric pressure and is exhausted through the outlet port.

The ultimate pressure of the pump is determined by the dead volume near the exhaust port. Pumping stops when the inlet pressure is so low that even compression into the dead volume does not raise it above atmospheric pressure. Figure A.3 shows the pumping speed of the typical single-stage, rotary pump as a function of pressure. Notice the ultimate pressure attainable with a mechanical pump is on the order of a few millitorr. However, continuous pumping at low pressures can lead to oil backstreaming into the system, so it is not advisable to use mechanical pumps for long periods at pressures under 50 millitorr.

When pumping condensable vapors such as water vapor (or air), at low pressures, compression can lead to condensation of the vapor. To prevent this, nitrogen is often added to the exhaust side of the rotor as ballast. This forms a gas mixture outside the condensation range.

Mechanical pump oils can be broadly classified as hydrocarbon based and perfluoropolyether (PFPE) based. Whereas the hydrocarbon oils are cheaper, they can react violently with oxygen under compression. Where toxic and pyrophoric gases along with air are pumped from the same chamber, the more expensive PFPEs are recommended for safety.

Recently a class of pumps called oil-free mechanical pumps or dry pumps has become commercially available. These are designed with precise dimensional tolerance to avoid the use of the oil film. This is advantageous for eliminating oil backstreaming into the chamber, for disposal of toxic contaminated oil and for the general cleanliness of the pump. However, these pumps are more expensive and they lack the natural filtration of gases offered by the oil film. Hence the exhaust system for dry pumps needs to account for additional solid residues that get deposited on the exhaust pipes.

A.2.2. Roots Pumps

Roots pumps or blowers represent the next level of vacuum from the rotary vane pumps. A Roots pump consists of two rotary lobes, as shown in Figure A.4, which form a close-fitting pair rotating in opposite directions. The lobes do not actually touch each other. The small mechanical clearance between the two lobes determines both the upper and ultimate pressure limits of the pump. At the high end, very high gas volumes can cause excessive heating, so the pump has an upper pressure limit of a few hundred torr. The backstreaming of gas through the gap determines the lower pressure limit, which can be as low as 10^{-6} torr.

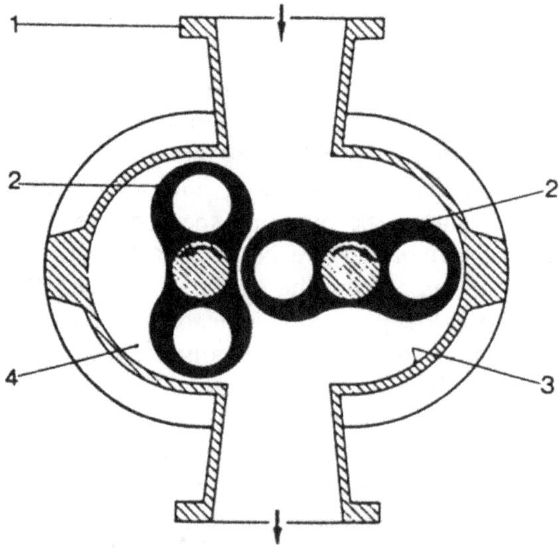

FIGURE A.4. Cross-section of a single-stage Roots blower from Leybold-Heraeus. The pump consists of lobes (2) rotating in a casing (3), compressing a volume of gas (4) to the exhaust. Reprinted with permission.

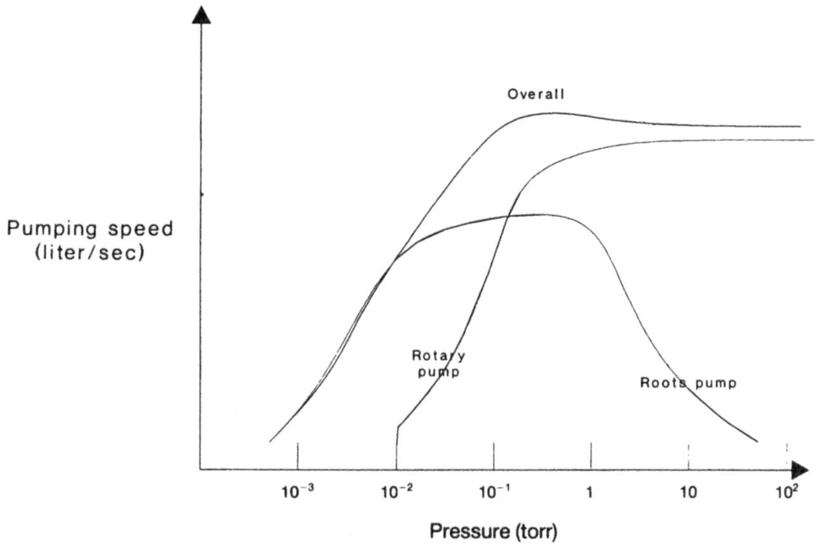

FIGURE A.5. A Roots blower–vane pump tandem achieves the ultimate pressure of the blower, while maintaining an overall higher pumping speed.

Since the two lobes do not physically touch each other, high speeds of rotation are possible, resulting in large pumping speeds. A Roots blower–vane pump combination is very popular as an LPCVD pumping package. Figure A.5 shows the pumping speed characteristics of this tandem as a function of pressure.

A.2.3. Turbomolecular Pumps

Turbomolecular pumps (TMP), axial flow molecular turbines, or molecular drag pumps have become increasingly popular for CVD applications in the last several years. TMPs (shown in Figure A.6) offer extremely high compression ratios, need little maintenance and can pump to pressures as low as 10^{-10} torr. However, in general, they are expensive due to the precision machining required.

TMPs act on the principle of imparting directional velocity to gas molecules in the molecular flow regime when they strike a fast-moving surface. They consist of a rotary turbine moving inside a stator housing at a very high rotation rate, as much as 50 000 rpm. The turbine blades are angled to direct the molecules toward stationary plates on the stator (Figure A.6), and are staged in nine or ten blade-stator combinations, each one having a

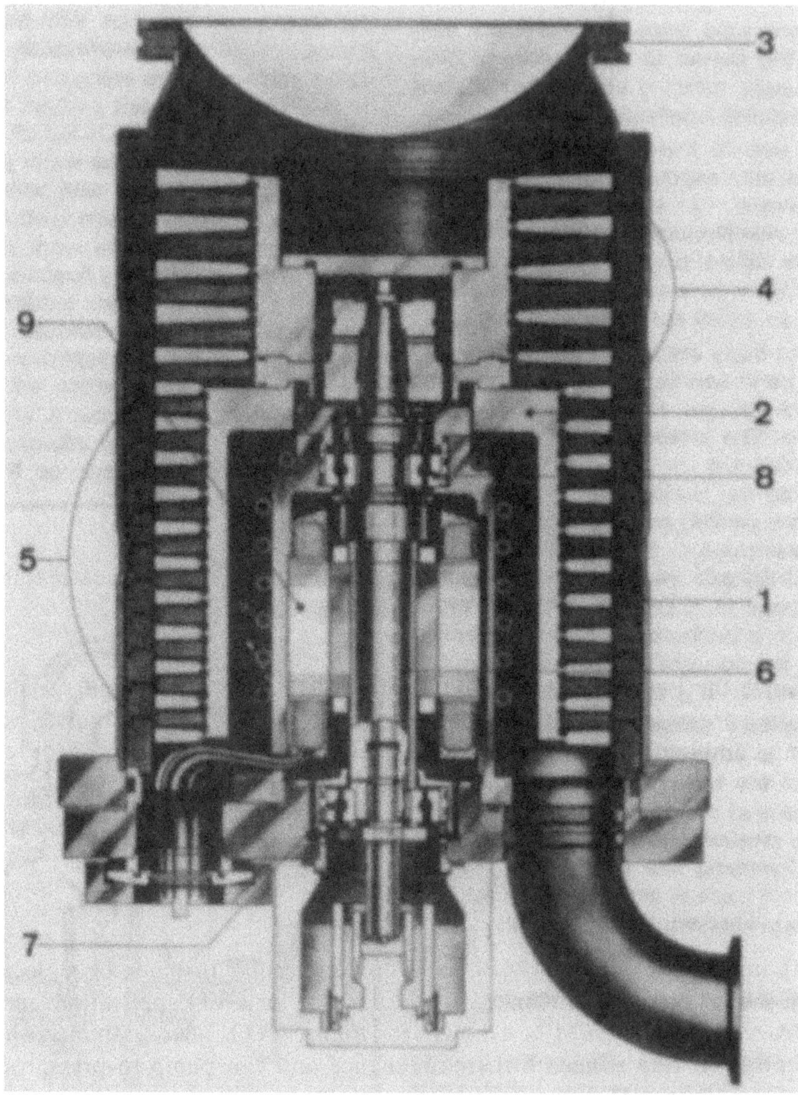

FIGURE A.6. Cross-section of a single-ended axial flow turbomechanical pump from Leybold-Heraeus. The pump consists of a stator (1) and a rotor (2); (3) is the inlet, (4) and (5) show the blades, (6) is the drive shaft mounted on bearings (7) and (8); (9) is the motor that drives the shaft. Reprinted with permission.

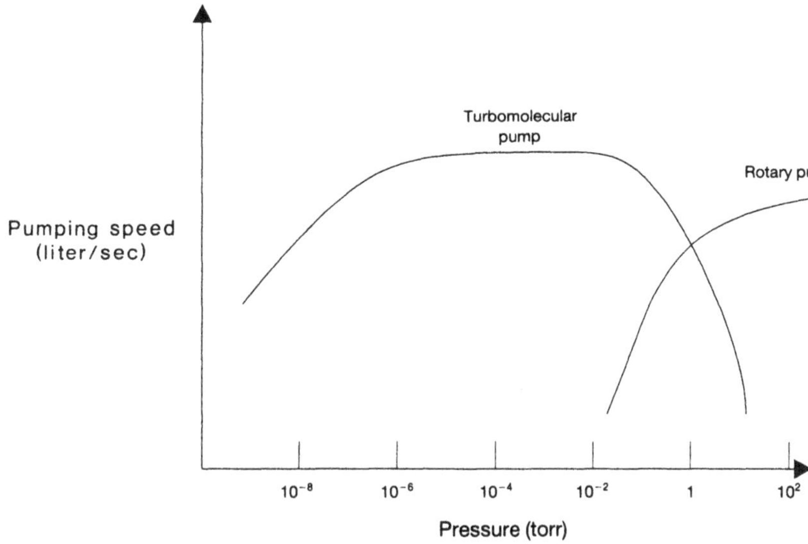

FIGURE A.7. The turbomechanical–vane pump tandem is commonly used to stage the turbopump to atmospheric pressure.

compression ratio of about 5. This results in an overall compression ratio of $5^9 = 2 \times 10^6$.

The fact that TMP does not use hydrocarbon or other oils in direct contact with the airstream helps it to be compatible with reactive gases while keeping it very clean. New magnetically levitated air bearings have enhanced the service life of TMPs and correspondingly increased their cost!

Since turbomolecular pumps work in the molecular flow regime, the chamber pressure needs to be reduced before switching over to them. Similarly, they cannot exhaust to the atmosphere and have to be backed by a mechanical pump. Figure A.7 shows the pumping speed characteristics of the TMP–vane pump combination.

A.2.4. Pressure Gauges

We will discuss three types of pressure gauges commonly seen in LPCVD/PECVD systems. Thermocouple gauges are used for measuring pressures in the roughing lines and in coarse pressure measurements from a vacuum up to a few millitorr. Capacitance manometers are used for precise measurements of processing pressures from a few millitorr to several hundred torr. Bayard and Alpert ionization gauges are used for measuring high vacuum pressures.

The thermocouple gauge (TC gauge) is the simplest and most rugged of

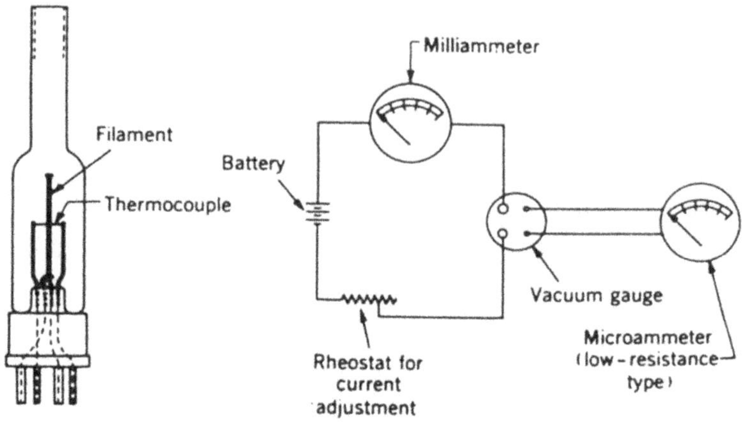

FIGURE A.8. Configuration of a thermocouple gauge and readout.

vacuum gauges. It senses the change in temperature of a heated resistor element as the pressure changes. Since pressure also determines the conductivity of the gas around the gauge, a high pressure leads to a larger temperature drop in the element and vice versa. A thermocouple embedded in the resistor, or separately mounted close to the heater element, senses the change in temperature, hence the name of the gauge. Figure A.8 shows a typical TC gauge configuration. The Pirani gauge also operates under the same principle.

The diaphragm manometer (such as the Baratron™ made by MKS Instruments) is used for the direct measurement of pressure over two or three orders of magnitude, from near atmospheric down to about 10^{-5} torr. Gauges for the different pressure ranges also differ in sensitivity. They consist of a thin diaphragm, often made of stainless steel, and measure the deflection of the diaphragm through capacitance bridges. The sensitivity and reproducibility of the gauge are excellent. They do suffer long-term drift due to aging of the diaphragm, corrosion of the diaphragm when exposed to reactive gases, and damage to the diaphragm when exposed to excessive pressure differences. Figure A.9 shows the internal construction of a diaphragm manometer.

For the measurement of high vacuum pressures, up to 10^{-12} torr or lower, ionization gauges have been commonly used. For CVD applications where such vacuum levels are not common, the use of high vacuum gauges are limited to the measurement of base pressures of the system. A commonly used ionization gauge configuration is the Bayard and Alpert (BA) gauge.

Ionization gauges operate under the principle of detecting the number

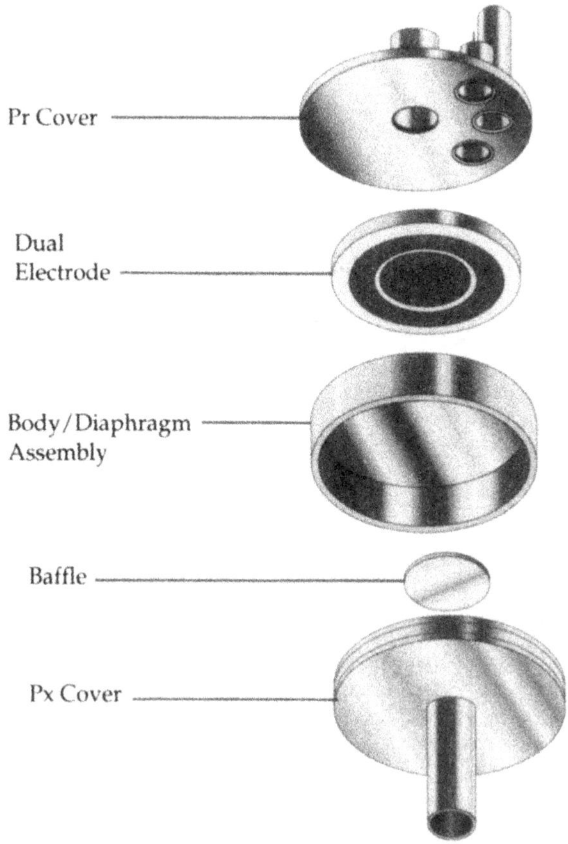

FIGURE A.9. Assembly drawing of a diaphragm manometer (Baratron TM) from MKS.

density of the gas by subjecting it to an ionizing electron stream and collecting the ion current. In structure, they resemble a triode, with an emitter, grid and collector. They consist of an electron emitter, such as a tungsten wire thermionically emitting electrons, an accelerating grid to provide energy to the electron for ionization and a thin wire biased to collect the ions. Figure A.10 shows the configuration of the BA gauge. Before a measurement is taken, the measurement surfaces are degassed. This is to desorb any adatoms on the measurement surfaces, otherwise the pressure reading would be too high. At high vacuum, the surfaces in the gauge can also provide adsorbing surfaces for gas molecules in the chamber, thereby acting as a pump. But this is not an issue in the CVD pressure range. Accelerated electrons that strike the grid can produce soft X rays which, in turn, can strike the collector,

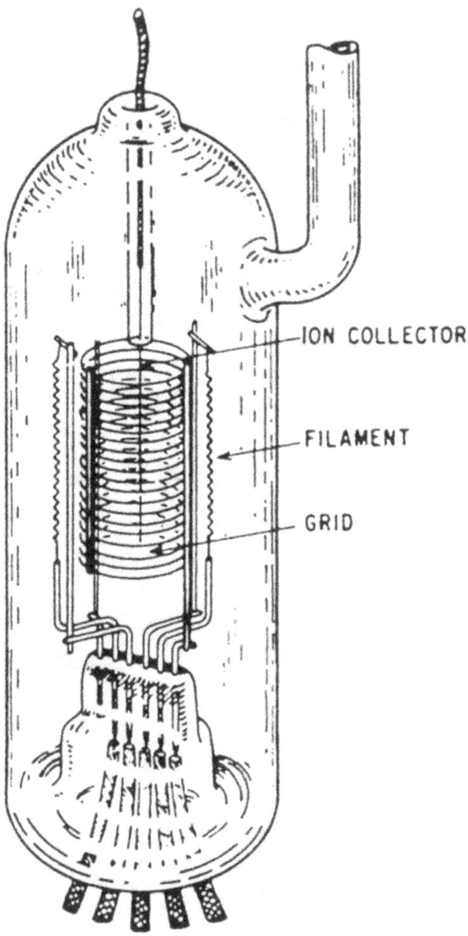

FIGURE A.10. Bayert-Alpert type ionization gauge used for the measurement of high vacuum.

sometimes producing a spurious secondary electron current. The collector is designed to be a wire so that it subtends a small solid angle to the X ray producing grid.

A cold cathode ion gauge, circumvents the X ray problem by using a Penning discharge (see Section 6.3.1) for ion production. Since the cold cathode gauge does not have the degassing requirements, it is starting to be popular in CVD equipment.

Index

Printed in the USA
CPSIA information can be obtained
at www.ICGtesting.com
LVHW020745150224
771920LV00001B/58

9 781475 747539